SYSTEMATIC BOTANY MONOGRAPHS

VOLUME 34

Systematics of Epilobium (Onagraceae) in China

Chia-jui Chen
Peter C. Hoch
Peter H. Raven

THE AMERICAN SOCIETY OF PLANT TAXONOMISTS
6 April 1992

SYSTEMATIC BOTANY MONOGRAPHS
ISSN 0737-8211

ISBN 0–912861–34–7

Printed in the United States of America

SYSTEMATICS OF EPILOBIUM (ONAGRACEAE) IN CHINA

Chia-jui Chen
Institute of Botany
Academia Sinica
Xiangshan, Beijing 100093, China

Peter C. Hoch and Peter H. Raven
Missouri Botanical Garden
P.O. Box 299
St. Louis, Missouri 63166

ABSTRACT. In this comprehensive revision, the taxonomy of the Chinese species of *Epilobium* is presented along with discussions about their phytogeography and intrageneric relationships. The study is based on examination of many new collections from China, on field work in Taiwan, northwest China, and parts of south-central China, and on detailed study of pollen, seeds, and trichomes. Thirty-seven species, including eight subspecies, are recognized. Three species are newly described (*E. fangii, E. pengii, E. taiwanianum*), and one new combinations is proposed (*E. amurense* subsp. *cephalostigma*). Nine taxa are reported from China for the first time. Four species are recognized in sect. *Chamaenerion,* a small, well-delimited group sometimes segregated as a separate genus. Data from cytology and distribution support the hypothesis of a Eurasian origin for the distinctive group. The remaining species belong to sect. *Epilobium,* a group of about 150 species of worldwide distribution. The Chinese species are arranged in groups of probable affinities, based on morphology. The major groups discussed are the circumboreal species, the Eurasian species (mostly in the Tian Shan-Altay region of northwestern Xinjiang), the Sino-Himalayan species (the large majority of the species treated), the Sino-Japan species (including several ranging into subtropical Vietnam and Luzon), and the Taiwan species. The last comprise seven taxa, including four endemics and two with striking range disjunctions to southwest China. Five species are endemic to mainland China: *E. blinii, E. fangii, E. kingdonii, E. sinense,* and *E. subcoriaceum. Epilobium ciliatum,* of northeast China, has its major distribution and affinities in North America.

INTRODUCTION

Epilobium is the largest genus in the Onagraceae, with about 165 species distributed mostly at high latitudes and altitudes worldwide (Raven 1976, 1988). Occurring most often in moist and often disturbed places, *Epilobium* is a common component of most temperate floras, and many species have large geographical ranges. Widespread sympatry, a tendency to form interspecific hybrids, and substantial vegetative plasticity often produce complex patterns of variation (Raven & Raven 1976), and have led to considerable confusion regarding taxonomic delimitation in most areas. This is notably true of *Epilobium* in China, where relatively few collections have been available until recently. No thorough revision of the genus in China exists, although Haussknecht (1884) included all Chinese collections extant at that time in his worldwide monograph. Recent treatments of *Epilobium* in Japan, the Soviet Union, and the Himalayan region have provided some insight, but also some confusion, since they often have employed different names for the same widespread taxa and, of course, have not included taxa endemic to China. Therefore, we undertook this study to provide a modern revision of the species of *Epilobium* in China, as a major step toward understanding the taxonomy and evolution of the genus in Eurasia and worldwide.

Epilobium is divided into six sections of very unequal size; sect. *Epilobium* comprises about 150 species, and the other five sections a total of 16 species (Raven 1976). Four sections are endemic to North America, although the monotypic sect. *Xerolobium* Raven is introduced in South America (Solomon 1982) and Europe (Izco 1983). Of the other two, sect. *Chamaenerion* (8 species) is a primarily Eurasian group, including two very widespread circumboreal species in North America as well, whereas sect. *Epilobium* occurs throughout the range of the genus. Systematic revisions of the genus are now complete for North America (36 native species, including those with the greatest diversity in morphology and chromosome number in the world; Hoch and Raven, unpubl.), South America (14 sp.; Solomon 1982), Africa (10 sp.; Raven 1967a), and Australasia (51 sp., including a remarkable assemblage of 37 endemic species in New Zealand; Raven & Raven 1976).

The Eurasian group of some 80 species has not been treated in a comprehensive fashion since the worldwide monograph of Haussknecht (1884), but recent regional treatments—for the U.S.S.R. (Steinberg 1949), the Himalayan region (Raven 1962a; Hara 1966, 1971, 1979), Turkey (Raven 1962b; Chamberlain & Raven 1972), Iran (Raven 1964), Japan (Hara 1965), Taiwan (Raven 1977), Malesia (Raven 1967b), Europe (Raven 1968), Pakistan (Hoch 1981), Bhutan (Hoch 1991) and others—provide the basis for a modern revision of this large complex of species. Many of the species that occur in China have been reviewed in these recent treatments, but for about half of the species, this is the first treatment in which they are clearly delimited, based on new collections, a knowledge of the genus outside of China, and a consistent taxonomic philosophy.

We examined more than 10,000 herbarium sheets of Chinese plants from 17 herbaria in China and from many other herbaria worldwide, thousands of additional specimens from areas adjacent to China, and type material for most relevant names. We received new collections from various recent botanical expeditions in Gansu, Hunan, Hubei, Sichuan, Guizhou, and Yunnan, and undertook our own fieldwork in Beijing, Gansu, Qinghai, Xinjiang, Zhejiang, Taiwan, Guangdong, Hunan, Hubei, and Sichuan. At the Missouri Botanical Garden, we grew more than 200 strains, representing most taxa, for detailed morphological and cytological study. We also made a limited survey of seed and pollen morphology by using light, scanning electron, and transmission electron microscopy as a supplement to earlier studies of these features (Seavey et al. 1977; Skvarla et al. 1975, 1976, 1978).

A Numerical List of Species (following the Literature Cited) is presented as an aid to understanding the introductory discussion; our reasons for proposing this arrangement will be discussed below. We recognize 37 species and 4 additional subspecies in China; these include three new species (*E. taiwanianum, E. fangii,* and *E. pengii*) and one new subspecific combination (*E. amurense* subsp. *cephalostigma*). All taxa are fully illustrated. As an aid to those unfamiliar with the provinces of China, we also include a map (Fig. 1); the numbers on the map refer to the sequential order of provinces in the listing of cited specimens and in Appendix 2.

FIG. 1. Map of China with province names and sequential numbering system used in lists of cited specimens and in Appendix 2.

TAXONOMIC HISTORY

Because *Epilobium* is so widespread and common, the history of the Eurasian species is extremely complex. In order to include the most relevant treatments of some of the more widespread taxa, it is necessary to consult studies of the Central Asian, Siberian, Japanese, and Himalayan floras, as well as the huge literature on plants of Europe and western Asia. The following survey includes only the major literature relevant to the history of the Chinese species of *Epilobium,* in both the Chinese literature and that of surrounding countries.

Epilobium was apparently first described in the Chinese literature in *Jiuhuang Bencao* (*Plants for Disaster Relief Food*), written in 1368 by Zhu Su. Calling the plant "Liuyecai," Zhu described what we recognize now as *E. hirsutum* as follows: "Liuyecai occurs in waste places in Zhongmou [probably in North China]. Plants are over two feet tall. Stems like those of 'Hao' [*Artemisia*]. Leaves are willow-like but shorter. Flowers are small, white, but silver and brown in the center. The leaves taste slightly hot. Young leaves can be used as tossed salad after boiling." The famous naturalist Li Shizhen described this plant again in his great scientific encyclopaedia, *Ben Cao Gang Mu* (*Compendium of Materia Medica;* S. Z. Li 1596), which consists of 52 volumes and includes 1195 plants illustrated by more than 1100 figures. Li described Liuyecai (*E. hirsutum*) as follows: "Liuyecai grows in valleys in Mt. Guyu, Zhengzhou. Plants are over two feet tall, light green. Leaves are willow-like but shorter and thicker, covered with long hairs. Flowers produce from axillae of the upper stems and branches, the petals 4, dark red. Fruits are long cylindrical. Leaves taste somewhat sweet, and can be used as salad after boiling."

Epilobium also has a long history in the literature of the West. In his *Monographie der Gattung Epilobium,* Haussknecht (1884) provided an exhaustive review of the pre-Linnaean literature, including, in the case of *E. angustifolium,* references from Theophrastus (371–287 B.C.). Linneaus (1753) included in *Species plantarum* four of the widespread and common species in China (*E. angustifolium, E. latifolium, E. hirsutum,* and *E. palustre*), although not based on any material from China. Schreber (1771) later described *E. roseum* from European material.

Maximowicz (1859) was the first western botanist to include Chinese material from Manchuria (Heilongjiang), when he included five taxa of *Epilobium* in his *Primitiae Florae Amurensis.* The only taxon he described as new, *E. origanifolium* var. *pubescens,* served as the basis for Haussknecht's (1884) *E. amurense.*

In a series of papers culminating in his classic *Monographie der Gattung Epilobium,* Haussknecht (1884) treated the genus worldwide in a comprehensive and consistent manner, and for the first time included collections from China, although the material available to him was limited. Haussknecht's exceptional knowledge of geography and all of the relevant literature, and his familiarity with the genus worldwide combined to make this work the standard treatment (Hoch & Raven 1990). The *Monographie* includes 32 species known from China, three of which (*E. tanguticum, E. subcoriaceum,* and *E. propinquum*) were described from Chinese collections. In the present work, we recognize 19 of the 32 species included in Haussknecht's work; we reduce two others to subspecies. His *Monographie* provided the foundation for all subsequent taxonomic work on *Epilobium* in China, and, considering the limited number of collections available to him, was an outstanding contribution.

Forbes and Hemsley (1887) wrote the first comprehensive taxonomic study of the entire Chinese flora, *Index Florae Sinensis,* and in three volumes they reported Chinese plants based on material available at Kew. They included only six species of *Epilobium* and an additional unnamed species (which we refer to *E. platystigmatosum*). Three names were correctly assigned, *E. japonicum* is now considered a synonym of *E. pyrricholophum* (Hara 1965), and the other two names were misapplied.

Komarov (1901, 1905) led a number of Russian expeditions to Mongolia, Manchuria, Korea, and the Soviet Far East. From his collections, he described several new *Epilobium* species, all of which are now considered synonyms of some of the more variable taxa of the region.

In a series of publications that appeared from 1907 to 1916, Léveillé proposed more than 24 new species of *Epilobium* from China and the surrounding regions. However, we recognize only two of these, *E. blinii* and *E. sinense,* and have placed the others in synonymy with various species (e.g., there are eight Léveillé names included in *E. pyrricholophum*). The uneven quality of Léveillé's taxonomic studies have been noted by Stearn (1938) and Lauener (1972); Raven (1962a) commented on the erratic quality of Léveillé's work on Onagraceae.

Handel-Mazzetti (1933), in another comprehensive work on the whole Chinese flora, *Symbolae Sinicae,* included ten species of *Epilobium* (incl. *Chamaenerion*). We recognize seven of these as valid species or subspecies. His unnamed species,

"*E. sp. nov.,*" based on *T. H. Wang 56* from Wugang, Hunan Province, was later validly described by Raven (1962a) as *E. kermodei,* based on a collection from northern Myanmar (Burma).

Kitagawa (1939) recorded eight species of *Epilobium* in Manchuria in his catalog, *Lineamenta Florae Manshuricae,* none of them new. We recognize only three of these as valid taxa. It appears that this book was primarily a compilation of previous work, and was not based on new collections or data.

The noted Japanese botanist Hiroshi Hara clarified the taxonomy of *Epilobium* in Japan, which includes at least eight taxa in common with China (Hara 1942, 1965). He also introduced a complex infraspecific taxonomy within several of the more complex and widespread taxa (incl. *E. palustre* and *E. pyrricholophum*), which he later abandoned (Hara 1965); based on our examination of ample Chinese collections, we largely concur with his revised species circumscriptions.

Since 1950, several studies have been undertaken of the flora of northeastern China, based on a large number of new collections from the region. These include *Illustrated Flora of Ligneous Plants of N. E. China* (Liou et al. 1955), *Claves Plantarum Chinae Boreali-Orientalis* (Liou et al. 1958), and a comprehensive herbaceous flora entitled *Flora Plantarum Herbacearum Chinae Boreali-Orientalis.* This latter work is projected to include 15 volumes, 8 of which have appeared; vol. 6 (Liou et al. 1977) includes the Onagraceae. This work treats 12 species of *Epilobium* (including *Chamaenerion*) from this region; we recognize six of these as valid taxa. Most of the other names were misapplied; for example, *E. angulatum, E. tenue, E. cylindrostigma,* and *E. glandulosum* were all applied to material we have determined as *E. ciliatum.*

Raven (1962a) provided a modern treatment for many of the Chinese species in "The Genus *Epilobium* in the Himalayan Region," in which he described 13 new species and one new subspecies, and proposed four new combinations. Although we have changed the rank of some taxa and have placed in synonymy at least four others described in that treatment, this work established the basic circumscription of 23 of the Chinese taxa. Hara (1966, 1971, 1979) also followed Raven's treatment for the majority of taxa he recognized in his summary of the results of expeditions to the eastern Himalaya organized by the University of Tokyo; 13 of these taxa also occur in China.

Iconographia Cormophytorum Sinicorum, compiled by the Institute of Botany, Academia Sinica, is a very comprehensive work, consisting of five volumes and two supplements in which more than 10,000 economic and ornamental species (with 9,082 figures) are reported. Because of the more limited focus of the work, only six Chinese species of *Epilobium* (including *Chamaenerion*) are treated in vol. 2 (Institute of Botany 1972).

Raven (1977) reported four species of *Epilobium* in his treatment in the *Flora of Taiwan,* whereas we now recognize seven species, based on availability of many more collections and results from new field work. These include *E. hohuanense,* a species first described by Ying (1975); because it appeared in an obscure journal after Raven's treatment was in press, however, it was not included in that treatment.

In *Flora Intramongolica,* C. J. Wu (1979) treated six species of *Epilobium* (including *Chamaenerion*). The circumscription of *E. palustre* includes collections referable to *E. fastigiatoramosum,* and that of *E. cephalostigma* includes material of *E. minutiflorum.*

Six species (incl. *Chamaenerion*) were reported in *Flora Tsinlingensis* (Northwest Institute of Botany 1981). Li (1980, 1986) recorded 15 species of *Epilobium* (incl. *Chamanerion*) from Xizang (Tibet) on the basis of material collected by many exploratory expeditions to Xizang during the past 34 years.

In summary, although several very useful treatments of *Epilobium* are available for the Himalayan region (Raven 1962a) and Japan (Hara 1965), the most comprehensive revision of the genus in China is that of Haussknecht (1884), now badly out of date in light of abundant new collections. The taxonomy of many species, especially those endemic to China, and even some widespread species for which regional treatments have failed to provide a satisfactory taxonomy, is confused or non-existent. Additional collections and careful fieldwork are still needed to clarify the circumscription and relationships of several species, especially from relatively inaccessible alpine regions of southwestern China.

In China, 43 species, 11 subspecies, one variety, and one forma were previously recorded. We recognize 41 taxa (37 species, including eight subspecies). We describe three new species (two from Taiwan and one from western Sichuan), propose one new combination (from species rank to subspecies), and report nine new records in China.

MORPHOLOGY

The sections *Epilobium* and *Chamaenerion* have sometimes been accorded separate generic status (e.g., Steinberg 1949), although most authors follow Raven (1976) by keeping them together, based on such shared characters as presence of a coma on the seeds, a basic chromosome number ($x = 18$) unique in the Onagraceae, and similarities in seeds (Seavey et al. 1977) and pollen (Skvarla et al. 1976, 1978). Section *Chamaenerion* (=*Chamerion* as a genus; Holub 1972) nevertheless differs sharply from sect. *Epilobium* in having leaves spirally arranged (vs. opposite, at least below the inflorescence), flowers slightly zygomorphic (vs. actinomorphic), pollen in monads (vs. tetrads), floral tube absent (vs. present), stamens in one subequal set of eight (vs. two unequal sets of four), style initially deflexed (vs. erect throughout), and petal notch absent (vs. present). These sectional characters do not differ within the sections, although there may be, for example, quantitative differences in length of floral tube or petal notch, etc. Most of the morphological variation we discuss is infrasectional.

Habit and perennating structures. All species of *Epilobium* in China are perennial herbs (only three species, all in North America, are annual; Raven 1976), although all flower normally in their first year. The species vary substantially in stature and degree of clumping. Exceptionally large individual plants of *E. angustifolium, E. hirsutum,* or *E. kermodei* may exceed 2 m in height and have rather woody bases, whereas individuals of *E. clarkeanum* or *E. nankotaizanense* rarely reach 0.2 m in height; most species fall between these extremes. Many of the larger species have erect primary stems, some form clumps of slender ascending stems (e.g., *E. fangii* and *E. wallichianum*) or of thicker stems (e.g., *E. latifolium* of sect. *Chamaenerion*), whereas others, like *E. hohuanense* and *E. williamsii,* form spreading clumps or mats without a primary stem. Stems may be simple or densely branched, but this sometimes varies within species, depending on the plant's age and habitat. Older plants, ones at the lower end of an elevational range, or in open, disturbed habitats tend to be well branched. On the other hand, some taxa, including those of sect. *Chamaenerion,* rarely tend to be branched.

A very useful taxonomic character in *Epilobium,* which unfortunately is overlooked by many collectors and thus often not preserved, is the perennating structure (Keating et al. 1982). Vegetative propagation is by basal axillary bud development, which takes place at the caudex or rarely higher on the stem. These buds provide the means by which plants overwinter, and may take the form of soboles, stolons, turions (sessile or at the ends of stolons), or sessile or pedicellate rosettes (Fig. 2; Keating et al. 1982). Starting with Haussknecht (1884), who clearly defined each of these types of structures and utilized them in his taxonomy, most subsequent taxonomists have used them as a major delimiting feature. However, as demonstrated by Keating et al. (1982), similar perennating structures may be analogous rather than homologous, and with few exceptions, cannot be used to construct phylogenies, because the types do not necessarily define monophyletic groups. The types also intergrade, depending on the degree of condensation or leafiness of the bud, which may depend on the harshness of the microclimate or other ecological factors. Finally, although most species can be characterized by one type of structure, others are more variable and may have two or more types.

Perennating structures nevertheless are useful in distinguishing many species, especially those that diverge from the basic type. The most basic type in China (found in 25 of the 41 taxa), and for the genus as a whole, is *soboles,* which are simple shoots with distinct internodes and loose scales or small fleshy leaves. They tend to be short (1–2 cm or less) and often are achlorophyllous when produced below ground, or more leafy and green when at ground level. We are including in the sobole type those structures characterized as "caespitose shoots" by Keating et al. (1982), and found in *E. clarkeanum, E. pengii,* and *E. anagallidifolium.* Although plants of this latter group tend to form many soboles at the caudex, giving rise to a clumped or matted habit, and to be somewhat leafier (as opposed to scaly), the soboles themselves do not differ fundamentally from those of other soboliferous plants.

A more diagnostic type of perennating structure is the *turion,* essentially a condensed sobole lacking internode elongation, which forms a strobiloid bud with thick, succulent, and imbricate cataphylls. The fleshy starch-filled leaves may enable plants to survive dormancy in particularly cold and/or dry conditions (Keating et al. 1982). One to several turions are usually formed below ground, some as much as 8 cm deep, and give rise to erect rhizomes with dark, coriaceous scales at the base (Fig. 2b). Many of the Chinese species with turions—*E. sikkimense, E. williamsii, E. subcoriaceum, E. gouldii,* and the newly described *E. fangii*—occur at higher elevations in southwest China. *Epilobium taiwanianum,* an endemic to the high mountains of Taiwan, also has fleshy turions. Most collections of *E. ciliatum* from China form rosettes, but a few from higher elevations in northeast China have more condensed turions, as also documented in the major part of its range in North America (Hoch 1978).

Rosettes, with leaves scarcely distinguishable from typical cauline leaves, may be sessile (Fig. 2d) or short-pedicelled (as in *E. parviflorum, E. cylindricum,* or *E. pannosum*), or terminating long (up to 10 cm) stolons (*E. kermodei;* Fig. 2e). The rosette leaves may shrivel and remain attached at the base of the stem, or fall off, leaving a scar. In general, rosuliferous plants grow in less harsh habitats.

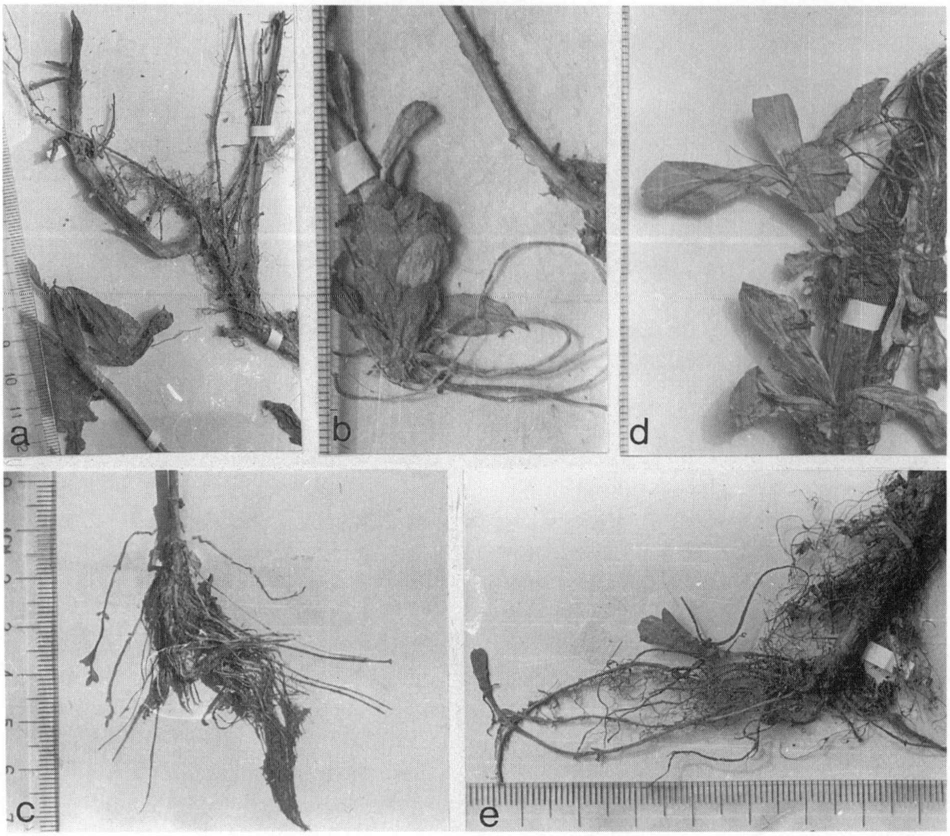

FIG. 2. Vegetative perennating forms in *Epilobium*. a. Fleshy sobole of *E. hirsutum* (*C. J. Chen 83007*). b. Plant arising from turion, with basal scales; *E. fangii* (*T. P. Soong 38435*). c. Filiform stolons of *E. palustre* (*C. G. Yang 1489*). d. Sessile rosettes of *E. wallichianum* (*C. J. Chen 83037*). e. Pedicillate rosettes of *E. kermodei* (*Raven et al. 26117*). Scales in mm.

The last major type of perennating structure in *Epilobium* is *stolons,* which are ground-level filiform structures with small decussate leaves or scales and a fleshy or leafy bud at the tip. Found in only a few Chinese species, and apparently associated with particularly wet, even boggy, habitats, they are notably well developed in *E. palustre,* which forms small succulent turions at the ends of stolons that can be 50 cm long (Fig. 2c), and in *E. pyrricholophum. Epilobium hirsutum,* one of the most common species of the genus in the world, uniquely forms thick, ropelike stolons (Fig. 2a), which radiate 1 m or more from the base of the plant and from which new stems arise directly or after formation of a small rosette.

The only major departure (in an anatomical sense) from the typical mode of perennation occurs in *E. angustifolium* and a few other species, in which shoots may also arise directly from horizontal underground roots, often a considerable distance from the main caudex (Keating et al. 1982). As a result, *E. angustifolium* generally forms large extended clones (Vodalazskij 1979; Flint 1980), and some authors have suggested that this species spreads primarily by vegetative means rather than by seed (Moss 1936; van Andel et al. 1978). This "shoot-from-root" method of vegetative growth may be partly responsible for the remarkable ecological "success" of *E. angustifolium* (Vodolazskij 1982), which has one of the largest ecological and geographical ranges of any species in the genus (Mosquin 1966; Chen et al. 1988). These species also form more typical axillary buds from the caudex.

Although one does not normally associate vegetative perennation with dispersal, in *Epilobium* we have observed that the small fleshy turions of some species detach easily from the parental caudex. A number of these species grow on shifting scree slopes, and the detached turions, equipped with their own roots, may be dispersed by normal rock movements. We also observed detached rosettes of *E. wallichianum* being transported in a stream (Yunnan, Kunming, *C. J. Chen 83037;* Fig. 3b); the dispersal of either axillary buds or seeds (Fig. 3a) by water may not be uncommon in the genus.

Pubescence. There is substantial variation in *Epilobium* in the type, quantity, and distribution of hairs on the plant, and this variation provides several useful taxonomic characters. Although a few completely glabrous plants have been detected (e.g., *E. nudicarpum* in *E. amurense* subsp. *cephalostigma,* and *E. royleanum* f. *glabrum*), these appear to be local or even infra-populational variants, and there are no glabrous taxa found among the Chinese species of *Epilobium.* There are four basic hair types found in the Chinese species:

1. Strigillose pubescence consists of relatively short (0.2–0.4 mm long), stiff, incurved or appressed hairs (Fig. 4a–c, e). It is the most common hair type in *Epilobium* but often is mixed with glandular hairs, especially on the inflorescence. Pure strigillose pubescence is found in the species of sect. *Chamaenerion* and a few species of sect. *Epilobium,* notably *E. minutiflorum* and *E. palustre,* both of which often have canescent inflorescences. Even on plants with strigillose pubescence all over the stem, hairs on the leaves are usually restricted to the margins and veins.

2. Glandular hairs may be short (0.1–0.2 mm; Fig. 4f, h) or somewhat longer (0.3–0.8 mm; Fig. 4c–d) and erect, with a thickened apex from which is exuded a small amount of viscid fluid. The surface of the apex is quite smooth (Fig. 4d, h) compared to that of other hair types. Although in a few species (e.g., *E. kermodei* and *E. pyrricholophum*) glandular hairs may predominate over most of the stem, in most species this type of hair is restricted to the inflorescence, where it is mixed with other types of hairs. Glandular hairs are characteristically absent from some species. The sticky fluid may act as a deterrent to aphids and other insects, since one frequently finds small dead insects entrapped in the glandular hairs.

3. Villous pubescence consists of long (0.5–1 mm), thin hairs that are lax and either suberect or spreading (Fig. 4f–g). Villous pubescence, frequently mixed with glandular hairs, is found mostly in *E. hirsutum* and *E. parviflorum,* imparting to those species one of their most characteristic aspects. In some populations of *E. hirsutum,* especially from the Tian Shan region of Xinjiang, the villous pubescence is extremely dense, grading into the tomentose type.

4. Tomentose pubescence also consists of relatively long (0.5–1.2 mm) and thick hairs that are more rigid and erect than those of the villous type (Fig. 4i–j). Tomentose pubescence is found only in *E. pannosum,* often mixed with glandular hairs. In most individuals, the hairs are very abundant, covering the entire stem and leaves.

The distribution of hairs on the stem is highly diagnostic for some species or

FIG. 3. Water as dispersal agent for seedlings and seeds in *Epilobium*. a. Seedlings of *E. amurense* subsp. *amurense*, which germinated in capsules, floating in stream (*C. J. Chen 83038*). b. Rosettes of *E. wallichianum*, detached from parent plant, growing at margin of stream (*C. J. Chen 83037*). Scales in mm.

groups of species. A broad and useful distinction is between species with hairs covering the entire stem (e.g., *E. brevifolium* subsp. *trichoneurum,* Fig. 43D) and species that are subglabrous below the inflorescence except for raised strigillose lines decurrent from the margins of the petioles (e.g., *E. amurense* subsp. *amurense,* Fig. 45A, D). Members of the latter group often have distinctively tetragonal or ridged stems, depending on whether there are 2 or 4 distinct lines through the internodes, whereas those of the former group are mostly terete. The Chinese species are roughly equally divided between these major patterns, although a few species (e.g., *E. cylindricum, E. tibetanum,* and *E. minutiflorum*) are either intermediate or mixed in this feature. Plants with either kind of cauline pattern have pubescent inflorescences, frequently with varying admixtures of glandular hairs even if these are absent elsewhere on the plants.

Leaves. A major difference in phyllotaxy exists between sect. *Chamaenerion,* which has 2/3 to 3/5 (Keating et al. 1982) to 3/8 (Shea & Watson 1989) spiral phyllotaxy (depending on the robustness of the apical meristem), and sect. *Epilobium,* which almost exclusively has opposite and decussate phyllotaxy below the inflorescence. The arrangement changes in the inflorescences of sect. *Epilobium* to alternate, and in sect. *Chamaenerion* to 5/13 or other arrangements (Shea & Watson 1989). In at least some collections of *E. sinense,* the leaves appear to be spiral throughout the stem, but the most basal leaves were not present on those individuals; this phenomenon clearly is exceptional in sect. *Epilobium.* In a few species, the leaves are characteristically either crowded (short internodes, e.g., *E. pannosum*) or well spaced (long internodes, e.g., *E. clarkeanum*).

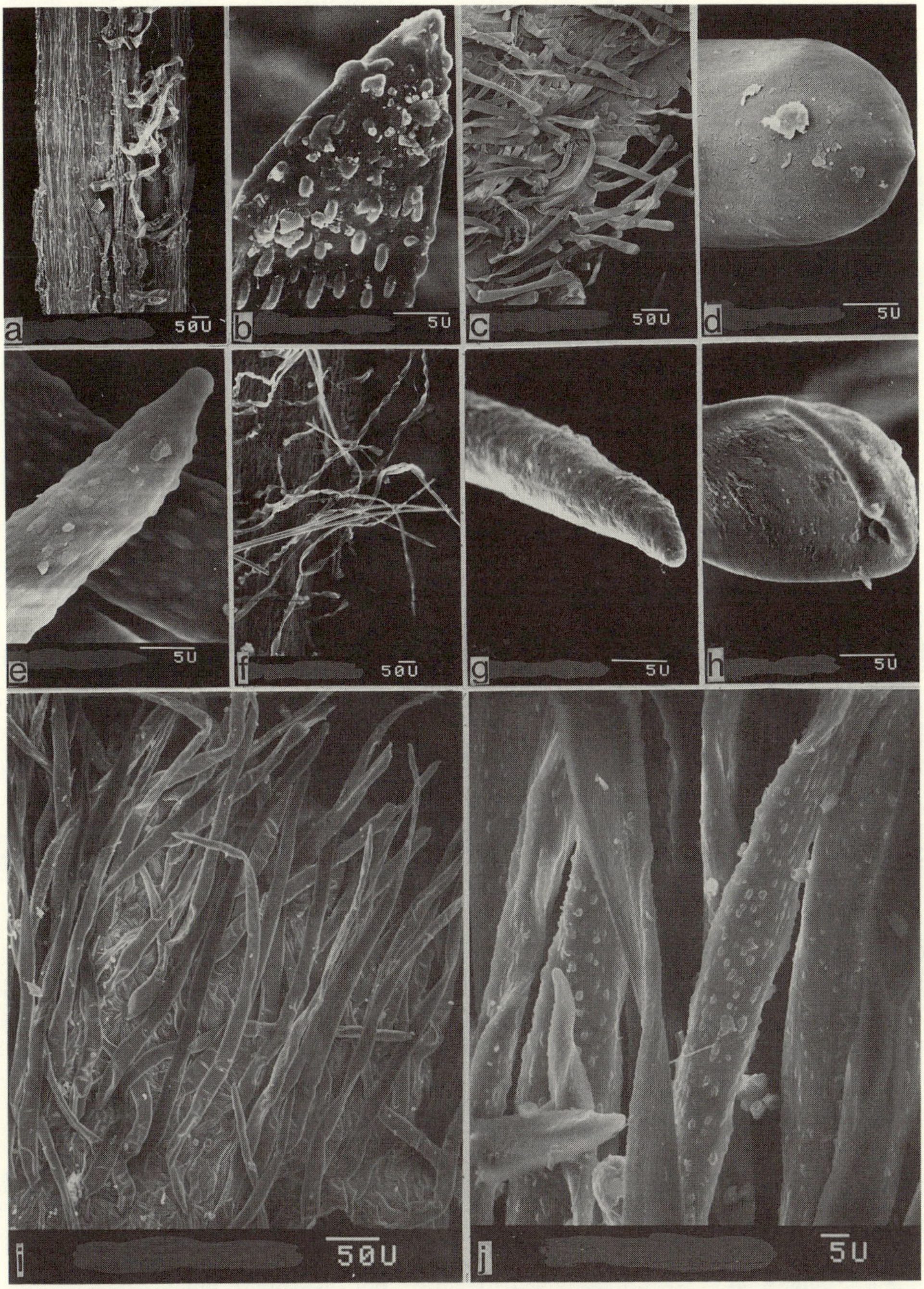

FIG. 4. SEM's of pubescence types of *Epilobium*. a. Strigillose pubescence of *E. amurense* subsp. *amurense* (*W. L. Chen & C. J. Chen 8549*). b. Detail of a, showing scattered tubercles on hair surface. c. Mixed glandular and strigillose pubescence of *E. pyrricholophum* (*Guizhou Bot. Exped. 3625*). d. Detail of c, showing tip of a glandular hair. e. Detail of c, showing tip of a short appressed hair of strigillose pubescence type. f. Villous pubescence mixed with short glandular hairs of *E. hirsutum* (*W. L. Chen & C. J. Chen 8715*). g. Detail of f, showing tip of a long thin suberect hair of villous pubescence type. h. Detail of f, showing tip a of glandular hair. i. Tomentose pubescence of *E. pannosum* (*T. N. Liou 14292*). j. Detail of i, showing scattered tubercles on surface of some hairs, none on others. Scales in microns.

Size of the leaves for the most part varies in proportion to the overall plant size, and also varies considerably within species or populations. However, some plants have disproportionately small (*E. blinii, E. clarkeanum*) or large (*E. kermodei*) leaves.

Leaf shape in Chinese species of *Epilobium* is variable from broadly ovate to sublinear, with the majority of the species having ovate to lanceolate to oblong leaves. In a few species, leaf shape is quite diagnostic (e.g., elliptic in *E. blinii* and *E. fangii,* narrowly spatulate in *E. sinense,* and sublinear in *E. platystigmatosum*). In most species, the lowest leaves tend to be smaller, obtuse, and obovate compared with cauline leaves, and the uppermost leaves and/or bracts tend to be smaller and much narrower, although this is not always the case. Leaf texture is normally herbaceous, but a few species (*E. nankotaizanense, E. sinense,* and *E. subcoriaceum*) have subcoriaceous leaves.

Leaf margins are denticulate in most species, with some more coarsely serrulate (*E. amurense* subsp. *cephalostigma* and *E. kermodei*), and others only remotely toothed (*E. blinii* and *E. platystigmatosum*). Some species have subentire leaves (*E. latifolium, E. palustre,* and *E. fastigiatoramosum*), and a few others are variable, but often have subentire margins (*E. angustifolium* subsp. *angustifolium, E. speciosum, E. clarkeanum,* and *E. anagallidifolium*).

Many species of *Epilobium* have sessile leaves, but others have leaves attached by petioles to 10 mm or more in length. Even in species with subsessile cauline leaves, the basal leaves may be petiolate; unless noted, the key and descriptions refer to the cauline leaves. About ⅔ of the Chinese species have subsessile leaves, which provides a useful diagnostic character for those with longer petioles (especially *E. roseum* subsp. *roseum, E. cylindricum, E. tibetanum, E. sinense,* and *E. royleanum*). The clasping leaf bases of *E. hirsutum* distinguish it from the otherwise similar *E. parviflorum.*

Inflorescence. All inflorescences in Chinese representatives of *Epilobium* are terminal, simple or branched racemes (polytelic botrya or spikes: cf. Weberling 1988). Some species have primarily simple inflorescences (e.g., all four species of sect. *Chamaenerion*), but in most species at least larger individuals are branched. Bracts are generally foliaceous and either progressively reduced in size up the axis or large and subequal to the capsules they subtend. The most extreme reduction occurs in *E. angustifolium,* which has minute cartilaginous bracts. In some collections from western Sichuan and Yunnan, however, the inflorescences are rather leafy, suggesting possible hybridization with *E. conspersum* in that region.

Some species (*E. pannosum, E. kingdonii, E. sikkimense, E. williamsii, E. gouldii,* and *E. clarkeanum*) characteristically have nodding inflorescences compared to the more typical suberect ones. When plants are fully in seed, however, the inflorescence of all species is nearly erect, so this character can be observed reliably only in younger individuals. Also, improper pressing of plants may obscure the character on some herbarium sheets.

Flowers. The flowers of sect. *Chamaenerion* differ sharply from those of sect. *Epilobium,* as previously mentioned. Within sect. *Chamaenerion,* flowers vary from 2–6 cm in diameter when fully open, but much of this variation is within the widespread species, *E. angustifolium* and *E. latifolium,* and overlaps among taxa. Indeed, there are few features of the petals, sepals, or androecium that can be used to distinguish among the taxa of sect. *Chamaenerion.*

Within sect. *Epilobium,* the petal, sepal, and androecium characters are also quite uniform, although some diagnostic size differences exist. Stigma morphology is a useful character, distinguishing species with 4-lobed stigmas (*E. hirsutum, E. parviflorum, E. blinii,* and *E. nankotaizanense*) from those with entire ones (all other taxa), and among the species with entire stigmas those with capitate vs. clavate or cylindrical ones. Some species have petals that are consistently white or rose-purple, but many are variable for color. Large flower size distinguishes a few species, but most taxa have funnel-form flowers 3–7 mm long; even among the small-flowered group, however, size differences distinguish some pairs of species. *Epilobium minutiflorum* is notable for its very small flowers (mostly 3–4 mm long).

Pollen characters are extremely useful in delimiting the Onagraceae within Myrtales (Raven 1988), notably by presence of viscin threads (Fig. 5f–g, i; Skvarla et al. 1978) and of 'paracrystalline beaded' pollen ektexine (Fig. 5h, j–l; Skvarla et al. 1975, 1976), but within *Epilobium* there are few characters of phylogenetic value. Pollen grains are roughly spheroidal, mostly triangular in polar view with three protruding apertures (rarely quadrangular and four-pored in some species). Although some differences in viscin thread morphology and other features exist within *Epilobium* (Skvarla et al. 1978), the only one useful in delimiting the Chinese species concerns pollen in monads (Fig. 5d, f; sect. *Chamaenerion*) vs. in tetrads (Fig. 5g; sect. *Epilobium;* Skvarla et al. 1975). Also, the two subspecies of *E. angustifolium* differ in pollen size and aperture number (Fig. 5a–b). Although variation exists in pollen surface sculpturing (Fig. 5e, h), it does not provide taxonomic characters for this group.

Pollen also provides a rapid method of corroborating suspected instances of interspecific hybridization; a solution of malachite green-acid fuchsin-Orange G differentially stains aborted and nonaborted pollen (Alexander 1969; Raven & Raven 1976).

Capsules and seeds. There is considerable variability in capsule length, but much of it is intraspecific; even on a single plant, there can be more than a two-fold difference between the earliest and late-season capsules. Nevertheless, certain species have notably long capsules (*E. speciosum* and *E. sikkimense,* 5–9 cm; *E. kermodei,* 7–11 cm), or very short capsules (those of *E. clarkeanum* and *E. anagallidifolium* do not exceed 4 cm), and this feature may be useful in distinguishing those taxa. Full development of capsules depends in large part on normal seed formation, and a small proportion of *Epilobium* collections have obviously small, shrivelled capsules with few or no seeds. In the case of obligate out-crossing species (primarily taxa with four-lobed stigmas, especially in sect. *Chamaenerion*), this occurs occasionally if cross-pollination fails to take place; however, most plants with shrivelled, undeveloped capsules also have low pollen fertility and are apparent interspecific hybrids.

Pedicel length in most species varies from 1 to 2 cm, but some species have particularly long pedicels (*E. conspersum* and *E. palustre* to 5 cm, and *E. sinense* to 4 cm) or short ones (*E. taiwanianum, E. laxum,* and *E. williamsii,* less than 1 cm). *Epilobium anagallidifolium,* a circumboreal species that barely reaches the Altay Mountains in northern Xinjiang, has pedicels (1–5 cm long) that equal or exceed the length of its capsules (1.7–3.6 cm).

Variation in seed size, surface morphology, and comas provides some of the most useful diagnostic characters in *Epilobium* (Figs. 6–8; Raven & Raven 1976;

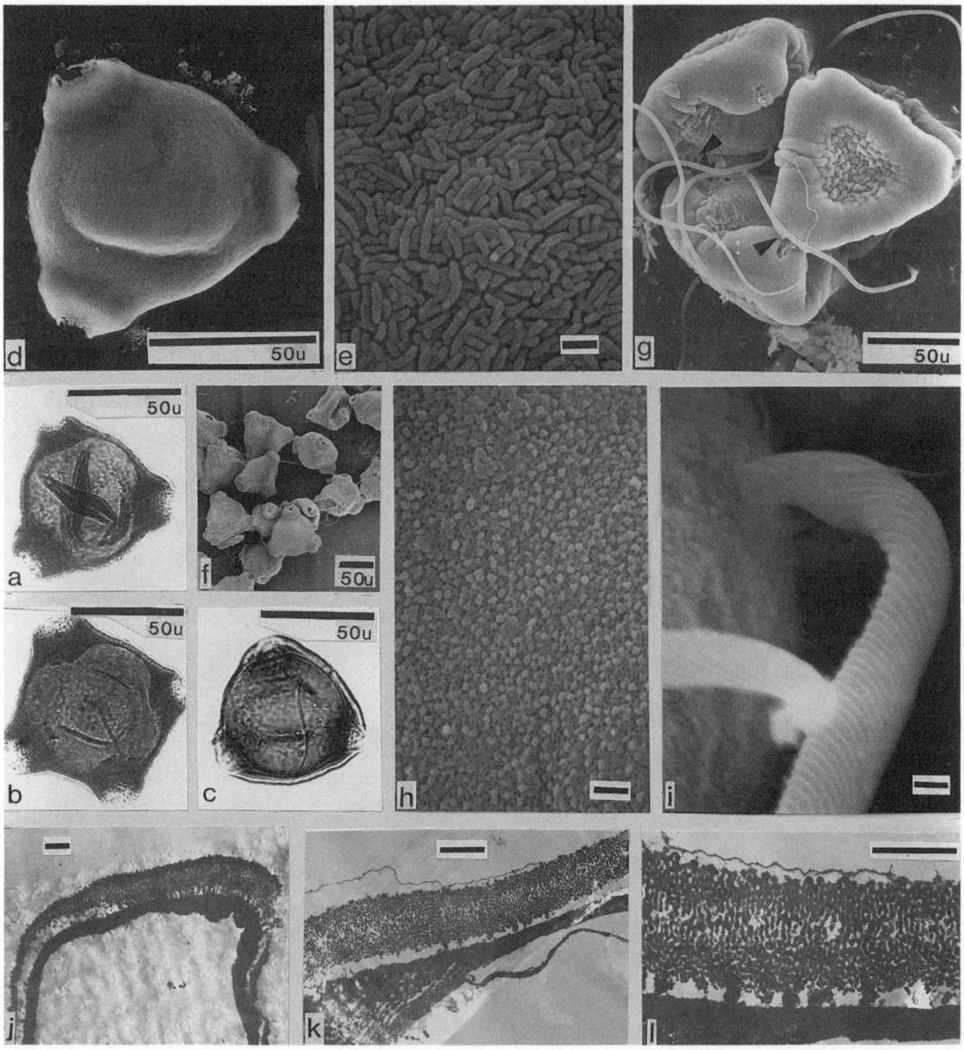

FIG. 5. Pollen of *Epilobium*. a–c. Light photomicrographs of acetolyzed pollen: a. Three-pored grain of *E. angustifolium* subsp. *angustifolium* (*R. C. Ching 1275*). b. Four-pored grain of *E. angustifolium* subsp. *circumvagum* (*G. H. Yang 57442*). c. Three-pored grain of *E. ciliatum* subsp. *ciliatum* (*W. Wang et al. 2393*). d–i. SEM's of acetolyzed pollen: d. Pollen of *E. conspersum* (*Xizang Med. Exped. 547*), shed as monads. e. Detail of d, showing surface of exine. f. Pollen grains with viscin threads of *E. angustifolium* subsp. *circumvagum* (*G. H. Yang 57442*). g. A pollen tetrad of *E. blinii* (*Handel-Mazzetti 6211*), showing bridge area (arrows) and viscin threads on the proximal pole. h. Detail of g, showing surface of exine. i. Detail of g, showing viscin threads. j–l. TEM's of acetolyzed pollen; detail sections of pollen wall of *E. pannosum* (*T. N. Liou 20664*). Except as noted, scale bars equal 1 μm.

Seavey et al. 1977). Even though the seeds of some species are highly variable in length (e.g., *E. latifolium,* 1.2–2.1 mm; *E. palustre,* 1.3–2.2 mm; and *E. anagallidifolium,* 0.7–1.4 mm), most have much more limited size ranges, and this character can be used to distinguish a number of species. The largest seeds are found in some of the variable species just noted, but other species with large seeds include

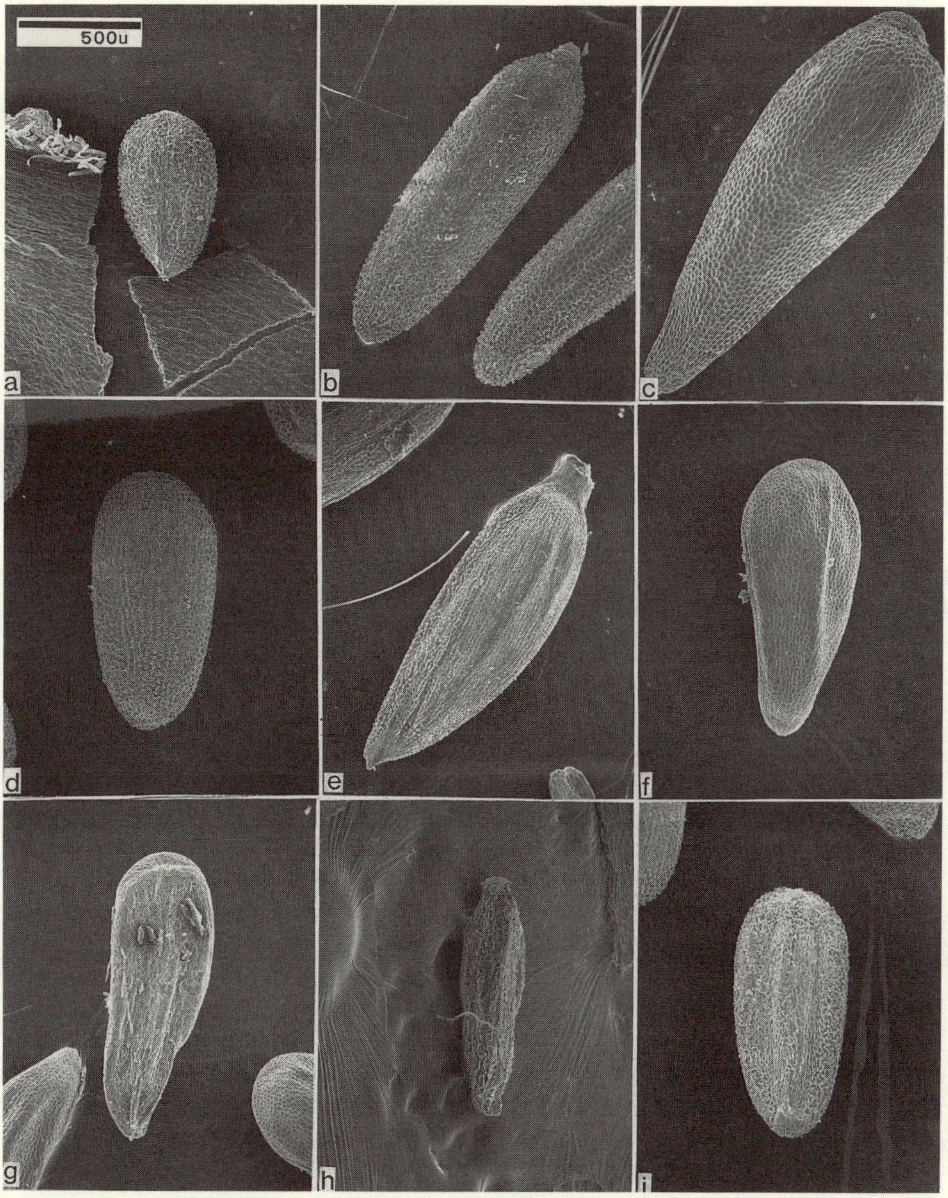

FIG. 6. SEM's of seeds of *Epilobium*. a. Obovoid seed, *E. kermodei* (*Raven et al. 26112*). b. Oblanceolate seed, *E. blinii* (*Ducloux 7573*). c. Oblanceolate seed, *E. nankotaizanense* (*Yamazaki et al. 321*). d. Oblong seed, *E. pannosum* (*T. N. Liou 14292*). e. Subfusiform seed, *E. palustre* (*S. W. Liu & Y. F. Huang 869*). f. Pyriform seed, *E. tibetanum* (*Wallich 6863*). g–i. Dorsal views of seeds: g. Same as f, showing almost flat abaxial surface. h. *E. conspersum* (*K. T. Fu 9564*). i. Same as d, showing abaxial surface.

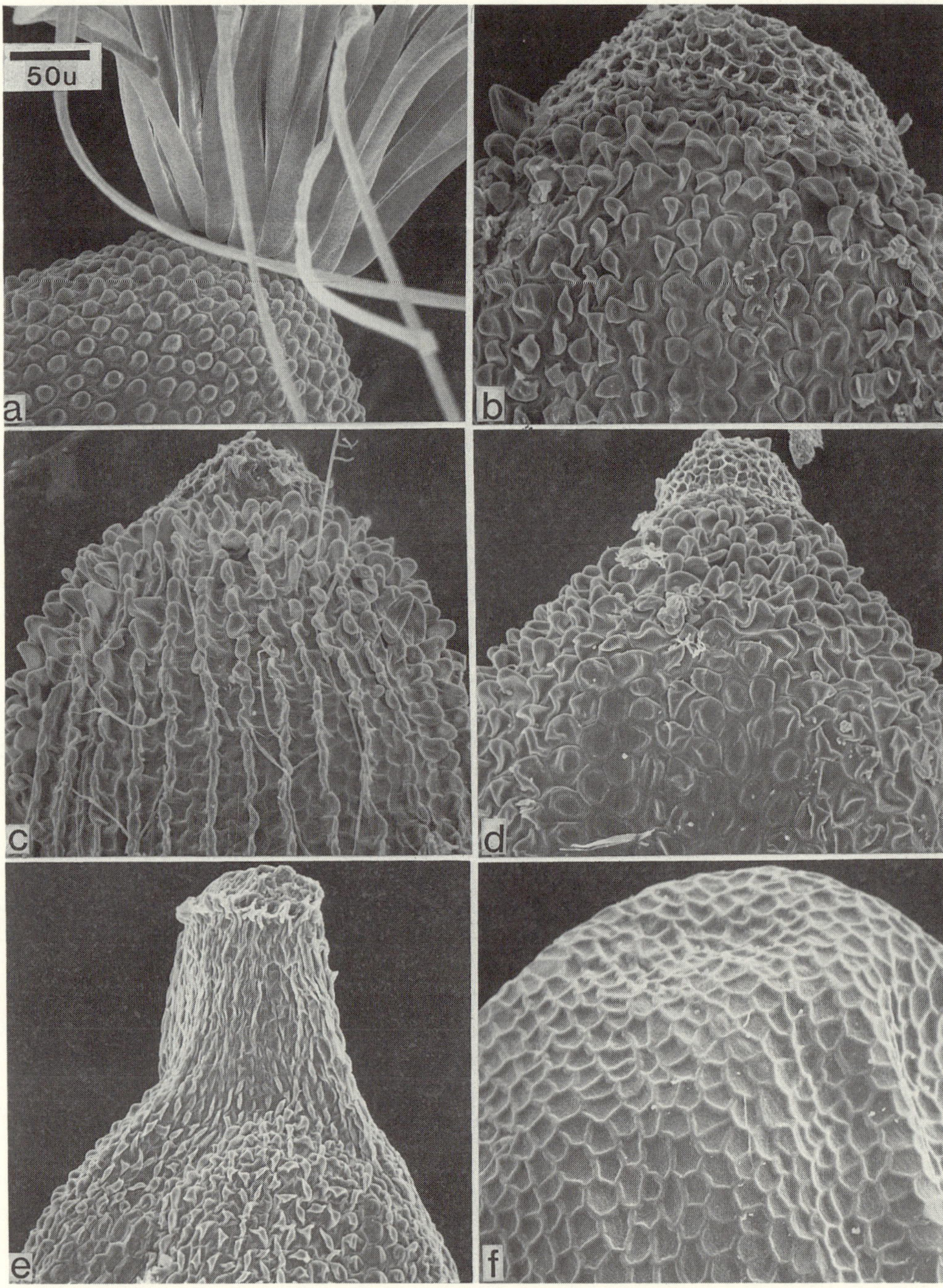

FIG. 7. SEM's of chalazal collar of seeds of *Epilobium*. a. *E. cylindricum*, showing attachment of coma at chalazal neck (*Shennongjia Exped. 31514*). b. *E. pyrricholophum* (*C. S. Fan & Y. Y. Li 627*). c. *E. ciliatum* subsp. *ciliatum* (*T. N. Liou 4103*). d. *E. blinii* (*Handel-Mazzetti 6211*). e. Long, very conspicuous chalazal collar of *E. palustre* (*S. W. Liu & Y. F. Huang 869*). f. Rounded, inconspicuous chalazal collar of *E. tibetanum* (*Wallich 6863*).

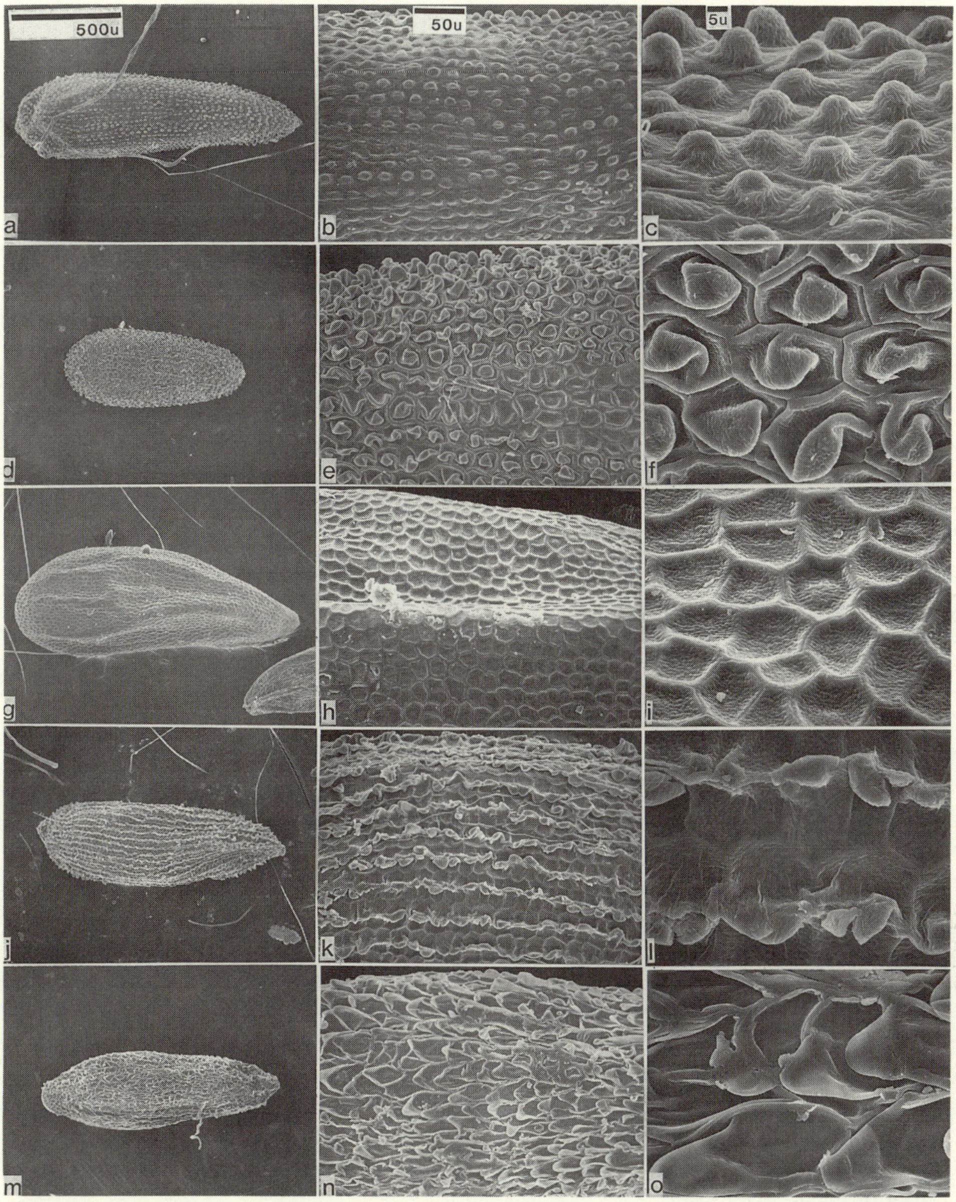

FIG. 8. SEM's of seed surface morphology of *Epilobium*. a–f. Papillose type: a. *E. wallichianum* (*Smith 2849*). b–c. Details of a. d. *E. kermodei* (*Raven et al. 26112*). e–f. Details of d. g. Reticulate type: *E. tibetanum* (*Wallich 6863*). h–i. Details of g. j. Ridged type: *E. ciliatum* subsp. *ciliatum* (*T. N. Liou 4103*). k–l. Details of j. m. Irregularly reticulate (or squamose) type: *E. conspersum* (*K. T. Fu 9564*). n–o. Details of m.

E. nankotaizanense (1.6–1.8 mm; Fig. 6c) and *E. pyrricholophum* (1.5–1.8 mm). Seeds of several other species rarely exceed 1 mm (Fig. 6a), and this gross size difference is useful in diagnosing some species even without microscopic measurements. Seed shape is also variable, but because it requires microscopic examination, we have not employed it in our taxonomy. Nevertheless, one can distinguish obovoid (Fig. 6a), oblanceolate (Fig. 6b–c), oblong (Fig. 6d), subfusiform (Fig. 6e), and pyriform (Fig. 6f) shapes in various species.

The coma is attached to the seed at the chalazal collar or neck (Fig. 7a; Seavey et al. 1977). In most species, this collar is low, often not protruding much beyond the rounded end of the seed, and more or less inconspicuous (e.g., Fig. 7a, *E. cylindricum;* Fig. 7f, *E. tibetanum*). However, in a few taxa the collar is long and prominent (Fig. 7e, *E. palustre;* also in sect. *Chamaenerion* excl. *E. angustifolium*). The coma may be easily detached, as in most species, or more persistent (e.g., Fig. 7a), particularly in most taxa of sect. *Chamaenerion.* The comas of most species are white or dull white, contrasting with a few species with tawny or reddish comas (*E. palustre, E. fastigiatoramosum,* and especially *E. pyrricholophum*).

Despite their microscopic nature, seed surface features provide some excellent diagnostic characters. Most Chinese species, and indeed most *Epilobium* species worldwide (Seavey et al. 1977), have papillose seeds (Fig. 8a–f), in which convex papillae arise centrally in cells that are distributed in more or less regular lines on the seed surface and separated from neighboring cells by a radial wall reticulum. The papillae may be more or less pronounced on the surface and in outline, and are noted in descriptions as low ("fine" or "rugose") or high ("coarse").

A previously unreported modification of this basic papillose type surface appears in *E. conspersum* (Fig. 8m–o), in which the papillae appear to be collapsed tangentially, making the surface appear squamose or ramentaceous. At low magnification, the outline of these seeds appears to be "smooth" or reticulate (Fig. 8m), but they are fundamentally different from actual reticulate seeds (Fig. 8g–i).

Departures from the basic papillose surface provide a strong distinguishing feature for certain species. Reticulate seeds (="foveolate" seeds of Seavey et al. 1977), in which the surface cells are well marked by radial wall reticula but lack raised central papillae, are found in *E. tibetanum* (Fig. 8g–i), providing the most consistent and certain character for delimiting it from *E. cylindricum;* the group of *E. clarkeanum, E. pengii,* and *E. anagallidifolium* (the latter at least in part); and most species of sect. *Chamaenerion.* As reported and illustrated by Seavey et al. (1977, figs. 25–30), however, the surface reticulation on *E. angustifolium* and *E. latifolium* is highly irregular compared to that in species of sect. *Epilobium,* and may not be homologous.

Perhaps the most specialized and diagnostic seed surface type is the ridged type found in *E. ciliatum* (Fig. 8j–l; Seavey et al. 1977, figs. 157–174). These seeds are marked by longitudinal rows or ridges of raised, laterally compressed, fused scales or flattened papillae. Because the ridges appear to be in line with the edges rather than the middles of the surface cells, it is unclear whether or not the scales or flattened papillae making up these ridges are homologous to the raised central papillae of papillose seeds. *Epilobium ciliatum* is a particularly widespread and weedy species (Raven & Raven 1976; Hoch 1978; Solomon 1982), although currently found in China only in the extreme northeast; this seed surface feature, visible without magnification, is most useful in distinguishing the species from all congeners in China.

CYTOLOGY AND CYTOGENETIC RELATIONSHIPS

The gametic chromosome number is known or reported for 32 of the 41 taxa of *Epilobium* in China; the remaining nine species have not been counted (*E. clarkeanum, E. gouldii, E. hohuanense, E. kingdonii, E. pengii, E. speciosum, E. subcoriaceum, E. taiwanianum,* and *E. williamsii*). Counts for 21 taxa are reported here (Appendix 1); we have found no previous chromosome reports for at least 15 of these taxa. Of the 21 reported counts, 14 are from material of Chinese origin, and most of the remaining counts are from areas adjacent to China. Six collections of *E. angustifolium* subsp. *circumvagum* yielded counts of n = 36 (Chen et al. 1988); all of the remaining taxa (20/21) for which we are reporting counts had n = 18.

In fact, n = 18 is the only gametic number known for any species of sect. *Epilobium,* for almost 85% of which (124 of 147 species) counts are available (Raven 1988). The base number for tribe Epilobieae is probably x = 9, which implies that all species of *Epilobium,* with reported counts of n = 12, 13, 15, 16, 18, 30, 36, and 54, are polyploids (Raven 1976); for this discussion, we use "polyploid" to refer to numbers higher than n = 18. Although only diploid numbers are known for four of the eight species in sect. *Chamaenerion* (two are unknown), *E. angustifolium* (n = 18, 36, 54) and *E. latifolium* (n = 18, 36) are two of the only three species in the entire genus that include polyploid entities (Mosquin 1966; Raven 1988). Chromosome numbers provide no information regarding relationships among the Chinese species of *Epilobium;* the same can be said for the whole genus except for the North American species (Raven 1976; Raven & Raven 1976).

Within sect. *Epilobium,* cytogenetic analysis has provided valuable information regarding relationships. The fact that almost all species in the section that have been tested are able to form hybrids has made it possible to observe chromosomal behavior and to detect translocation heterozygosity (Mosquin 1966; Seavey & Raven 1977a, 1977b, 1977c, 1978, unpubl.). Each taxon, by being crossed to species of known arrangements, can be characterized as having one, or very rarely more than one, arrangement. Most of the species characterized, including the very generalized *E. obcordatum* A. Gray of western North America, have what appears to be the ancestral arrangement, which was designated "BB" by Seavey and Raven (1977a). This arrangement is found in nine taxa that occur in China: *E. amurense* subsp. *amurense, E. hirsutum, E. minutiflorum, E. palustre, E. parviflorum, E. pyrricholophum, E. roseum* subsp. *subsessile, E. royleanum,* and *E. tibetanum.* Because BB is the plesiomorphic condition, this information does not suggest anything certain about relationships among these species.

The AA arrangement, which differs from BB by one translocation (hybrids form a ring or chain of 4 chromosomes at meiosis), characterizes a probably monophyletic group in North America that includes *E. ciliatum* (the only AA species in China). However, the AA arrangement is found also in several unrelated species in Europe, which suggests that this arrangement is polyphyletic. No other Chinese species is known to have the AA arrangement. The CC arrangement, which also differs from BB by one translocation and from AA by two, characterizes the so-called Alpinae group, which includes *E. anagallidifolium* in China. Despite uncertainty about the homology of this arrangement in all species in which it occurs (Seavey & Solomon 1984, unpubl.), presence of CC in all taxa of Alpinae supports the morphological hypothesis of monophyly for that group; using that information, it may be possible to test our hypothesis that the Chinese species *E. clarkeanum* and *E. pengii* may also be directly related to the Alpinae group.

Two taxa from eastern Asia appear to have arrangements different from any others. *Epilobium platystigmatosum* and *E. amurense* subsp. *cephalostigma* both differ from each of the three known arrangements by one or more translocations; however, it has not been possible to establish unambiguously how or whether they may differ from each other (Seavey and Raven, unpubl.). If these character states are autapomorphies, they demonstrate nothing about relationships. If these unique arrangements occur in other species, however, it may be possible to use them to test for certain relationships hypothesized from morphology, e.g., between *E. platystigmatosum* and *E. sinense* or *E. hohuanense*. Also, if the arrangement in *E. amurense* subsp. *cephalostigma* is shared with other taxa, and supported by other data, it may require revision of the taxonomy for this species, since subsp. *amurense* has the BB arrangement. Despite the difficulty in establishing the monophyletic nature of these translocation arrangements, the presence in East Asia of several novel arrangements suggests that a program of selective hybridization and cytogenetic analysis might provide valuable clues to the phylogeny of at least some species or groups.

It should be emphasized that even hybrids involving different chromosome translocation arrangements are viable, and many have moderate levels of pollen fertility (Seavey & Raven 1977a, 1977b, 1978). Not surprisingly, the highest fertility levels (to 94%) are within-arrangement crosses, but fertility of crosses involving one reciprocal translocation was 10–59%, and those involving two translocations, 4–46%. Thus, even relatively distant hybrids can be nearly 50% fertile; however, a few within-arrangement crosses were totally or nearly sterile, indicating that fertility is controlled by more than cytogenetic factors. In *Epilobium,* for example, Michaelis (1954), Schmitz (1988), and others have demonstrated the strong effects on the phenotype, and particularly on male sterility, of genetic factors in the cytoplasm. We have no data regarding the occurrence of these factors in the Chinese species.

REPRODUCTIVE BIOLOGY

Most species of *Epilobium* are genetically self-compatible (Raven 1979), although at least partial self-incompatibility has been detected in *E. angustifolium, E. latifolium* (both sect. *Chamaenerion;* Mosquin 1966; Flint 1980), and *E. obcordatum* (sect. *Epilobium;* Seavey & Bawa 1986). In *E. angustifolium,* Flint (1980) reported significant reduction in seed set in controlled self-pollinations vs. outcrosses in both diploid and tetraploid individuals. However, she found that selfing in at least some clones of tetraploids produced as much fertile seed as did outcrossing; this was not true of diploids. Further studies of the phenomenon may be useful, especially in clarifying its occurrence elsewhere in sect. *Chamaenerion.* Whether the mechanism is similar to the "late-acting" self-incompatibility described in *E. obcordatum* by Seavey and Bawa (1986) is not known.

In parallel with the floral morphology that so clearly marks it, sect. *Chamaenerion* has a highly distinctive flowering phenology (e.g., Knuth 1908; Benham 1969). In fact, the first concise report of pollination in plants dealt with *E. angustifolium*

(Sprengel 1793). The buds droop just prior to anthesis; as anthesis begins, the pedicel straightens, the perianth opens flat, and the erect stamens dehisce promptly, while the style is sharply deflexed below the stamens. As flowering continues, the staminal filaments gradually senesce and reflex as the style straightens into the central axis of the flower, the stigmatic lobes spread open and become receptive, and the nectary at the base of the style increases nectar production. Individual flowers are open for five or more days. Because of this marked protandry, all taxa of the section are predominantly outcrossing.

The flowering phenology of sect. *Epilobium* is much simpler: the anthers and stigma generally mature simultaneously as the funnel-shaped or campanulate flower opens and remain erect throughout anthesis. Nectar is produced, but in small quantities. In those species with four-lobed stigmas, and a few species with entire stigmas (*E. brevifolium* subsp. *trichoneurum, E. laxum, E. pyrricholophum,* and *E. wallichianum*), the stigma is exserted beyond the anthers; these taxa are more or less modally outcrossing, depending on the degree of separation of stigma from anthers. The large majority of species of sect. *Epilobium,* and most Chinese taxa (28 of 41), are modally autogamous; in them, the anthers are subequal with the stigma and shed pollen directly onto it, often before the flower opens (Raven & Raven 1976; Raven 1979). Even in species of this section with four-lobed or normally exserted stigmas, the probability of selfing increases in smaller late-season flowers with shorter stigmas.

This modal autogamy produces a pattern in many *Epilobium* species of well-differentiated local populations that vary moderately from populations in other areas or habitat types (Raven & Raven 1976; Hoch 1978). In some widespread species (in China, esp. *E. pyrricholophum, E. amurense* subsp. *cephalostigma,* and *E. wallichianum*), this has resulted in the formal recognition of many entities that are better viewed as local variants in the broad variation within the species.

Despite this tendency toward selfing in most taxa, outcrossing occurs at least occasionally in all species of *Epilobium* that have been studied. Because many species occur sympatrically (discussed below), are relatively undifferentiated florally and share generalized pollinators, and have poorly developed internal barriers to hybridization, there is a certain probability that when outcrossing occurs, it results in interspecific hybridization. We have documented examples of more than 20 hybrid combinations involving 26 taxa; these are proposed based on intermediate morphology, reduced seed/capsule set, and reduced pollen fertility. Most hybrid combinations involve at least one species in which some degree of cross-pollination is the rule. This estimate for China is lower than the frequency of hybridization among *Epilobium* species in Australasia (Raven & Raven 1976; Raven 1980) or Europe (Stace 1975), but it is almost certainly an underestimate due primarily to our less precise knowledge of *Epilobium* in China relative to those other regions. Also, the morphological similarity of certain pairs of species (e.g., *E. cylindricum–E. tibetanum*) and the moderately high fertility of many hybrids means that some combinations may be overlooked.

We have limited, mostly anecdotal information regarding pollinators in China, but the situation appears to be comparable to that in other parts of the world. Species of sect. *Chamaenerion* are visited intensively by bees (*Apis* and *Bombus*), both in China and elsewhere in their ranges; *E. angustifolium* is a major source of honey in many areas (Mosquin 1966; Pellett 1976; Howes 1979), and its pollination biology has been the subject of many studies (see Benham 1969). Other species of sect. *Chamaenerion* are less well-known. In sect. *Epilobium,* butterflies, skippers, flies, and bees have been observed on some species, especially those with larger flowers (Raven & Raven 1976; Hoch, unpubl.). Visits by non-specific pollinators probably provide the means by which sympatric species of *Epilobium* can hybridize.

In summary, two very different reproductive modes are found in Chinese *Epilobium.* The first characterizes sect. *Chamaenerion,* a group of large-flowered plants in which modal outcrossing is achieved by pronounced protandry, reinforced in some species by genetic self-incompatibility (Flint 1980). These species fail to set seed in insect-free greenhouses (pers. obs.), and appear to require insect pollination, mostly by large bees, as amply documented for *E. angustifolium.* The species are well differentiated, in part by habitat specialization, but some hybrids have been reported (Mosquin 1966).

A second reproductive mode characterizes the species of sect. *Epilobium,* virtually all of which are self-compatible and either occasionally or modally autogamous. Outcrossing is rare in most species, and primarily involves small, non-specific insects. A few species of sect. *Epilobium* have relatively large flowers and four-lobed and/or exserted stigmas, and although primarily outcrossing, they are not protandrous and in all other aspects are typical species of sect. *Epilobium.* In terms of reproductive biology, they represent an intergrading series between the outcrossing and selfing extremes, and as such these taxa are commonly involved in hybridization events.

HABITATS

In China, most of the species of *Epilobium* grow in habitats like those in which species of the genus occur elsewhere in the world. A majority of taxa (24 of 41) occur in moist, open or semi-shaded, often disturbed places like streambanks, roadside ditches, margins of lakes, marshes or bogs, and along edges of forests or cultivated areas where there is sufficient moisture. Some of these taxa occur primarily at lower elevations (below 2000 m) in southwestern and southern China in warm temperate, broadleaf or mixed broadleaf-conifer communities (*E. amurense* subsp. *cephalostigma, E. kermodei, E. pannosum, E. platystigmatosum, E. pyrricholophum,* and *E. sinense*). Several other species grow in similarly low but cooler, coniferous regions in northeastern China (*E. ciliatum, E. fastigiatoramosum,* and *E. minutiflorum;* the last also extends far west into the Central Asian region). A group of species occurs in this type of moist, open habitats in coniferous (mostly *Abies-Picea*) communities primarily at middle elevations (mostly 2000–3000 m) in the Hengduanshan-Himalayan region of southwest China (*E. amurense* subsp. *amurense, E. brevifolium* subsp. *brevifolium* and subsp. *trichoneurum, E. cylindricum, E. fangii, E. royleanum, E. subcoriaceum, E. tibetanum,* and *E. wallichianum*), although many of them occur at higher and lower elevations, depending on conditions. In the drier and colder mountain-and-desert region of Xinjiang in northwestern China, *E. roseum* subsp. *roseum* and subsp. *subsessile* and *E. tianschanicum* grow along open streambanks and ditches in the coniferous forest zone. Three species are widespread altitudinally throughout most of China in open, disturbed areas but excluding the desert and high mountains (*E. hirsutum, E. palustre,* and *E. parviflorum*).

The remaining Chinese species of *Epilobium* grow in more specialized or restricted habitats. For example, *E. williamsii* and the Taiwanese endemics *E. nankotaizanense* and *E. taiwanianum* occur primarily on loose scree slopes or rocky areas in alpine regions, mostly well above 3000 m. Another group occurs at similarly high alpine areas, but along streams, seeps, and grassy swales (*E. anagallidifolium, E. clarkeanum, E. gouldii, E. hohuanense, E. kingdonii, E. laxum, E. pengii,* and *E. sikkimense*). *Epilobium blinii* grows in sphagnum bogs and other very wet places at lower elevations; because these habitats have largely been destroyed through drainage for agriculture, this species is rare and may be endangered.

The species of sect. *Chamaenerion* grow in rather unusual habitats relative to the rest of the genus. The three primarily Himalayan taxa (*E. conspersum, E. latifolium,* and *E. speciosum*) occur on gravel beds, outwashes of glacial till, or rocky open ridges, mostly at high elevations (3000–5000 m), or (*E. latifolium*) lower in the Tian Shan-Altay region. *Epilobium angustifolium* occurs in a very wide range of habitats both altitudinally and ecologically, and tends to grow in drier areas than most species. This species is best known for its ability to colonize rapidly large areas following forest fires and other types of disturbances, and these areas often are relatively dry. It also occurs in disturbed places like roadsides, but more often in meadows or even moderately shaded forests. In shaded areas it may fail to produce flowering stems, but can maintain itself indefinitely by vegetative stems (van Andel 1975; Flint 1980); this may also explain its sudden appearance after fires (Holm 1922).

As noted, the Chinese species of *Epilobium* occur over a wide altitudinal range, mostly 500–2500 m in northern and eastern China, and about 1000–4500+ m in southwestern China. Some pairs or groups of taxa are well differentiated by altitude. The best-documented case is in *E. angustifolium,* in which subsp. *angustifolium* occupies higher elevations than does subsp. *circumvagum* in areas where they overlap (Chen et al. 1988). For example, the two taxa occur geographically sympatrically in the Hengduan Mountains, but subsp. *angustifolium* is found at 2900–4700 m, whereas subsp. *circumvagum* occurs almost exclusively from 1000 to 3000 m in that region. On the other hand, some taxa (*E. amurense* subsp. *amurense, E. palustre,* and *E. royleanum*) have enormous altitudinal ranges, the most extreme is that of *E. palustre:* 200–4500 (–4950) m.

Flowering and fruiting periods in Chinese species of *Epilobium* also vary somewhat, with the typical flowering season being July to early September, and fruiting season August to late September or October. Taxa growing at lower elevations may start blooming in May (e.g., *E. amurense* subsp. *amurense, E. fangii,* and *E. wallichianum*) or even earlier (e.g., *E. blinii* and *E. kermodei*), but may still fruit into September or later. Some alpine species may not bloom until August, and have a short season (e.g., *E. kingdonii*). Not surprisingly, species with wide geographical and altitudinal ranges also have longer flowering and fruiting periods (e.g., *E. amurense* subsp. *amurense, E. angustifolium,* and *E. kermodei*).

DISTRIBUTION

Geography and Climate

China occupies the southeastern portion of the Eurasian continent, between 18°N (southern Hainan Island) and almost 54°N and between 73°30′E and almost

135°E. It is extremely diverse in climate, topography, and vegetation, home to one of the richest floras in the world with approximately 30,000 species (Wu & Hou 1980). Much of the climatic and floristic diversity is due to the extremes in topography; nearly 50% of the land surface is mountains and plateaus and only about 10% forms river basins and the coastal plain (the main agricultural regions). The remainder consists of low to mid-elevation hill country (Lee & Chang 1953). The topography of China is often described as consisting of a three-step west-east "staircase" (Hou 1983). The first stair is the famous 'roof of the world', comprising the Qinghai-Xizang (Tibet) Plateau, which is mostly above 4000 m and bounded by the Kunlun and Qilian Mountains on the north and the Himalaya and Hengduan Mountains on the south and east. The land slopes from there to the second stair, mostly 1000–2000 m, which consists of the northern plateaus (including the Mongolia Plateau, Loess Plateau, Junggar Basin, and Tarim Basin) and the Yunnan-Guizhou Plateau. The third stair comprises the northeastern and the eastern plains and the hill country of southeastern China, with elevations below 1000 m.

The major mountain ranges in China, particularly in terms of the distribution of *Epilobium,* are the Changbaishan (eastern Jilin), Da Hinnganling (western Heilongjiang-northeastern Nei Monggol), Yinshan (central Nei Monggol-Shanxi), Qinlingshan (southern Shaanxi-southern Gansu), Tian Shan (west-central Xinjiang), Kunlunshan (southern Xinjiang-Xizang), Himalaya (southern Xizang), and Hengduanshan (Sichuan-Yunnan-Xizang).

The climate of China is rather sharply divided into two broad zones: the northwestern and western part, an arid to semi-arid region with less than 500 mm of annual precipitation, where temperature variations are extreme; and the more humid regions of the south and southeast, where the climate is much less variable. The climate pattern is largely dictated by the southeast (Pacific) and southwest (Indian) monsoons, which bring moisture in the summer and fall. These moisture-bearing air masses are mostly blocked from the west and northwest portions of China, where the cold and dry Siberian and Central Asian high pressure systems predominate, and where the small amount of precipitation is distributed fairly evenly throughout the year. The net effect of these predominating high and low pressure systems, overlain on the diverse topography of China, is a stratification of the country into different climate regions from northwest to southeast: cold desert to cool-temperate, temperate, warm-temperate, subtropical and tropical zones (Hou 1983).

This climatic variation strongly affects the floristic diversity of China, as does its recent geological history. The vegetation of China was less devastated than other temperate floras, because continental glaciation during the Quaternary period was slight, restricted mostly to northeastern China, the Himalayan region, and the higher mountains stretching from northeast to southwest. Also, the biota were able to migrate far to the south, moving north again as the climate warmed (Hsü 1983). West-central China has served as a refugium for plants in the Northern Hemisphere since the Tertiary period; the present flora includes very isolated Tertiary relicts (e.g., *Metasequoia, Cathaya,* etc.), at least three endemic families, and about 200 endemic genera (Wu & Wang 1983). Many of these relict species were once part of a widespread north temperate (Holarctic) flora, and there are strong floristic connections between the floras of China, eastern North America, and Europe (e.g., Boufford & Spongberg 1983; Hsü 1983; Cheng 1983). Likewise, there are floristic connections between the tropical and subtropical flora of China and the Indo-Malaysian Paleotropical flora, and to some degree to the Pantropical flora.

Patterns of Distribution

Chinese species of *Epilobium* occur widely in the mountain and temperate regions of China, and are absent only from the southern tropical region including Hainan Island, and from the extreme desert regions of northern and western China. Appendix 2 lists distributions of all taxa by province. Those provinces with the greatest concentration of taxa include: Sichuan (25), Yunnan (25), Xizang (21), Gansu (17), Shaanxi (16), Guizhou (14), Hubei (13), Qinghai (12), Xinjiang (12), Hebei (10), and Henan (10). Thirty-one of the 41 taxa occur in the provinces of Sichuan, Yunnan, and southeastern Xizang; of the remainder, four occur only in Taiwan, five in the Tian Shan-Altay region, and one, *E. ciliatum,* only in northeast China. All taxa are part of the Holarctic floristic element, even though several extend into high elevations in the subtropical areas. For the sake of our discussion, we have divided the taxa into groups according to their primary floristic affinities.

1. Circumboreal. Five taxa have essentially circumboreal and/or circumarctic distributions: *E. anagallidifolium, E. angustifolium* subsp. *angustifolium* and subsp. *circumvagum, E. latifolium,* and *E. palustre.* With the exception of *E. anagallidifolium,* which barely reaches China in the Altay Mountains of Xinjiang, all of the other taxa are widespread in China. There are no compelling data regarding areas of origin for most of the taxa, although it seems likely that sect. *Chamaenerion,* including *E. angustifolium, E. latifolium,* and six other species endemic in Eurasia, originated there rather than in North America. The distribution of *E. latifolium* in China shows an unexpected and unusual disjunction from the Tian Shan-Altay region in the northwest, probably connected more or less continuously to the stations in western Xizang and the western Himalaya (Raven 1962a), to a few localities in eastern Xizang-northern Yunnan and one in east-central Qinghai. We have no data to determine if it reached southwestern China recently or if its occurrence there is part of an earlier, broader distribution across the Himalaya, with subsequent disappearance from the central part of the range.

2. Eurasian. This group includes taxa whose native distribution is in Eurasia, with ranges substantially west of China. *Epilobium hirsutum, E. parviflorum,* and *E. roseum* subsp. *roseum* all occur from Europe more or less continuously to China; indeed, *E. hirsutum* extends to Japan and the Soviet Far East, and as an adventive to North America (Stuckey 1970). *Epilobium parviflorum* also occurs as an adventive in Australasia (Raven & Raven 1976). In addition, two other taxa, *E. minutiflorum* and *E. roseum* subsp. *subsessile,* occur from Turkey and the Caucasus region through Central Asia to China, although the latter taxon extends no further than the Tian Shan-Altay region of China.

Epilobium tianschanicum is morphologically similar to *E. roseum* and *E. minutiflorum,* and may have a distribution well to the west of our documented localities in Xinjiang. However, we lack precise data regarding its distribution in Soviet Central Asia, and at present are treating it as an endemic in the Tian Shan region.

3. Himalayan. This group includes species whose ranges fall primarily in the Himalaya Mountains, although some extend beyond the Himalaya in their distribu-

tion in China, and might alternatively have been included in group 4. The Himalayan group includes 10 species, two from sect. *Chamaenerion* (*E. conspersum* and *E. speciosum*), the rest from sect. *Epilobium* (*E. brevifolium* subsp. *brevifolium, E. cylindricum, E. gouldii, E. laxum, E. royleanum, E. sikkimense, E. tibetanum,* and *E. williamsii*). The very beautiful *E. speciosum* occurs in China only in extreme eastern Xizang, with a general distribution confined to the southern slope of the western Himalaya, whereas *E. conspersum* occurs from Nepal eastward to the Qinlingshan in western Shaanxi. *Epilobium gouldii* is endemic to a very small region of Sikkim and adjacent Xizang. We include *E. laxum* with this group, despite the fact that the only collection in China is from the Tian Shan region; it is a characteristic species of the western Himalaya, and has affinities with species from that region (Raven 1962a; Hoch 1981). The only Chinese collections of *E. brevifolium* subsp. *brevifolium* are from eastern Xizang; these plants appear to be outliers from the main range of this taxon in the western Himalaya from Himachal Pradesh (India) to central Nepal. The other five species in this group have distributions that extend into central China either from the central Himalaya (*E. sikkimense, E. williamsii*) or from Afghanistan along the length of the Himalaya (*E. cylindricum, E. royleanum, E. tibetanum*).

4. Southwestern China (Hengduanshan region). This group includes 11 species, three of which (*E. clarkeanum, E. kingdonii,* and *E. wallichianum*) might also have been put into the Himalayan group, since they occur almost equally in the two areas, the former two based on very few collections. *Epilobium amurense* subsp. *amurense* has a very extensive range throughout the Himalaya to Kashmir (Raven 1962a), but also occurs across China to Jilin in the northeast and to Hokkaido, Japan, and Sakhalin (Hara 1965); for that reason it might fit with either groups 3 or 5. Several taxa that are fairly widespread in southwestern China also extend into Myanmar (Burma) (*E. kermodei*); Assam, Myanmar, and Vietnam (*E. pannosum*); or Bhutan, Myanmar, Vietnam, and northern Luzon (*E. brevifolium* subsp. *trichoneurum*). *Epilobium sinense* has a range from this region to the northeast, more like those of the East Asian group (5); its affinities, however, seem to be with taxa of the Himalayan and southwest Asian groups, so we include it here. The remaining three species (*E. blinii, E. fangii,* and *E. subcoriaceum*) are Chinese endemics in this region.

5. East Asian (Sino-Japan). These species are grouped because they share a primary distribution from Japan to central China. Many taxa of this and the previous group reach the limits of their distributions in a transition area defined by the Qinlingshan of southern Shaanxi and Gansu and the eastern extent of the Hengduanshan at about 105°E, a line noted by C. Y. Wu (1979) that roughly divides the Sino-Japan and Sino-Himalayan floras. *Epilobium amurense* subsp. *cephalostigma* occurs widely in eastern China from Heilongjiang to Guangdong and Yunnan. *Epilobium fastigiatoramosum* ranges from northeast China to Qinghai and northern Sichuan. The very characteristic, narrow-leaved *E. platystigmatosum* occurs in central China north to Hebei, and disjunctly to Taiwan. *Epilobium pyrricholophum* has an anomalous distribution at relatively low elevations in southeast China, although it is unexpectedly absent from Taiwan.

We include *E. ciliatum* with this East Asian group, because it occurs in northeastern China (Heilongjiang and Jilin), Japan, Korea, and the Soviet Far East. Its primary native range and phylogenetic affinities are in North America (Hoch 1978), and its occurrence in eastern Asia might represent a secondary range extension. This species also occurs rather widely in southern South America (Solomon 1982), Australasia (Raven & Raven 1976), and Europe (Raven 1968), introduced recently in at least the last two. The weedy nature of this species, its invasive spread in other temperate regions, and its rather restricted occurrence in East Asia suggest that it arrived there relatively recently.

6. Taiwan. In view of the broad nature of the other groups, the restriction of this group to species on Taiwan may be unexpected. However, there are four endemic species of *Epilobium* on Taiwan (out of nine in China), including two newly described here. The affinities of two of these endemics (*E. pengii* and *E. taiwanianum*) appear to be with different taxa in southwest China, but for the other two (*E. hohuanense* and especially *E. nankotaizanense*), the relationships are more obscure. In addition to these endemics, two other taxa of *Epilobium* on Taiwan are widely disjunct from the rest of their ranges (*E. amurense* subsp. *amurense* and *E. platystigmatosum*). This disjunctive pattern between the mountains of Taiwan and southwest China, and sometimes even the northern mainland, can also be found in other groups, such as *Taiwania* (Taxodiaceae; Hu 1980), *Fagus* (Fagaceae; Shen & Boufford 1988), *Rubiteucris* (Lamiaceae; Wu & Li 1977), *Hemiphragma* (Scrophulariaceae; Tsoong & Yang 1979), *Whytockia* (Gesneriaceae; Weber 1982), *Pouzolzia elegans* Weddell (Urticaceae; Institute of Botany, Academia Sinica 1972), *Elatostema obtusum* Weddell (Urticaceae; Wang 1980), and, in Onagraceae, *Circaea glabrescens* (Pampanini) Handel-Mazzetti (Boufford 1982).

Taiwan and the Chinese mainland were connected directly through much of the Tertiary, and at least periodically during the Quaternary glacial cycles; the present opening of the Taiwan Haixia Straits occurred only recently (Wu & Wang 1983). Most of these endemic species of *Epilobium* (or their predecessors) probably reached Taiwan during cooler periods of the Quaternary and were subsequently restricted to higher elevations in Taiwan (the mountains reach nearly 4000 m, the highest in eastern Asia) as the temperature warmed in the Pleistocene and Holocene. Because no comparably high areas exist on the mainland within 1200 km, the alpine flora on Taiwan is effectively isolated.

In summary, all species of *Epilobium* in China are probably native, and almost all have circumboreal, Himalayan, or East Asian affinities. There appear to be no close phylogenetic relationships between the Chinese species and those of Australasia or North America, except for *E. ciliatum,* which has the major part of its native range and its closest affinities in North America (Hoch 1978).

The group of species of *Epilobium* in the Sino-Himalayan region, including 31 taxa treated here and 10 additional species from the western Himalaya (Raven 1962a), is the richest concentration of the genus in Eurasia. This finding is consistent with the richness of the entire flora of that region, which was estimated at over 20,000 species by C. Y. Wu (1979).

CRITERIA FOR ASSIGNMENT OF TAXONOMIC RANK

This treatment follows the criteria set forth most clearly by Raven and Raven (1976) in their monograph of the Australasian species of *Epilobium.* These guidelines, first organized by Lewis and Lewis (1955) in their revision of *Clarkia* Pursh, have also been applied consistently to the rest of Onagraceae; this consistency provides the basis for direct comparison of the patterns of variation and relationship between different genera and within genera between different areas.

As noted above, many species of *Epilobium* primarily self-pollinate, which results in series of local populations that may be more or less morphologically distinct. In attempting to delimit units, we benefitted enormously by the availability of many new collections and by opportunities for fieldwork in various parts of China. The reticulate nature of the variation in *Epilobium* makes a conservative taxonomy most appropriate (Raven & Raven 1976), therefore we have defined species by major morphological discontinuities. Where units are discrete in some areas of their ranges and intergrading in others, they have been designated as subspecies (as in *E. amurense, E. angustifolium, E. brevifolium,* and *E. roseum*). We view subspecies as important biological units, composed of series of populations sharing a suite of character states and occupying a discrete range. It does not seem useful to subdivide these units further in a formal sense.

TAXONOMY

In view of the limited geographical scope of this paper, we include sectional descriptions of only those sections that occur in China; descriptions of other sections are available in Raven (1976).

As established by Raven and Raven (1976), and in compliance with the International Code of Botanical Nomenclature (1988 edition), we have not recognized the "formae" listed by Haussknecht (1884) as validly published unless they have been validated formally by subsequent authors. Haussknecht used these formae as a device to organize his discussion of patterns of diversity, purposely providing them with feminine endings to indicate his intent (Hoch & Raven 1990).

The synonymies include names and information from treatments of the genus outside of China. However, the synonymies presented for some very widespread species, especially those occurring in Europe and North America, are incomplete, including only names relevant to the taxon in eastern Asia. The lists of specimens examined include every available collection from China for all but the most common species (*E. angustifolium* and *E. hirsutum*); for these few widespread taxa, we cite most collections, but include only a selection from well-collected areas. Although non-Chinese collections were always consulted in the preparation of descriptions, several rare taxa (especially *E. clarkeanum* and *E. gouldii*) could not be described adequately without using specimens from outside of China. Because the nomenclatural histories of many of the species are complex, and have never before been treated in a comprehensive fashion, we have prepared a summary of these histories as Appendix 3, which charts the treatment of the Chinese species of *Epilobium* in most of the floras that include China and/or those species.

A Chinese "common name" is provided for almost all taxa. Unlike common names in English, these are very standardized in Chinese usage, and may be useful in reference to the Chinese systematic literature.

Epilobium Linnaeus, Sp. pl. 347. 1753.—LECTOTYPE, designated by Britton and Brown, 1913: *Epilobium hirsutum* Linnaeus. (For generic synonymy, see Raven 1976.)

Perennial or rarely annual (3 species, not in China) herbs, sometimes suffrutescent, sometimes woody near the base, with erect, ascending or decumbent stems; with fibrous roots (taproot in annuals), perennials wintering and multiplying vegetatively by leafy rosettes, stolons or soboles (shoots), turions (subterranean globose or ovoid winter buds with fleshy scales), or buds (gemmae) in the leaf axils, rarely with spreading lateral roots. Stems glabrous to densely pubescent, often with raised lines of hairs decurrent from margins of the petiole. Leaves opposite below the inflorescence and alternate above in most species, or completely alternate (in sect. *Chamaenerion*), subentire to denticulate or serrulate; stipules absent; bracteoles absent. Flowers borne singly in axils, usually clustered to form spikes, racemes, panicles, or corymbs, actinomorphic to somewhat zygomorphic in a few species, perfect, 4-merous, the stamens and pistil maturing simultaneously, or sometimes protandrous or very rarely protogynous; floral tube evident or (in sect. *Chamaenerion*) absent, when present deciduous soon after anthesis; petals purplish or pink to white, rarely (in 2 species) orange-red or (in 2 species) cream to yellow, emarginate or (in sect. *Chamaenerion*) entire, obovate or obcordate; stamens 8, subequal in one whorl or unequal in two whorls, the episepalous ones longer; pollen yellow or in some species bluish grey, shed in tetrads in most species, singly in a few, connected by viscin threads; stigma clavate or capitate, receptive on all surfaces, or deeply 4-lobed, the commissural lobes then opening to expose receptive inner surfaces. Fruit an elongate, slender, 4-loculate, loculicidal capsule, pluriovulate or in 2 species with a single seed in each locule, the capsule then fusiform. Seeds many (4 in 2 species), in all but 2 species with a terminal tuft of silky hairs, the coma. Chromosome numbers: n = 12, 13, 15, 16, 18, 30, 36, 54.

The genus *Epilobium* comprises some 165 species assigned to six sections (Raven 1976). Only the two most widespread sections, sect. *Chamaenerion* and *sect. Epilobium,* occur in China; they are differentiated in the first couplet of the following key. Raven (1976) provides a key to all sections.

KEY TO THE SPECIES AND SUBSPECIES OF EPILOBIUM IN CHINA

1. Leaves spirally arranged or alternate (only rarely opposite at stem base); floral tube absent; petals entire; stamens in a single whorl; stigma deeply 4-lobed; flowers slightly zygomorphic. I. *Epilobium* sect. *Chamaenerion*.
   2. Bracts much smaller than cauline leaves, coriaceous, sublinear; leaves linear to lanceolate, with distinct submarginal vein; seeds with inconspicuous chalazal collar (<0.05 mm long).
      3. Leaves glabrous on abaxial midrib, 7–14 (–18.5) cm long, 0.3–1.3 (–2.5) cm wide, base obtuse or subrounded, margins subentire, subsessile; stems subglabrous; pollen usually 3-porate and less than 75 μm in diameter; petals 9–15 (–19) mm long, 3–8 (–11) mm wide. 4a. *E. angustifolium* subsp. *angustifolium*.
      3. Leaves usually pubescent on abaxial midrib, (6–) 9–23 cm long, (0.7–) 1.5–3.4 cm wide, base cuneate, margins denticulate, with petioles 2–7 mm long; stems strigillose, at least above; pollen mixed 3- and 4-porate (rarely 5-porate) and mostly more than 75 μm in diameter; petals 12–23 mm long, 7–13 mm wide. 4b. *E. angustifolium* subsp. *circumvagum*.

2. Bracts about half as long as cauline leaves, foliaceous, lanceolate to elliptic; leaves narrowly ovate, elliptic to lance-elliptic, lacking submarginal vein; seeds with distinct chalazal collar ($\geq$0.08 mm long).
   4. Style pubescent on lower half; stems densely strigillose; seeds 1–1.3 mm long; primary leaf veins distinct.
      5. Secondary leaf veins conspicuously reticulate or anastomosing; plants 30–120 cm tall; sepals 11–15 mm long; petals 8–14 mm long; fruiting pedicels 1.5–5 cm long. 3. *E. conspersum.*
      5. Secondary leaf veins obscure, not conspicuously anastomosing; plants 20–45 cm tall; sepals 15–20 mm long; petals 17–25 mm long; fruiting pedicels 1–3 cm long. 2. *E. speciosum.*
   4. Style glabrous; stems subglabrous to sparsely strigillose, only rarely densely strigillose; seeds 1.2–2.1 mm long; primary leaf veins obscure. 1. *E. latifolium.*

1. Leaves opposite at least below the inflorescence; floral tube present, 0.6–2.8 mm long; petals cleft or emarginate; stamens in 2 unequal whorls; stigma entire or shallowly emarginate to 4-lobed; flowers actinomorphic. II. *Epilobium* sect. *Epilobium.*
   6. Stigma deeply 4-lobed (rarely shallowly so in *E. blinii*).
      7. Stems (18–) 30–120 (–250) cm tall, covered with long spreading villous and short erect glandular pubescence; leaves 3–12 cm long, subacute or acuminate, with (15–) 20–60 teeth on each side; seeds 0.8–1.2 mm long, coarsely papillose.
         8. Leaves distinctly clasping; petals (7–) 10–20 mm long; stigma exserted beyond anthers at anthesis. 5. *E. hirsutum.*
         8. Leaves subsessile but not clasping; petals 5–8 mm long; stigma surrounded by anthers of longer stamens at anthesis. 6. *E. parviflorum.*
      7. Stems 3–25 (–45) cm tall, strigillose; leaves 0.8–3 cm long, obtuse, with 3–7 indistinct teeth on each side; seeds 1.2–1.8 mm long, finely papillose or subreticulate.
         9. Loosely matted herb, stems 3–18 cm long, branched; leaf petioles 1–3 mm long; petals 16–31 mm long; fruiting pedicels 0.4–0.7 cm long; seeds reticulate; Taiwan. 8. *E. nankotaizanense.*
         9. Ascending or erect herb, stems 10–25 (–45) cm long, usually simple; leaves subsessile; petals 10–15 mm long; fruiting pedicels 1.5–3.5 cm long; seeds finely papillose; southwestern China. 7. *E. blinii.*
   6. Stigma entire or shallowly emarginate.
      10. Stems pubescent all over, either lacking raised pubescent lines decurrent from margins of the petioles, or with inconspicuous, sparsely pubescent lines.
         11. Plants forming short basal soboles (shoots), often clumped; stems occasionally with persistent basal scales.
            12. Inflorescence variously pubescent but without glandular hairs.
               13. Ovaries white-canescent; seeds with conspicuous chalazal collar; leaves 2–7 cm long, oblong-lanceolate to narrowly ovate. 33. *E. minutiflorum.*
               13. Ovaries strigillose, not white-canescent; seeds with inconspicuous chalazal collar; leaves 1–4.5 cm long, sublinear to elliptic or lanceolate.
                  14. Plants erect, loosely or not clumped; stems mostly well branched throughout; leaves 1–4.5 cm long, linear to narrowly lanceolate. 14. *E. platystigmatosum.*
                  14. Plants ascending, forming distinct clumps or mats; stems simple or branched only from caudex; leaves 1–2 cm long, elliptic to lanceolate. 15. *E. hohuanense.*
            12. Inflorescence variously pubescent but always with glandular hairs.
               15. Mid-cauline and upper leaves with distinct petioles (2–7 mm long) and narrowly cuneate bases. 17. *E. royleanum.*
               15. Mid-cauline and upper leaves subsessile (petioles 0–2 mm long), with subcordate, rounded, or broadly cuneate bases.
                  16. Leaves subentire, lance-elliptic; seed coma tawny. 32. *E. fastigiatoramosum.*
                  16. Leaves denticulate, ovate to lanceolate or oblong-lanceolate; seed coma white or dingy.

17. Petals 3–4.3 (–5) mm long; ovaries white-canescent; seeds with conspicuous chalazal collar. 33. *E. minutiflorum.*
17. Petals 7–11 mm long; ovaries pubescent but not white-canescent; seeds with inconspicuous chalazal collar.
18. Leaves ovate to broadly ovate with subcordate base, usually herbaceous; stigma broadly clavate to clavate. 20a. *E. brevifolium* subsp. *brevifolium.*
18. Leaves lanceolate to narrowly ovate or elliptic with rounded or cuneate base, subcoriaceous; stigma clavate. 20b. *E. brevifolium* subsp. *trichoneurum.*
11. Plants with basal rosettes, turions or stolons, not soboles, rarely clumped; stems often with persistent basal scales.
19. Plants covered all over with dense appressed tomentose pubescence; plants with leafy rosettes; petals 8–16 mm long. 19. *E. pannosum.*
19. Plants strigillose, villous, and/or glandular-pubescent, but not tomentose; plants forming stolons or turions, not rosettes; petals 4–8 mm long (except up to 15 mm long in *E. kermodei*).
20. Plants forming fleshy turions, leaving dense coriaceous basal scales; fruiting pedicels 0.5–1 cm long; plants 7–25 cm tall; Taiwan. 16. *E. taiwanianum.*
20. Plants forming stolons; fruiting pedicels 0.7–5 cm long; plants 25–120 cm tall (except in *E. palustre* sometimes less than 25 cm); widespread, but not in Taiwan.
21. Seeds 0.8–1.2 mm long; capsules 7–11 cm long; plants forming thick ropy stolons with fleshy terminal buds. 18. *E. kermodei.*
21. Seeds 1.3–2.2 mm long; capsules 3–9 cm long; plants forming filiform stolons, sometimes with terminal buds.
22. Leaves sublinear to narrowly lanceolate, subentire to obscurely denticulate; stolons terminating in fleshy turions; fruiting pedicels 1–5 cm long; seed comas dingy white. 31. *E. palustre.*
22. Leaves ovate or broadly oblong to lanceolate, sharply serrulate; stolons not terminating in buds; fruiting pedicels 0.7–1.5 cm long; seed comas reddish. 21. *E. pyrricholophum.*
10. Stems subglabrous below inflorescence except for 2 or 4 raised strigillose lines decurrent from margins of the petioles.
23. Inflorescence glabrous or rarely with scattered hairs on ovaries and sepals.
24. Plants low, mat-forming; stems 3–20 cm tall; leaves 0.8–2.5 cm long; petals 2.5–6.5 mm long; capsules 1.7–3.6 cm long. 37. *E. anagallidifolium.*
24. Plants erect, clumped; stems (5–) 10–25 (–60) cm tall; leaves (0.8–) 1.5–7.5 cm long; petals 7–14 mm long; capsules 5–9 cm long. 26. *E. sikkimense.*
23. Inflorescence generally pubescent.
25. Inflorescence variously pubescent, but entirely lacking short, erect glandular hairs.
26. Leaves subsessile (lower ones with petioles 0–2 mm long).
27. Seeds papillose; stems 10–50 cm tall, erect; leaves 3.4–6 cm long. 9b. *E. roseum* subsp. *subsessile.*
27. Seeds reticulate; stems 7–25 cm tall, ascending; leaves 1–2.5 cm long.
28. Leaves 1.1–1.6 cm long; seeds 0.8–0.9 mm long; capsules 3.5–4 cm long, on pedicels 0.6–0.7 cm long; southwestern China. 35. *E. clarkeanum.*
28. Leaves 1.5–2.5 cm long; seeds 1–1.1 mm long; capsules 4.5–5 cm long, on pedicels 1.5–2 cm long; Taiwan. 36. *E. pengii.*
26. Leaves with distinct petioles 2–11 mm long.
29. Leaves narrowly spatulate to rarely broader, with prominent pale midrib beneath, often crowded and spirally arranged on mid-stem; petioles 2–11 mm; fruiting pedicels 1.3–4 cm long. 13. *E. sinense.*

29. Leaves sublinear to narrowly ovate, without prominent pale midrib, always opposite below inflorescence; petioles 2–7 (–10) mm; fruiting pedicels 0.5–3 cm long.
30. Leaves narrowly lanceolate to sublinear, on petioles 3–7 (–10) mm long; seeds 0.8–1 mm long, papillose. 11. *E. cylindricum.*
30. Leaves lanceolate to narrowly ovate, on petioles 2–5 mm long; seeds 1–1.3 mm long, reticulate or papillose.
31. Seeds reticulate; petals 5–8 mm long; plants well branched, not much clumped; stigma capitate to broadly clavate; Himalaya and southwestern China. 12. *E. tibetanum.*
31. Seeds coarsely papillose; petals 5.5–6.5 mm long; plants mostly simple, forming clumps; stigma clavate or rarely subcapitate; Tian Shan region. 10. *E. tianschanicum.*
25. Inflorescence variously pubescent, always with some short, erect glandular hairs.
32. Plants forming fleshy basal turions, which leave coriaceous thick brown scales.
33. Leaves elliptic or oblong to elliptic-lanceolate, with mostly cuneate bases and distinct petioles (1–) 2–15 mm long.
34. Petioles 3–10 (–15) mm long; seeds 1–1.2 mm long, 0.45–0.55 mm in diameter, obovoid; stigma broadly clavate to subcapitate; Xinjiang and western Asia. 9a. *E. roseum* subsp. *roseum.*
34. Petioles 1–4 (–6) mm long; seeds 1.1–1.4 mm long, 0.3–0.45 mm in diameter, narrowly obovoid; stigma capitate; southwestern China. 30. *E. fangii.*
33. Leaves ovate to lanceolate, with mostly rounded bases and obscure petioles 0–3 mm long (except in *E. laxum,* to 8 mm long in some basal leaves).
35. Petals 4.5–5.5 mm long; seeds 0.9–1 mm long; stigma clavate to subcapitate. 29. *E. gouldii.*
35. Petals 5.5–16 mm long; seeds 1.1–1.5 mm long; stigma capitate to subcapitate.
36. Fruiting pedicels 1–7 (–10) mm; petals (7–) 10–16 mm; leaves 2–7 cm long, 1.2–2.6 cm wide, ovate to narrowly ovate, crowded, usually longer than internodes; western Himalaya and Tian Shan. 25. *E. laxum.*
36. Fruiting pedicels 4–12 mm; petals 5.5–11 mm; leaves 1.5–5.5 mm long, 0.5–1.5 cm wide, narrowly ovate to lanceolate, not crowded, usually shorter than internodes; endemic in southwest China. 28. *E. subcoriaceum.*
32. Plants forming soboles, stolons, or rosettes, but not fleshy turions, with rather loose, more or less herbaceous scales, or scales absent.
37. Seed surface with conspicuous longitudinal ridges of flattened fused papillae; plants forming leafy rosettes; northeastern China. 34. *E. ciliatum.*
37. Seed surface papillose or reticulate, without longitudinal ridges; plants forming soboles or stolons.
38. Stems ascending, forming clumps or mats; seeds reticulate; capsules 1.7–3.6 cm long, on pedicels 1–3.5 (–5) cm long; leaves subentire to very remotely denticulate. 37. *E. anagallidifolium.*
38. Stems mostly erect, loosely or not clumped; seeds papillose; capsules 1.5–11 cm, mostly longer than 3.5 cm, on pedicels 0.3–2.5 cm long; leaves denticulate or serrulate.
39. Plants forming thick ropy stolons 1–12 cm long with fleshy terminal bud; robust stems 40–120 (–200) cm tall; capsules 7–11 cm long. 18. *E. kermodei.*
39. Plants forming soboles, not stolons; stems 4–80 (–150) cm tall, mostly less than 50 cm; capsules 1.5–7.5 cm long (rarely to 9 cm in *E. sikkimense*).

40. Stem bases with crowded brown scales; stems mostly less than 25 cm tall.
    41. Petals 7–14 mm long; capsules 5–9 cm long, on pedicels 0.6–2 (–2.5) cm long; leaves (0.8–) 1.5–7.5 cm long; stems mostly erect. 26. *E. sikkimense.*
    41. Petals 5–6.5 mm long; capsules 3.5–5 (–6) cm long, on pedicels 0.4–1 cm long; leaves 0.7–2.2 cm long; stems mostly ascending. 27. *E. williamsii.*
40. Stem bases without scales; stems mostly more than 25 cm tall (except with stems 8–25 cm tall in *E. kingdonii*).
    42. Seeds 1.4–1.6 mm long, very finely papillose; plants 8–25 cm tall, ascending; floral tube with scattered pubescence. 23. *E. kingdonii.*
    42. Seeds 0.8–1 mm long, coarsely papillose; plants (15–) 20–100 cm tall, erect; floral tube with tufts of hair on the costae and at insertion of the sepals, or scattered throughout.
        43. Fruiting pedicels 1–2.5 cm long; petals 5.5–13 mm long; stems 15–80 cm tall, 4-angled, with 4 (or 2) strigillose lines on the internodes; leaves oblong, oblong-ovate, or elliptic, apex obtuse or rarely acute. 24. *E. wallichianum.*
        43. Fruiting pedicels 0.3–1.3 cm long; petals 4.5–8 mm long; stems (10–) 20–150 cm tall, subterete or somewhat 2-edged, with 2 strigillose lines on the internodes; leaves ovate to oblong-lanceolate, apex acute or acuminate.
            44. Stems with prominent strigillose lines decurrent from margins of petioles; inflorescence with scattered glandular hairs; floral tube with noticeable tufts of hair at insertion of sepals; leaves mostly ovate. 22a. *E. amurense* subsp. *amurense.*
            44. Stems with weak, barely elevated strigillose lines, often incomplete through internodes; inflorescence pubescent but eglandular; floral tube evenly pubescent, without tufts of hair at insertion of sepals; leaves oblong-lanceolate to narrowly ovate. 22b. *E. amurense* subsp. *cephalostigma.*

I. Epilobium section **Chamaenerion** Tausch, Hort. Canal. fasc. 1. 1823. *Chamaenerion* Seguier, Pl. Veron. 3: 168. 1754; nom. illeg. *Chamaenerion* S. F. Gray, Natur. Arrang. Brit. Pl. 559. 1821; nom. illeg. *Chamaenerion* Spach, Hist. Nat. Veg. 4: 346. 1835; nom. illeg. *Chamaenerion* Kostel, India Pl. Hort. Bot. Prag. 34. 1844; nom. illeg. *Chamaenerion* (Tausch) Schur, Sertum Fl. Transsilv. 25. 1853; nom. illeg.—Lectotype, designated by Raven, 1976: *Epilobium angustifolium* Linnaeus.

Chamerion (Rafinesque) Rafinesque, Herb. Raf. 51. 1833. *Epilobium* subg. *Chamerion* Rafinesque, Amer. Monthly Mag. & Crit. Rev. 2: 266. 1818.—Lectotype, designated by Holub, 1972: *Epilobium amenum* Rafinesque [=*Epilobium angustifolium* Linnaeus].

Erect perennial herbs, often clumped, sometimes basally woody. Stems terete, simple or rarely sparsely branched, pubescent to subglabrous, if pubescent the hairs always eglandular. Leaves spirally arranged or very rarely subopposite or subverticillate, scalelike and minute below ground, the basal ones 0.5–1 cm long, coriaceous to submembranous, triangular-ovate to lanceolate; the cauline blades 3–23 cm long, linear to lanceolate, elliptic, or rarely ovate, often subcoriaceous. Inflorescence suberect, simple or rarely with a few branches. Flowers slightly zygomorphic, the lower petals somewhat narrower than upper ones and the style deflexed at onset of anthesis, opening on axis nearly perpendicular to stem axis, strongly protandrous; floral tube absent, instead with disc at base of style and stamens, the disc green, fleshy and translucent, producing nectar mostly coincident with stigma maturation; petals entire; stamens subequal, in a single whorl, erect at onset of anthesis, recurved later; filament bases forming a chamber around disc; pollen bluish to yellow, shed singly; style deflexed at onset of anthesis, gradually becoming erect as stamens recurve; stigma deeply 4-lobed, exserted beyond anthers, the lobes opening and spreading 2–3 days after onset of anthesis, becoming revolute as style becomes erect, commissural, receptive only on inner surfaces. Chromosome numbers: n = 18, 36, 54.

Raven (1976) divided this section into two subsections, one of which, subsect. *Rosmarinifolium* (T. Tacik) Raven, occurs in western Asia and Europe; no taxa of that subsection reach China. The other subsection, subsect. *Leiostylae* (Steinberg) Raven, includes all and only the four species treated here in sect. *Chamaenerion.* [Raven (1976) included only three species in this group, because he treated *E. speciosum* as a subspecies of *E. latifolium* (Raven 1962a).]

1. Epilobium latifolium Linnaeus, Sp. Pl. 1: 347. 1753. *Chamaenerion latifolium* (Linnaeus) Fries & Lange, Fl. Dan. xvii, 49: 7. 1877. *Chamerion latifolium* (Linnaeus) Holub, Folia Geobot. Phytotax. 7: 86. 1972.—TYPE: U.S.S.R. "Siberia, ♃," [possibly Kamchatka], 1733–43, *Gmelin* (holotype: LINN 486.2!).

Epilobium changaicum Grubov, Bot. Mater. Gerb. Bot. Inst. Komarova Akad. Nauk SSSR 17: 20. 1955.—TYPE: MONGOLIA. Changai, (Cecerleg, 47°25′N, 100°17′E), at springs of the Urdu-Tamir River, 20 km above Tzetzerlik, in swales along the riverbanks, 7 Jul 1949, *Junatov 11905* (holotype: LE!).

Epilobium kesamitsui Yamazaki, J. Jap. Bot. 65: 141. 1990.—TYPE: JAPAN. Pref. Yamanashi, Mt. Kitadake, in herbages on stony places, 2800 m, Aug 1986, *Fukazawa 5381* (holotype: TI; isotypes: KYO, TI!).

Common name: Kuanye Liulan (Li 1986).

Clumped erect perennial herb, with thick rhizome up to 10 mm in diameter, somewhat woody, with wiry mass of roots, forming rather fleshy shoots from root crown. Stems several, 12–35 cm tall, 1.5–5 mm in diameter, simple or sparingly branched, glabrous below, subglabrous to sparsely strigillose all over upper stem and inflorescence, or rarely densely strigillose. Leaves spirally arranged throughout, or rarely subopposite near base, with subcoriaceous scales below ground, the basal leaves less than 1 cm long, brown, submembranous, triangular-ovate and sessile, the others subsessile or with a petiole 1–2 mm long; cauline blades 2–5 (–8) cm long, 0.6–1.7 (–2.6) cm wide, subcoriaceous, elliptic or ovate to lance-elliptic, green or pale green on both surfaces, subglabrous or strigillose, especially on veins, apex obtuse or acuminate, base cuneate or sometimes subobtuse, margin subentire to remotely punctate-denticulate with 4–7 teeth, the lateral veins obscure, 3–4 on each side of the midrib. Inflorescence an erect raceme, sparsely to moderately strigillose; bracts usually about half as long as cauline leaves, foliaceous. Flowers erect in bud, then nodding individually at early anthesis, opening laterally, eventually about 2–6 cm in diameter; buds 10–17 mm long, 5–7 mm in diameter, oblong-obovoid, acute; ovaries 1–2 cm long, purplish, densely canescent; pedicels 0.7–2 cm long; floral tube absent, instead with a disc 0.5–1 mm high, 3–4 mm in diameter, glabrous; sepals 10–16 mm long, 1.5–3.5 mm wide, oblong-lanceolate, acute, purplish, subglabrous or sparsely strigillose; petals 10–24 (–32) mm long, 7–15 (–23) mm wide, rose-purple or pink, usually slightly unequal, the lower two narrower than the upper two, obovate or oblong-obovate, entire and rounded or rarely retuse; filaments 6–11 mm long, subequal; anthers 1.2–4 mm long, 0.7–1.5 mm wide, oblong or elliptic-oblong; style 3.5–8 mm long, sharply deflexed at onset of anthesis, gradually becoming erect, glabrous, pinkish to rose; stigma deeply 4-lobed, the lobes 2–3.5 mm long and recurved at maturity, the surface papillose, white. Capsules 2.5–8 cm long, strigillose, borne on pedicels 1.2–2.5 cm long. Seeds 1.2–2.1 mm long, 0.4–0.6 mm in diameter, fusiform, acuminate, with a distinct chalazal collar 0.1–0.12 mm long, the surface appearing nearly smooth but inconspicuously and irregularly reticulate; coma 9–15 mm long, tawny or dingy, not readily deciduous. Chromosome numbers: n = 18, 36. Fig. 9.

Phenology. Collected in flower from June to August, in fruit from August to October.

Distribution (Fig. 10). In Asia from Kamchatka, eastern Siberia and Japan (rare in central Honshu) to western China, the Tian Shan and Altay regions, and the western Himalaya in Kashmir, Pakistan, and Afghanistan, across arctic Siberia to Novaya Zemlya, Norway, and Iceland; in North America from Alaska, arctic Canada, and Greenland south to California, Colorado, South Dakota, and Québec; in China found in northern Qinghai, the Tian Shan, Altay, and Pamir regions of Xinjiang, northwestern Yunnan, and eastern Xizang (Tibet); river gravel beds and grassy slopes in the northern part of the range at 1600–2500 m, and in Yunnan and Xizang on alluvial glacial soil in canyon bottoms, sandy or grassy places by rivers in alpine areas, or extending upward to the snow line on screes slopes, at (3480–) 4000–5000 m.

ADDITIONAL SPECIMENS EXAMINED. **China.** QINGHAI: Haixi, *Z. T. Yang 48* (HNWP).—XINJIANG: Urumqi, *Bayi Coll. Agr. 12360* (KUN, PE, YNU); Tian Shan, *K. C. Kuan 1767* (PE), *2095* (PE, XJBI), *Ludlow 663* (RSA), *Merzbacher 1029* (B, M); Shawan, *T. Y. Chou et al. 651105* (NAS, PE), *G. Z. Fu & Y. S. Zhang 4025* (XJBI), *K. C. Kuan 1288* (PE), *Xinjiang Exped. 3624* (XJBI), *3625* (XJBI), *4995* (XJBI); Altay, Qinghe, *R. C. Ching 819* (PE), *1023* (PE), *1099* (PE); 60 km SW Wenquan, *C. J. Chen & Hoch 86275* (MO, PE); Taxkorgan Tajik, Pamir, *S. Z. Hu T255* (PE), *Xinjiang Exped. 1171* (PE).—YUNNAN: Gongshan, Beima Shan, *Forrest 14443* (E).—XIZANG: Baxoi, *Qinghai-Xizang Exped. 73-1141* (PE); Nyingchi, *Qinghai-Xizang Exped. 751061* (PE); Lhari, *Ludlow et al. 15632* (BM, E); Gongbog-yamda, *Ludlow et al. 14014* (BM, E, UPS).

This circumpolar and circumboreal species occurs in the Tian Shan, Altay, and Pamir Mountains of Xinjiang in northwestern China, at a single locality in Qinghai, and in eastern Xizang and adjacent Yunnan. Populations from both regions are

FIG. 9. *Epilobium latifolium*. A. Habit: upper stem and inflorescence showing flowers erect in bud, nodding at initiation of anthesis. A[1]. Mid-stem. A[2]. Basal stem and caudex. B. Flower: style glabrous and initially sharply deflexed. C. Leaf (bract) from A. C[1]. Mid-cauline leaf from A[1]. C[2]. Leaf from lower stem. D. Axils from inflorescence. a. Capsule. b. Seed. (A–D based on *Xinjiang Exped. 4995*; a–b based on *Xinjiang Exped. 3625*.)

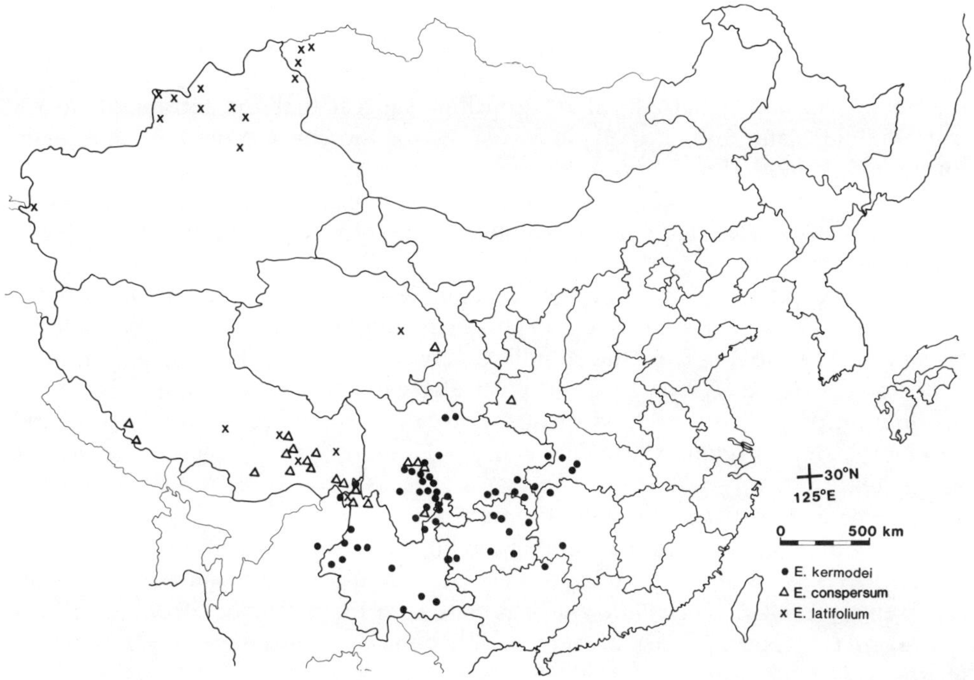

FIG. 10. Distribution in China of *Epilobium conspersum*, *E. kermodei*, and *E. latifolium*.

morphologically variable, but those from the Tian Shan-Altay region tend to have subglabrous stems and leaves, with subentire or remotely denticulate leaf margins and inconspicuous lateral veins, and a majority (72%) of the pollen grains 3-porate (the others 4–5-porate), with mean diameter of 65 μm. In contrast, the plants from Xizang and Yunnan tend to have strigillose stems and leaves, with more prominently toothed leaf margins and rather conspicuous lateral veins, and only 16.2% of the pollen grains 3-porate (83.8% 4–5-porate), with mean diameter of 80.5 μm. Small (1968) found a comparable pattern in *E. latifolium* in North America, although the variation in most characters varied more broadly; only pollen pore number showed a clear disjunction. Small also was able to determine chromosome counts on many populations and reported that diploids (n = 18), found in western North America, had smaller, 3-pored pollen, whereas tetraploids (n = 36), found in eastern North America, Greenland, and Iceland, had larger, 4–5-pored pollen. No chromosome counts for *E. latifolium* are available from China. If there is a comparable relationship between ploidy level and pore number in China, we hypothesize that populations in the Tian Shan-Altay and Pamir regions and Qinghai are diploid (n = 18), and those from southwestern China are tetraploid (n = 36). Small (1968) pointed out that the pattern may be more complex in Asia than in North America, and a more careful study of this species in Asia, including chromosome counts, is needed.

The variation pattern of *E. latifolium* has been discussed by Haussknecht (1884), Böcher (1962), Raven (1962a), Munz (1965), and Small (1968). In the most comprehensive recent analysis, Small (1968) concluded that populations of this species with n = 18 should not be separated formally from those with n = 36, because the morphological differences between them are inconsistent and mostly quantitative; many individuals appear to be intermediate. According to Small, the morphological and geographical relationships suggest that the tetraploids arose from diploid plants by autopolyploidy: 57.4% of the chromosomes at diakinesis form quadrivalents (Small 1968). However, similar results would probably be obtained from any tetraploid individual in sect. *Chamaenerion,* so these results in themselves do not exclude the possibility of alloploid origin for the tetraploid populations.

The geographical disjunction between southeastern Xizang and adjacent Yunnan and the Tian Shan-Altay region of northwestern China is striking, and the absence of *E. latifolium* from the Himalaya between Himachal Pradesh in India and easternmost Xizang (Raven 1962a) is particularly difficult to explain in view of the distribution of other species. Most of the area of southeastern Xinjiang and the Xizang-Qinghai plateau is extremely dry and inhospitable to most species of *Epilobium.*

The collection from Yunnan (Beima Shan, *Forrest 14443,* E) resembles *E. speciosum* in general appearance and in having larger flowers, more conspicuously veined leaves, and more densely strigillose stems and leaves than occur in most representatives of *E. latifolium,* but it is similar to the collections from Xizang in this regard (cf. Raven 1962a); all of these collections have the glabrous style characteristic of *E. latifolium.*

2. Epilobium speciosum Decaisne in Jacquemont, Voy. India IV, Bot.: 57, t. 69. 1844. *Epilobium latifolium* subsp. *speciosum* (Decaisne) Raven, Bull. Brit. Mus. (Nat. Hist.) Bot. 2: 349. 1962. *Chamerion speciosum* (Decaisne) Holub, Folia Geobot. Phytotax. 7: 86. 1972.—TYPE: INDIA. Himachal Pradesh, near Yurpo, 3800 m, 1830–31, *Jacquemont 1739* (holotype: P, not located; isotypes: G! K!).

Common name: Himalaya Liulan.

Clumped erect perennial herb, with a thick rhizome up to 15 mm in diameter, somewhat woody, with fleshy soboles arising from the crown. Stems several, 20–45 cm tall, 3–5 mm in diameter, simple or sparingly branched on the upper part, densely strigillose throughout, especially on the inflorescence. Leaves spirally arranged, the basal ones 1–2 cm long, brownish, submembranous, triangular-ovate to lanceolate and sessile, the others sessile or with petiole 1–2 mm long; cauline blades 3.5–7.5 (–9) cm long, 0.7–1.8 (–3) cm wide, subcoriaceous, narrowly ovate to lance-elliptic, pale green or brownish when dry, strigillose on both surfaces, apex acute, acuminate, or sometimes subobtuse, base cuneate, margin remotely punctate-denticulate with 7–15 teeth on each side, the lateral veins distinct, 3–5 on each side of the midrib, disappearing or sometimes anastomosing before reaching the margin. Inflorescence a loose, erect raceme, densely strigillose; bracts about half as long as cauline leaves. Flowers erect in bud, nodding individually at early anthesis; buds 14–20 mm long, 7–10 mm in diameter, oblong-obovoid, apiculate; ovaries 1.7–2.5 cm long, purplish red, densely canescent; pedicels 0.8–2.8 cm long; floral tube absent, instead with a disc 0.8–1 mm high, 3.5–5.5 mm in diameter, glabrous; sepals 15–20 mm long, 3–5

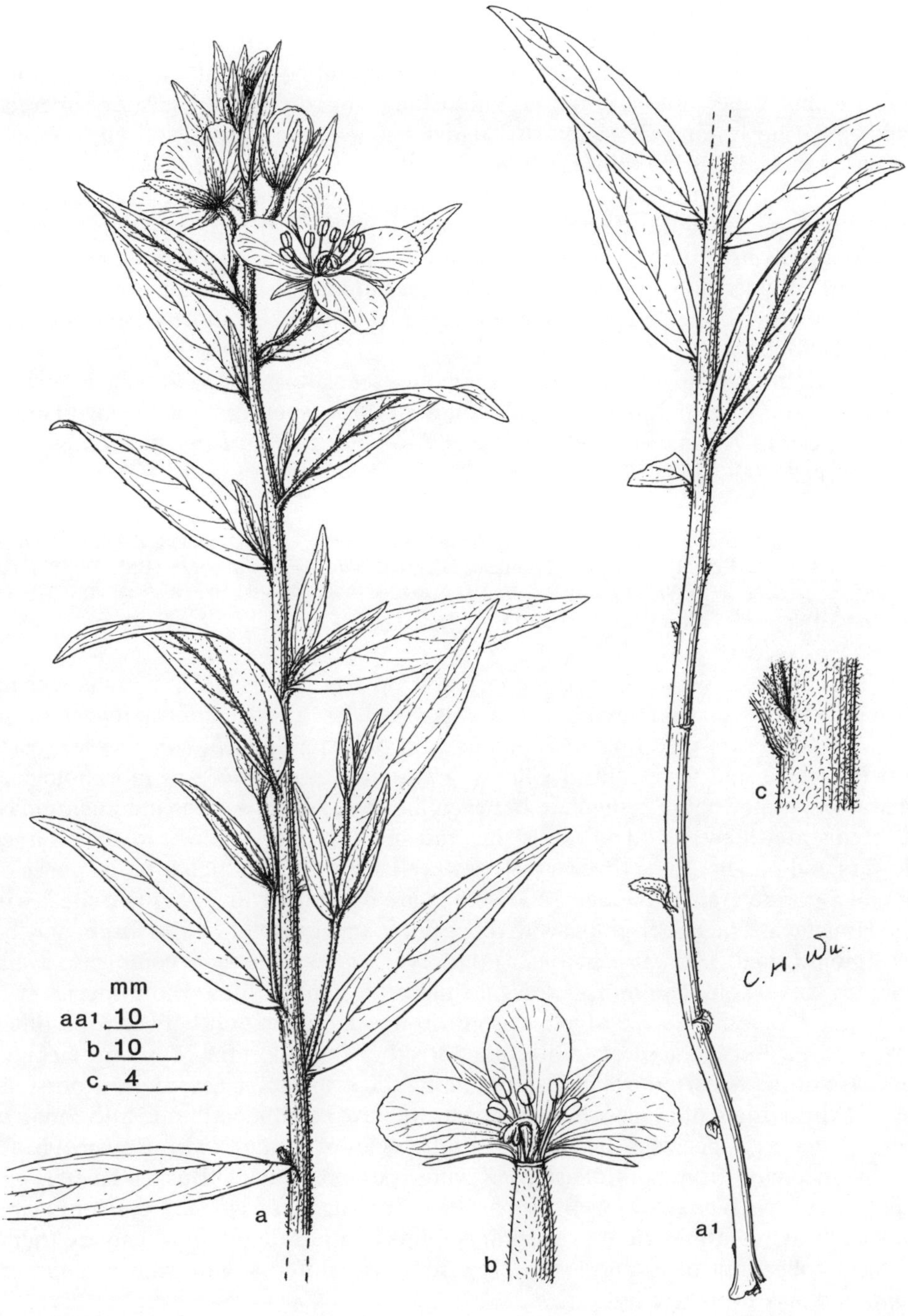

FIG. 11. *Epilobium speciosum*. a. Habit: upper stem and inflorescence. a[1]. Lower stem. b. Flower showing style deflexed; also note hairs near base of style (petal and three stamens removed). c. Axil from inflorescence. (Based on *Xizang Exped. 4113*.)

mm wide, linear to linear-lanceolate, acuminate, strigillose; petals 17–25 mm long, 12–22 mm wide, purplish red or rose, suborbicular to orbicular-oblong, entire, rounded or rarely retuse; filaments 8–10 mm long, subequal; anthers 3–4 mm long, 0.8–1.4 mm wide, oblong; style 5–7 mm long, initially deflexed, later suberect, villous on the lower half, purplish; stigma 4-lobed, the lobes 3–4 mm long and recurved at anthesis, papillose, white. Capsules 5–9 cm long, strigillose, borne on pedicels 1–3 cm long. Seeds 1.2–1.3 mm long, 0.4–0.6 mm in diameter, narrowly obovoid, acuminate, the chalazal end rounded, with a conspicuous collar 0.1–0.12 mm long, inconspicuously and irregularly reticulate; coma 10–13 mm long, tawny, not markedly deciduous. Chromosome number unknown. Fig. 11.

Phenology. Collected in flower in August to September, in fruit from September to October.

Distribution (Fig. 12). Himalayan region, especially in the west-central section from Kashmir to Nepal; in China only found in Zayu, southeastern Xizang (Tibet), and in western Xizang; in sandy or gravelly soil at the foot of boulder screes or in wet places; 3900–4500 m.

ADDITIONAL SPECIMENS EXAMINED. **China.** XIZANG: Southern flanks of Demulha, Zayu, *Pl. Sources Exped. 3944* (PE, XZ); Burma-Tibet Frontier (97°55′E), *Kingdon-Ward 10118* (BM, F); Burang, *Xizang Exped. 4113* (HNWP); "Zanskar," *Hay 124* (OXF); "Gnari Khorsum," *Wallich 6353* (HBG); Kamaon, *Wallich 9852* (P).

This species, which is here reported from China for the first time, occurs primarily in the central Himalaya, where its range lies approximately between those of *E. latifolium* (in the Himalaya, mainly west of Nepal) and *E. conspersum* (eastern Himalaya and west-central China). *Epilobium speciosum* has morphological characters somewhat intermediate between these two species in the indumentum of its stems and leaves, and in leaf veins and margins, but it tends to have larger flowers and pollen grains (average 85 μm in diameter), with higher percentages of 4- and 5-porate grains (average 52.3%) than are normal for the other two species in the Himalaya. The larger grains with more pores suggests that *E. speciosum* may be tetraploid (Small 1968; Raven 1962a); however, no chromosome counts are available for this taxon. The morphological similarities among these three species suggest that *E. speciosum* may have arisen through hybridization between the other two species, but cytological, molecular, or other corroborating data are lacking. Because of the uncertain relationships among these three species, and the possibility of hybrid origin of at least *E. speciosum,* we have concluded that *E. speciosum* is best treated as a species rather than as a subspecies of *E. latifolium* (Raven 1962a).

A specimen from south-eastern Xizang (perhaps Zayu, "Burma-Tibet Frontier," *Kingdon-Ward 10118,* BM) has long fruiting pedicels (2.5 cm long) and broader leaves with reticulate venation and less denticulate margins than are found in most collections of *E. speciosum.* It is somewhat intermediate with *E. conspersum,* and may even be a hybrid.

3. Epilobium conspersum Haussknecht, Oesterr. Bot. Zeitschr. 29: 51. Feb 1879. *Epilobium reticulatum* C. B. Clarke in J. D. Hooker, Fl. Brit. India 2: 583. May 1879, nom. superfl. *Chamaenerion reticulatum* (C. B. Clarke) Kitamura in Kihara, Fauna & Fl. Nepal Himal. 1: 185. 1953. *Chamaenerion*

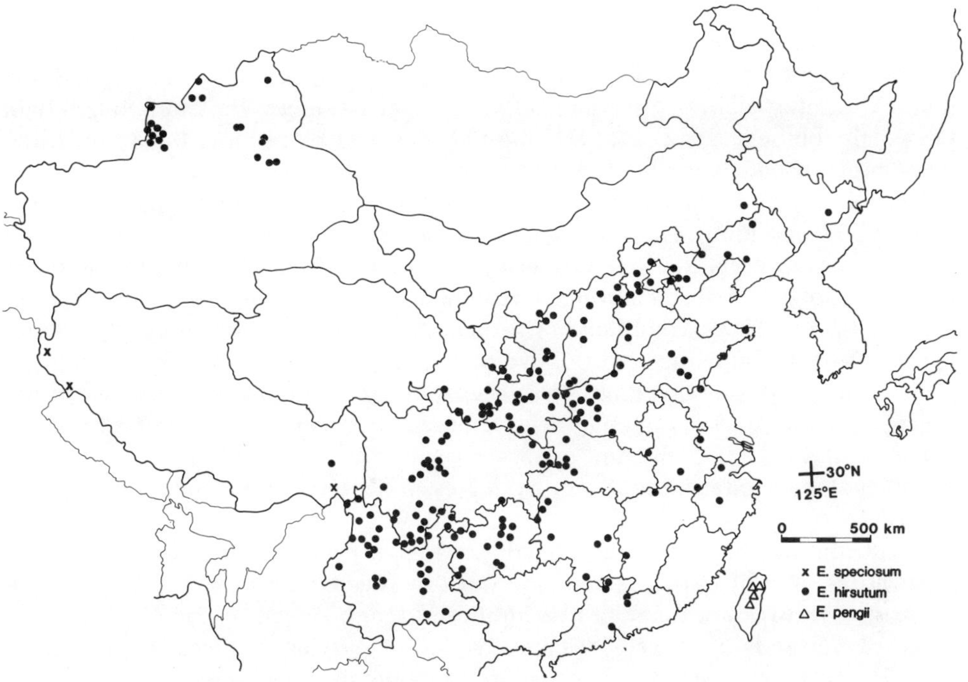

Fig. 12. Distribution in China of *Epilobium hirsutum*, *E. pengii*, and *E. speciosum*.

conspersum (Haussknecht) Kitamura, Results Kyoto Univ. Sci. Exped. Karak. & Hinduk. 1955, 8: 110. 1966. *Chamerion conspersum* (Haussknecht) Holub, Folia Geobot. Phytotax. 7: 86. 1972.—TYPE: SIKKIM. "Lama Konyr" [Kongra Lama], 4250 m, 27 Jul 1849, *Hooker s.n.* (holotype: K!; isotype?: BM!).

Common name: Wangmai Liulan (Li 1986).

Robust perennial herb, with somewhat woody rhizome up to 15 mm in diameter, forming short soboles at or just below root crown. Stems 30–120 cm tall, 2.5–5 mm in diameter, simple or few-branched, brown when dry, densely strigillose throughout. Leaves spirally arranged, the subterranean ones minute, scalelike, coriaceous, triangular and clasping, the basal ones less than 1 cm long, submembranous, narrowly triangular to narrowly lanceolate, subsessile, the others subsessile or with petioles 1–3 mm long; cauline blades 4.5–11 cm long, 0.7–1.4 cm wide, subcoriaceous, narrowly oblong- or elliptic-lanceolate, dark green above, pale green beneath, strigillose on both surfaces, especially on veins, apex acuminate, base cuneate, margin often more or less revolute, denticulate with 15–35 teeth, the lateral veins distinct, 4–5 on each side of the midrib reaching to the teeth, secondary veins conspicuous, finely reticulate or anastomosing, prominent on abaxial surface, impressed above. Inflorescence an erect raceme, densely strigillose; bracts less than half as long as cauline leaves, submembranous. Flowers suberect in bud, becoming pendulous individually at early anthesis; buds 9–12 mm long, 7–10 mm in diameter, broad-obovoid to subglobose, apiculate or mucronate; ovaries 1–2 cm long, purplish, densely canescent; pedicels 1.5–4 cm long, sometimes with basal portion fused to stem; floral tube absent, instead with disc 0.7–1 mm high, 3–4.5 mm in diameter, glabrous; sepals 11–15 mm long, 3–5 mm wide, often longer than the petals, oblong, acuminate, strigillose; petals 8–14 mm long, 6–13 mm wide, rose-purple, usually unequal, the lower two slightly narrower, subcordate to suborbicular, entire and rounded or rarely retuse; filaments 5–8 mm long, subequal; anthers 3–4 mm long, 1–1.3 mm wide, oblong; style 5–8 mm long, recurved initially, becoming erect, densely villous on the lower half, purplish; stigma deeply 4-lobed, the lobes 2.5–3.5 mm long, spreading at maturity, papillose, whitish. Capsules 2.5–7.5 cm long, strigillose, borne on pedicels 1.5–5 cm long. Seeds 1–1.2 mm long, 0.38–0.48 mm in diameter, obovoid-oblong, narrowed to more or less distinct chalazal neck 0.08–0.1 mm long, the surface squamose-papillose to inconspicuously reticulate; coma 10–12 mm long, tawny, not readily deciduous. Chromosome number: n = 18. Figs. 5d, e, 6h, 8m–o, 13.

Phenology. Collected in flower from July to September, in fruit from September to October.

Distribution (Fig. 10). Sino-Himalayan region from Shaanxi in central China westward to central Nepal; in China in western Shaanxi, eastern Qinghai, western Sichuan, northwestern Yunnan to southern Xizang (Tibet); in open moist areas, especially in sandy or gravelly streambeds, along margins of ditches or roadsides, on grassy or rocky open slopes, or ridges of mountains; 2300–3800 m, or in some areas in the south to 3800–4700 m.

Additional Specimens Examined. **China.** Shaanxi: Taibai Shan, *Z. X. Peng & X. Z. Zhang 1250* (LZU), *K. T. Fu 9564* (WUK).—Qinghai: Minghe, *S. W. Liu & C. J. Chen 2726* (HNWP, PE).—Sichuan: Erlang Shan, *Sichuan Econ. Pl. Exped. 894* (CDBI); Luding, *Z. A. Liu 22582* (CDBI), *G. Y. Xu 23477* (CDBI); Jinyan, *Z. S. Liu 189* (CDBI); Kangding, *Cunningham 224* (E), *H. S. Kung 6009* (MO, PE), *6296* (MO, PE), *McLaren AC249* (E, LD), *Smith 10524* (UPS), *10982* (UPS), *H. L. Tsiang & C. H. Tsiung 36510* (PE, SZ).—Yunnan: Zhongdian, *Zhongdian Exped. 791* (PE); Gongshan, *X. F. Deng 791311* (KUN), *K. M. Feng 7635* (KUN, PE), *Handel-Mazzetti 9863* (E); Deqen, *K. M. Feng 5416* (KUN, PE), *C. W. Wang 64845* (GH, NAS, PE, TAI), *64851* (NAS), *70150* (PE), *T. T. Yu 7890* (E), *19712* (KUN, PE), *27407* (KUN, PE).—Xizang: Mainkung, *T. T. Yu 22571* (E, GH, KUN, PE); Zayu, *Qinghai-Xizang Exped. 10266* (KUN, PE), *82-7744* (PE); Medog, *Ecol. Group Xizang Exped. 14717* (PE), *Qinghai-Xizang Exped. 74-3905* (PE); Bomi, *T. S. Ying & D. Y. Hong 906* (PE), *J. Y. Zhuang 1105* (PE); Mt. Nanjiabawa, *S. Z. Cheng & P. S. Li 74* (PE), *Plateau Group 14717* (PE); Nyingchi, *Xizang Exped. 670* (PE), *751081* (PE); Mainling, *P. S. Li et al. 5651* (PE), *Xizang Med. Exped. 4055* (PE); "Pangkar, Drukla Chu," *Ludlow et al. 6873* (BM, E); "Deyang La, Kongbo," *Ludlow et al. 14257* (BM, E); Lhunze, *Ludlow et al. 2083* (BM, E); Nyalam, *S. Jiang & L. Zheng 104* (PE), *Xizang Exped. 1551* (PE); Gyirong, *Xizang Med. Exped. 547* (PE); S. E. Tibet, Praigchu, *Ribu & Rhomoo 6606* (E); "Chumbi & Pari," 1879, *Dungboo s.n.* (K); Mt. Everest Expedition 1921, *Wollaston 99* (K); Yadong, 1897, *Hobson s.n.* (K).

Epilobium conspersum is an attractive and distinctive element of the Sino-Himalayan alpine flora. Although clearly related to *E. angustifolium* and *E. speciosum,* it differs sharply from them in the striking reticulate venation of its leaves, and in its pedicels, which may be 5 cm long. These three species appear to hybridize in some areas where their ranges overlap. In the discussion of *E. angustifolium,* we mention possible hybrids between that species and *E. conspersum* in southwestern China, some of them discussed earlier by Raven (1962a). We found another possible hybrid from western Shaanxi (Taibai Shan, *Z. X. Peng & X. Z.*

FIG. 13. *Epilobium conspersum*. A. Habit: inflorescence. A^1. Mid-stem. A^2. Basal stem showing thick caudex. B. Flower. a. Capsule. b. Seed. c. Axil from inflorescence. (A–B based on *T. T. Yu 22571*; a–c based on *S. Jiang & L. Zheng 104*.)

Zhang 1250, LZU), with intermediate morphology and pollen stainability of only 25%. Raven (1962a) discussed an apparent hybrid between *E. conspersum* and *E. speciosum* from Nepal, and we report a specimen of similarly intermediate morphology from Xizang (Tibet) in the discussion under *E. speciosum* (no. 2).

4. Epilobium angustifolium Linnaeus, Sp. Pl. 1: 347. 1753. *Chamaenerion angustifolium* (Linnaeus) Scopoli, Fl. Carniol. ed. 2, 1: 271, 1772. *Chamerion angustifolium* (Linnaeus) Holub, Folia Geobot. Phytotax. 7: 86. 1972.—TYPE: Probably from Sweden (holotype: LINN 486.1!).

Common name: Liulan ["willow orchid"] (Li 1986).

Robust erect perennial herb that forms large clones by vigorous soboles from a woody caudex or by long (up to 5 m), thick (up to 2 cm in diameter) lateral roots. Stems several to many, 20–200 cm tall, 2–15 mm in diameter, simple or rarely branched above, terete, glabrous to densely strigillose especially on inflorescence, basal part often woody and sometimes with exfoliating epidermis. Leaves spirally arranged or very rarely opposite near base, the lower ones 0.5–2 cm long, lance-oblong to obovate, submembranous, usually brown and shrivelled, subsessile, the others sessile or with petioles 2–7 mm long; cauline blades 3–23 cm long, 0.3–3.4 cm wide, linear to lanceolate, green or pale green, glabrous throughout or strigillose on the abaxial midrib, apex attenuate-acute, base obtuse or cuneate to attenuate, margin entire or obscurely denticulate, lateral veins 10–25 on each side of the midrib, diverging nearly at right angles from the midrib, confluent into a more or less prominent submarginal vein. Inflorescence a simple, lax, terminal raceme, unusually elongate, 5–50 cm long, glabrous or strigillose; lowest bracts foliaceous, the upper ones linear, coriaceous and much smaller than cauline leaves. Flowers nodding in bud, suberect at anthesis; buds 6–18 mm long, 4–7 mm in diameter, somewhat zygomorphic, obovoid, apiculate to cuspidate; ovaries 0.6–2.5 cm long, densely appressed-canescent; pedicels 0.5–3 cm long; floral tube absent, instead with raised disc 0.5–1 mm high, 2–4 mm in diameter, glabrous; sepals 6–19 mm long, 1.5–3 mm wide, oblong-lanceolate or upper two slightly oblanceolate, acuminate or attenuate, sometimes slightly clawed, canescent on abaxial surface, purplish; petals 9–25 mm long, 3–15 mm wide, pale pink to purple or very rarely white, slightly unequal, with upper ones longer and broader, obovate to narrowly obovate to suborbicular, entire to shallowly emarginate; filaments 7–15 mm long, subequal, reddish; anthers 2–3 mm long, 1–1.7 mm wide, oblong, initially red, purplish at dehiscence, bearing bluish pollen; style 8–16 mm long, sharply deflexed at onset of anthesis, becoming erect later, villous on lower part; stigma deeply 4-lobed, the lobes 3–6 mm long, 0.6–1 mm wide, oblong-lanceolate, with densely papillose surfaces, white. Capsules 4–9.5 cm long, densely appressed-canescent, borne on pedicels 0.5–3 cm long. Seeds 0.9–1.3 mm long, 0.3–0.45 mm in diameter, narrowly obovoid, brown, the chalazal end subacuminate with very short neck, the surface appearing smooth and rather shiny but irregularly reticulate; coma 10–17 mm long, dingy or white, dense, not easily detaching. Chromosome numbers: n = 18, 36, 54.

Epilobium angustifolium, an extremely variable species, is widespread over large areas of Eurasia and North America from about 25° to 70° N latitude. It is common throughout much of cool temperate China, especially in the northern half and at higher elevations in the south; it is absent from the deserts and highest mountain plateaus. Although its nomenclatural history is quite complex and many infraspecific taxa have been recognized, the basic delimitation of the species has remained relatively stable and uncontroversial since this taxon was first described by Linnaeus in 1753.

Epilobium angustifolium includes populations with three different chromosome levels ($n = 18$, 36, and 54; Mulligan 1957; Löve & Löve 1961; Mosquin 1963a, 1963b, 1966). Mosquin (1966) analyzed the morphological diversity of the species in relation to the chromosome numbers and geography, and recognized two broadly circumscribed subspecies. He reported original chromosome counts from nearly 100 populations, mostly from North America, and found that diploid ($n = 18$) populations could be distinguished from tetraploids ($n = 36$) by a combination of pollen size and pore number, leaf width, and pubescence of the abaxial leaf midrib. In North America, the taxa so delimited by morphology and ploidy level have different distributions, with diploids in the colder northern regions and tetraploids more widespread farther south. Although Mosquin attempted to apply this taxonomy to Eurasian material, there were too few collections with known chromosome numbers from Asia, particularly from China and the Himalayan region, to assess its validity in that region. In the absence of a comprehensive study of the Eurasian populations, Hoch (1981), e.g., chose to treat this species in Pakistan without infraspecific subdivision.

As reported previously (Chen et al. 1988), we found in our detailed study of collections from China that the characters used by Mosquin (1966) to delimit the two subspecies also separate the Chinese material into two groups. In addition, shape of the leaf base and petiole length also distinguish them, and upper stem pubescence, leaf margin, and petal size differ somewhat between the two taxa, although less consistently. The presumed diploid subsp. *angustifolium,* found mostly in northern China and at higher elevations in the south, has smaller, 3-porate pollen grains; glabrous, smaller, and subsessile leaves, usually with obtuse bases and revolute margins; glabrous stems; and smaller, narrower petals (mean area 91.2 mm^2; range 30–209 mm^2). The presumed tetraploid subsp. *circumvagum,* found mostly in southwestern China and at lower elevations, has larger pollen with a substantial proportion of 4- and 5-porate grains; usually strigillose abaxial leaf midrib and upper stem; broader, usually petiolate leaves with subcuneate bases and flat margins; and larger, broader petals (mean area 154 mm^2; range 72–299 mm^2).

Pollen size has long been used as an indicator of ploidy levels in plants (Stebbins 1950), and it proved very useful in this analysis in distinguishing the two groups of populations. In collections of *E. angustifolium* from China, increase in pollen size is closely correlated with pubescence on the abaxial leaf midribs (Chen et al. 1988). Individuals that we classified as subsp. *angustifolium* on the basis of leaf characteristics and especially of degree of pubescence on the abaxial leaf midvein had a mean pollen diameter of 67.6 μm (range 56.2–79.9 μm); 97% of the pollen grains were (2-)3-pored and only 3% were 4-pored. In contrast, those we classified as subsp. *circumvagum* had a mean pollen diameter of 85.1 μm (range 75.6–94.5 μm), and 73% of the grains were 3-pored and 27% were 4–5-pored. Increased pollen diameter in this species is correlated with increased pore number. Even in the absence of cytological data, pollen and leaf characters serve to distinguish the two subspecies, although there is some overlap between them, as also described by Mosquin (1966). More chromosome counts from Asia would probably make the pattern even clearer.

Because of the orographic complexity and the topographic extremes of southern and eastern Asia, correlation between geographic distribution of the two subspecies of *E. angustifolium* in China and their morphology is not as obvious as it is in North America. In North America, the diploid subsp. *angustifolium* rarely occurs south of the southern limits of the boreal forest, and then only at high elevations along the Rocky Mountains to Wyoming, whereas the tetraploid subsp. *circumvagum* is widespread south of those limits, and extends north along the Pacific coast to Alaska (Mosquin 1966). In the wide zone of overlap, populations of subsp. *angustifolium* usually grow at higher elevations than those of subsp. *circumvagum*. No such clear latitudinal demarcation occurs in eastern Asia. It is possible, however, to demonstrate a similar relationship among specimens from China, by assigning a combined altitudinal/latitudinal value to each. Using the rough approximation that 100 m increase in altitude is equivalent to 1° of latitudinal movement northward, and taking 25°N and sea-level as 0, we assigned each collection a value (units = degrees of latitude + hundreds of meters of elevation). The values for subsp. *angustifolium* range from 24 to 50 units, whereas those for subsp. *circumvagum* range from 10 to 38.5; they demonstrate an eco-geographical difference in the distribution of the two subspecies (Chen et al. 1988).

The distributions of diploid and tetraploid populations of *E. angustifolium* overlap broadly in China, as indicated in Figs. 15 and 17. In these areas of overlap, the populations of the two subspecies can generally be distinguished by the characters discussed above; relatively few individuals are morphologically intermediate. Some plants with pubescent leaf midribs have smaller pollen with a relatively high percentage of 3-porate and infertile grains and narrower subsessile leaves. Likewise, some glabrous-leaved plants have larger pollen with a higher percentage of 4-porate grains, and often broader leaves that have essentially flat margins. We studied some collections of plants from the Altai Shan and Tian Shan areas of Xinjiang in northwestern China that had very broad leaves, but that otherwise were characteristic of subsp. *angustifolium*. Careful field studies are needed in areas where diploid and tetraploid populations occur to determine the exact degree of local sympatry and the degree to which they actually interbreed, if at all.

Plants with n = 54 have been reported from only one population from the Nikko Mountains in Japan (Mosquin 1966). However, Dr. T. Hoshino of Okayama University has recently detected other populations with n = 54 (Hoshino, pers. comm.), all restricted to the same region of central Honshu. Until the morphology and geography of these plants have been more thoroughly studied, the significance of these findings cannot be determined.

Raven (1962a) suggested that *E. angustifolium* occasionally hybridizes with *E. conspersum* in the eastern Himalaya, Yunnan, and Sichuan. We have seen about 15 collections from these areas and from southeastern Gansu, which may have had such an origin. Although they resemble *E. angustifolium* overall, about ½ to ⅔ of the inflorescences are leafy, their pedicels are relatively long (2–3.5 cm), their flowers are more or less nodding, their leaves are elliptic-lanceolate with an attenuate base and distinctly petiolate; and their upper stems and abaxial midribs resemble those of *E. conspersum* in being densely pubescent. On the other hand, they resemble *E. angustifolium* in their nodding buds and in that the lateral veins of their leaves do not reach the margin but instead join by arching together.

A few collections in China have very high percentages of non-stainable pollen; for example, Xizang, Nagqu (31°30′N, 92°E), *D. D. Tao 10936* (PE; subsp. *angustifolium;* 3% fertile pollen); Sichuan, Li Xian (31°29′N, 103°15′E), *D. P. He 46527* (SZ; subsp. *circumvagum;* 9% fertile pollen); Xizang, Zhamu (28°5′N, 103°40′E), *Y. Chen 7133* (NAS; subsp. *circumvagum;* 9% fertile pollen). The cause of such sterility is unknown.

The young shoots and leaves of *E. angustifolium,* which are used as a vegetable or tea, or as pig feed in the U.S.S.R. (Steinberg, 1949; also herbarium label data), are used in rural areas of China as a medicine for dropsy and indigestion.

4a. Epilobium angustifolium subsp. **angustifolium.**

Epilobium spicatum Lamarck, Fl. Franc. 3: 482. 1778. *Epilobium neriifolium* H. Léveillé, Monde Pl. 6: 2. 1896, nom. superfl.—TYPE: FRANCE (not located).

Stems 20–130 cm tall, 2–10 mm in diameter, subglabrous. Leaves subsessile; cauline blades (3–) 7–14 (–18.5) cm long, (0.3–) 0.7–1.3 (–2.5) cm wide, usually subcoriaceous, linear-lanceolate or narrowly lanceolate, glabrous throughout, apex attenuate-acute, base obtuse to cuneate, margin subentire to obscurely denticulate, somewhat revolute, the lateral veins often indistinct. Inflorescence an erect raceme 5–40 cm long, subglabrous; bracts 5–20 mm long. Floral buds 6–12 mm long, 4–6 mm in diameter; ovaries 0.6–2 cm long; pedicels 0.5–1.8 cm long; sepals 6–15 mm long, 1.5–2.5 mm wide, attenuate or acuminate; petals 9–15 (–19) mm long, 3–9 (–11) mm wide, obovate or narrowly obovate, entire or shallowly emarginate; filaments 7–14 mm long; anthers 2–2.5 mm long; pollen 3-porate, rarely 4-porate, 57–75 (–80) μm in diameter; style 8–14 mm long. Capsules 4–8 cm long, borne on pedicels 0.5–1.9 cm long. Seeds 0.9–1 mm long, 0.35–0.45 mm wide. Chromosome number: n = 18. Figs. 5a, 14.

Phenology. Collected in flower from June to September, in fruit from August to October.

Distribution (Fig. 15). Circumpolar and circumboreal, throughout Eurasia, northern North America and Greenland, also found southward at higher elevations of the Himalaya and the western cordillera of North America; in China in Heilongjiang, Jilin, Nei Monggol, Hebei, Shanxi, Ningxia, Gansu, Qinghai, Xinjiang, Sichuan, northwestern Yunnan, and northern and eastern Xizang (Tibet); in moist, often disturbed places in the mountains; 500–3100 m in northern China, (2300–) 2900–4700 m in southwestern China.

ADDITIONAL SPECIMENS EXAMINED. **China.** HEILONGJIANG: Jixian, *Y. L. Zhang 1683* (IFP, PE); "Sachalien," *Komarov 1124* (K, P); along Wusuli River, *Wilford 335* (GH); Hulin, 1952, *G. Z. Wang et al. s.n.* (PE); Mishan, *S. X. Li et al. 2940* (IFP), *3037* (IFP); coast of Manchuria, 1859, *Wilford s.n.* (GH, K); Hegan, *P. Y. Fu & Z. H. Zhang 18* (IFP, PE); Shangzhi, *T. N. Liou et al. 7204* (PE), *7321* (PE), *Y. C. Zhu et al. 148* (PE); Maoershan, *T. Ye 1657* (IFP); Da Hinggan Ling, *Sino-German Exped. 8827* (PE); Gen He, *T. N. Liou et al. 8827* (PE); "Haritum," *Stotzner 55* (S); no locality, s.d., *Maximovitch s.n.* (NY), 1922, *Skvortzov s.n.* (GH).—JILIN: Helong, *Yanbian Exped. 2687* (PE); Antu, *Changbai Shan Exped. 667* (PE); Changbai Shan, *Yanbian Exped. 1160* (PE), 1886, *James s.n.* (K), *Changbai Shan Exped. 150* (PE), *222* (PE), *Fenzel 183* (G), *277* (W); Fusong, *Y. L. Zhang et al. 546* (PE), *643*

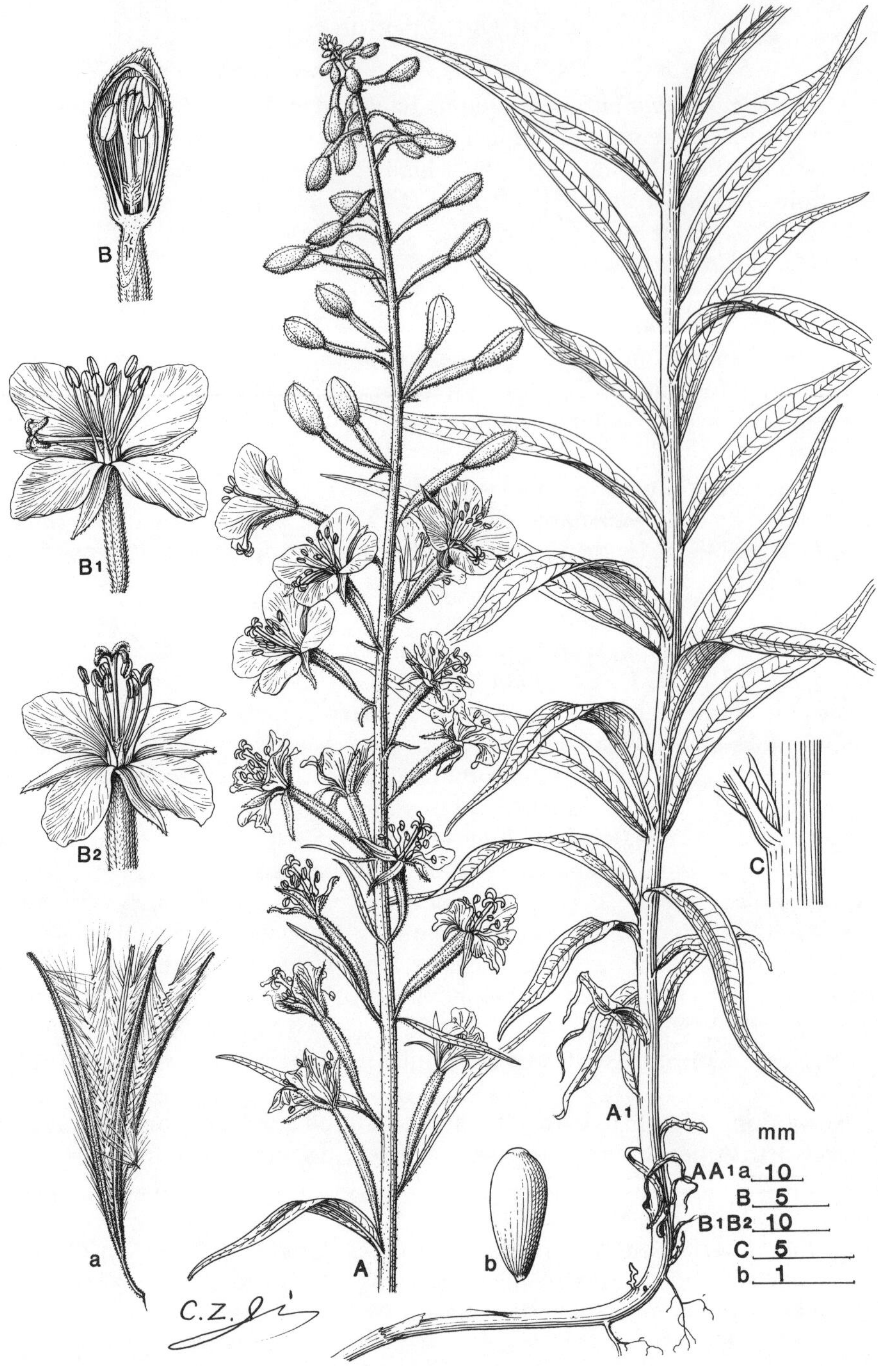

FIG. 14. *Epilobium angustifolium* subsp. *angustifolium*. A. Habit: inflorescence with flowers nodding in bud, erect at anthesis. A[1]. Lower stem and base. B. Flower: longitudinal section of floral bud showing erect style and commissural stigma lobes. B[1]. Open flower showing initially deflexed style and spreading lobes of stigma. B[2]. Open flower showing erect style and revolute lobes of stigma. C. Midcauline axil. a. Capsule. b. Seed. (A–C based on *S. W. Liu 2445*; a–b based on *C. Y. Yang & Z. G. Li 36431*).

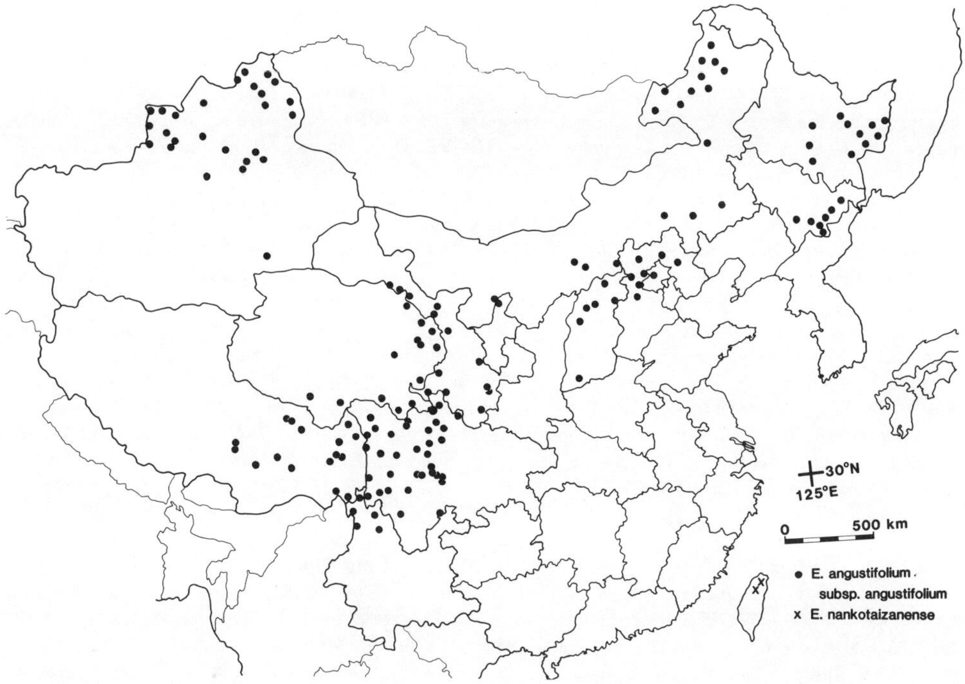

FIG. 15. Distribution in China of *Epilobium angustifolium* subsp. *angustifolium* and *E. nankotaizanense*.

(IFP); Horqin Youyi Qianqi, *Z. Wang et al. 1413* (IFP, PE); no locality, *F. H. Chen 144* (PE), *430* (PE), *2462* (PE); "Jalatun," *Dorsett & Dorsett 3470* (NAS, PE, US).—Nei Monggol: Erguu Youqi, *C. Q. Ling et al. 1884* (IFP), *Z. Wang 1182* (NAS, PE), *1413* (IFP, PE); Hailar, 1956, *Y. S. Liu & C. X. Fu s.n.* (NMU), *T. N. Liou et al. 8087* (NAS, PE), *Z. Wang 767* (IFP, PE), *2126* (IFP); Manzhouli, *Z. Wang 897* (IFP, PE); Huma, 1954, *Forest Exped. s.n.* (PE); Hulun Buir Meng, *Hulun Buir Meng Exped. 325* (IFP), *Y. Z. Liu 1* (NMU), *Nei Monggol Coll. Exped. 1176* (NMU); Xinghe, *Y. Z. Zhao 77-14* (NMU); Ju Ud Meng, *P. Y. Fu 4415* (IFP); *Ju Ud Meng Exped. 159* (NMU), *Y. Z. Liu 8* (NMU), *Y. C. Yang & Y. L. Zhang 384* (PE); Hohhot, *Nei Monggol Univ. Exped. 56* (NMU); Daqing Shan, *Nei Monggol Univ. Exped. 20* (NMU, PE); Helan Shan, *Nei Monggol Univ. Exped. 262* (NMU); "Clara- gol," 1903, *Molleson s.n.* (G); Hinggan Meng, *H. J. Meng 75* (NMU); no locality, *J. Zhou 108* (PE).—Hebei: Chengde, *Nankai Univ. 203* (PE); Beijing, 1862, *David s.n.* (P), *David 553* (P), *W. Y. Hsia 2029* (PE), *2957* (PE), *T. F. King 364* (PE), 1896, *Provost s.n.* (P), *W. T. Wang & S. J. Han 676* (PE), *679* (PE); Huailai, *Hebei Pl. Exped. 1674* (PE), *2095* (PE); Xiaowutai Shan, *Herb. Beijing Inst. Bot. 397* (PE), *787* (PE), *1457* (PE), *1474* (PE), *1555* (PE), *1806* (PE), *2336* (PE), *5236* (PE), *10178* (PE), *H. W. Kung 930* (PE), *C. F. Li 1154* (NY), *Y. Liu 10601* (PE), *10773* (PE); *10961* (PE), *12506* (PE), *Smith 1050* (UPS), *T. P. Wang 462* (PE), *Wu & Yang 2884* (PE), *Xiaowutai Shan Exped. 74078* (PE), 1906, *Y. Yabe s.n.* "*A*" (NAS), *C. Y. Yang & Z. G. Li 36431* (PE); Yi Xian, *Herb. Beijing Inst. Bot. 915* (PE); "Yang Ling Kie," *Licent 6603* (BM); no locality, *C. G. Yang 1667* (NAS), *C. F. Li 11054* (NAS, PE); *Y. W. Tsui 2378* (PE).—Shanxi: Mt. Wutai, *Y. L. Chen 1488* (PE), *Hancock 1876* (K), *W. Y. He 2957* (NY), *Shanxi Exped. 618* (PE), *T. Tang 1060* (GH, NAS, PE); Ningwu, *Shanxi Exped. 241* (NAS, PE); Luya Shan, *Smith 8297* (UPS); Lishan, *Huanghe Exped. 2544* (PE).—Ningxia: Helan Shan, *R. C. Ching 1064* (E, UC, US), *Helan Shan Exped. 262* (PE).—Gansu: Tianshui, *Y. L. Liu et al. 341* (LZU); Xinglong Shan, 1974, *J. T. Dong s.n.* (LZU); Tewo, *Z. X. Peng 219* (LZU); Xiahe, *Y. He 264* (LZU); Ping Fan, *R. C. Ching 401* (E, US); Maqu, *Z. X. Peng et al. 282* (LZU); Tianzhu, *Y. Q. He 4507* (PE), *S. Jian 5301* (PE), 1956, *Lanzhou Univ. Exped. s.n.* (LZU), *H. D. Zhao & Z. F. Jian 672* (LZU); Huangcheng, *Lanzhou Univ. Exped. 129* (LZU); Bei He, *Youngmongshan Exped. 522* (PE); central Gansu, *Farrer & Purdom 717* (E, MICH), *Rock 12755* (E); western Gansu, *Licent 4578* (BR, P); Sunan, 1957, *Z. X. Cao s.n.* (PE), *Y. Q. He 3653* (PE), *Qinghai-Gansu Exped. 3320* (PE), *Western River Exped. 473* (PE), *Q. Z. Yang 185* (LZU), 1957, *W. B. Yang s.n.* (PE), *Y. M. Zhang 30* (PE); "Labuwu," 1937, *K. T. Fu s.n.* (PE); "Liangchow," *Trippner 229* (M); border between Gansu and Qinghai, s.d., *Y. C. Wu s.n.* (NAS); no locality, *Licent 5472* (P).—QINGHAI: Semnyi, *S. Q. Lo 1334* (HNWP); Fuzhu, *B. Z. Gou 6643* (PE); Menyuan, *K. M. Liou 6885* (PE), *Qilian Shan Exped. 150* (PE), *P. C. Tsoong 8936* (PE), *Qinghai Exped. 1363* (PE); Qilian, *W. Z. Zhang 2549* (HNWP, PE); Datong, *B. W. Li 72-109* (HNWP), *Qinghai-Gansu Exped. 2913* (PE), *P. C. Tsoong 8529* (PE); Haibei, *B. Z. Go & W. Y. Wang 12261* (HNWP); Xinghai, *K. S. Hao 1069* (P, PE); Henan, *S. W. Liu 2003* (HNWP), *L. H. Zhou & L. N. Sun 2310* (HNWP); Jigzhi, *Golog Exped. 553* (HNWP), *Xizang Med. Pl. Exped. 720* (HNWP); Baima, *B. Z. Go 468* (HNWP), *D. S. Lo & Z. C. Zuo 801405* (PE), *801447* (PE); Chindu, Yushu District, *S. W. Liu 2445* (HNWP), *Qinghai Exped. 330* (PE); Nangqen, *D. S. Lo & Z. C. Zuo 800445* (PE); Angjiu, *Xizang Med. Pl. Exped. 1156* (PE), *Y. C. Yang 1493* (PE).—XINJIANG: Bogda Shan, *T. N. Liou 3286* (PE), 1931, *Ridley s.n.* (K); Fukang, *T. Y. Chou et al. 651180* (PE), *651185* (NAS, XJBI); Urumqi, *Bayi Coll. Agr. 18206* (PE, YNU), *Chaeden 15* (P), *G. Z. Fu & Y. S. Zhang 3972* (XJBI), *Y. R. Lin 74038* (PE), *T. N. Liou 3054* (PE), *Z. X. Peng 4810* (LZU), *Q. R. Wang et al. 3540* (PE), *Reymong 16* (P, W), *Xinjiang Exped. 5459* (XJBI), *5479* (XJBI), *Z. W. Zhang 3088* (XJBI), *K. Y. Zhou 69* (PE); Bayingolin, *G. J. Liu et al 593* (LZU); Tian Shan, *C. J. Chen & Hoch 86208* (MO, PE), *K. C. Kuan 1853* (PE), *2247* (XJBI), *3971* (PE); Hom, *R. C. Ching 3279a* (PE); Shawan, *T. Y. Chou et al. 651037* (NAS, PE), *Xinjiang Branch Acad. Sin. 3577* (XJBI); Zhongqinhe, *R. C. Ching 1456* (CDBI, PE); Qinghe, Altay, *R. C. Ching 871* (PE), *877* (KUN, PE), *1275* (PE), *1310* (PE), *1853* (PE); Altay, *R. C. Ching 3029* (PE), *3325* (PE), *T. Y. Chou et al. 652073* (PE, XJBI), *A. R. Li & J. N. Zhu 1310* (XJBI), *G. L. Liu et al. 172* (XJBI), *260* (XJBI), *454* (XJBI), *618* (XJBI), *685* (XJBI), *Xinjiang Exped. 2944* (XJBI), *Z. W. Zhang et al. 4443* (XJBI), *G. L. Zhu 5883* (NAS, PE), *G. L. Zhu et al. 7204* (NAS, PE); Fuyun, Altay, *Xinjiang Exped. 1854* (PE), *2169* (PE); Jaishihanagai, Altay, *C. J. Chen & Hoch 86139* (MO, PE), *86157* (MO, PE); Xiaodonggou, *C. J. Chen & Hoch 86065* (MO, PE), *86099* (MO, PE); Jeminay, *Xinjiang Exped. 7324* (XJBI); Fuhai, *B. L. Li & J. Z. He 507* (PE); Toli, *K. C. Kuan 2574* (PE, XJBI), *A. R. Li & J. N. Zhu 10931* (KUN, PE); Bole, *Xinjiang Exped. 3773* (PE, XJBI); Wenquan, *C. J. Chen & Hoch 86276* (MO, PE), *86329* (MO, PE), *Xinjiang Exped. 3711* (XJBI); Heqing to Xinyuan, *Xinjiang Exped. 2512* (PE); Gongliu, *Y. R. Lin 74399* (PE), *Z. X. Peng 5112* (LZU), *Q. R. Wang et al. 2589* (PE), *Z. W. Zhang 3509* (XJBI); Tekes, *K. C. Kuan 3358* (PE, XJBI), *Merzbacher 953* (M); Ili, *T. Y. Chou et al. 650907* (NAS, PE), *Ili Exped. 25* (XJBI); Zhaosu, *Bayi Coll. Agr. 107* (PE), *18206* (PE), *Xinjiang Exped. 2938* (PE), *2970* (PE, WUK); "Chen hua," *C. Liu 288* (NAS); "Xingoutai," *Xinjiang Med. Pl. Exped. 432* (XJBI); no locality, *Bayi Coll. Agric. 1945* (YNU).—SICHUAN: Zoige *Sichuan Med. Exped. 20145* (NAS); Hongyuan, *B. Z. Ni 351* (CDBI), *Sichuan Veget. Exped. 10411* (CDBI); Nanping, *W. L. Chen & C. J. Chen 8635* (PE); Nanping to Songpan, *W. L. Chen & C. J. Chen 8849* (MO, PE); Songpan, *W. L. Chen & C. J. Chen 8313* (MO, PE), *8477* (MO, PE), *W. P. Fang 4309* (CDBI), *K. T. Fu 1318* (PE), *1970* (HNWP, PE), *Smith 2847* (UPS); Zoige, *Sichuan Exped. 10379* (PE); Longriba, 1957, *X. Li 72100* (PE, SZ); Shuajingsi, *X. Li 74831* (PE, SZ), *74847* (PE, SZ), *74866* (PE, SZ), *Z. Y. Zhang & H. F. Zhou 23514* (PE, SZ), *23576* (SZ), *23763* (PE, SZ); Barkam, *P. X. Li 10250* (PE), *10747* (PE), *10517* (PE), *Sichuan Testing Ground of Forestry 146* (CDBI), *Sichuan Med. Exped. 20145a* (NAS), *22329* (NAS); Mou Xian, *Z. He & Z. L. Zhou 14241* (PE); Xiaojin, *S. S. Chang 5970* (CDBI, SZ), *S. S. Chang & Y. X. Ren 6050* (PE, SZ), *6170* (CDBI), *6172* (SZ); Aba, *W. L. Chen & C. J. Chen 7935* (PE), *7977* (MO, PE); Zamtang, *W. L. Chen & C. J. Chen 7503* (MO, PE), *Sichuan Veget. Exped. 9362* (CDBI); Mt. Emei, *Y. H. Dao 50442* (SZ), *J. L. Chen & B. K. Zhou 9* (CDBI), *W. P. Fang 20434* (NAS, PE, SZ), *G. H. Yang 56560* (NAS, PE, SZ), *Y. B. Yang 104* (CDBI); Ebian, *S. L. Sun 1135* (US); Yaan, *Rock 23817* (BM, E, F, GH, K, NY, UC, US); Baoxing, *T. P. Soong 39018* (PE, SZ), *T. H. Tu 2351* (PE), *4495* (PE); Tianquan, *W. P. Fang & W. K. Hu 51479* (SZ), *Sichuan Econ. Pl. Exped. 894* (PE), *F. C. Tai & C. M. Teng 5202* (SZ), *H. L. Tsiang & J. H. Xiong 34609* (PE, SZ), *34798* (PE, SZ), *H. L. Tsiang 35383* (SZ), *35421* (SZ), *37718* (SZ); Kangding, *W. L. Chen & C. J. Chen 6305* (PE), *6551* (MO, PE), *Cunningham 167* (E), *W. P. Fang 3662* (PE), *3667* (PE), *W. G. Hu & Z. He 10460* (PE, SZ), *10611* (PE, SZ), *10818* (SZ), *B. S. Gao & Y. T. Wu 111539* (CDBI, PE), *Z. P. Huan et al. 780* (KUN, PE), *1619* (KUN, PE), *1661* (KUN, PE), *X. L. Jiang 36317* (IBG, PE, SZ), *K. C. Kuan et al. 729* (PE), *H. S. Kung 6008* (MO), *6050* (MO), *Z. S. Liu 810* (PE), *846* (PE), *849* (PE), 1938, *McLaren s.n.* (DAO), *AC86* (E), *AC216* (E), *Mussot 155A* (P), *Pratt 527* (P), *Sichuan Econ. Exped. 348* (CDBI), *Stevens 326* (F, GH, NY), *496* (F); *H. L. Tsiang 36404* (PE, SZ), *Smith 10825* (UPS), *X. L. Jiang 36404* (PE); Qianning, *F. D. Fu et al. 9-115* (CDBI), *S. Jiang & C. L. Jing 2115* (KUN, PE), *Sichuan Veget. Exped. 5775* (CDBI, PE); Litang, 1890, *Bonvalot & Henri s.n.* (P); Dawu, *Dawu Exped. 106* (SZ), *192* (SZ), *W.*

G. Hu & Z. He 10981 (SZ), *10984* (PE), *Smith 11801* (UPS); Jiulong, *Q. Q. Wang 20295* (CDBI); Zogqensi, *Y. W. Tsui 5041* (PE); Garze, *Sichuan Veget. Exped. 7708* (CDBI, PE); Luhuo, *W. L. Chen & C. J. Chen 7357* (MO, PE); Jiulong, *W. L. Chen & C. J. Chen 6343* (PE); Dege, *Y. W. Tsui 4461* (PE), *5165* (PE); Baiyu, *W. L. Chen & C. J. Chen 6789* (MO, PE), *J. T. Peng 2073* (PE); Leibo, *T. T. Yu 3432* (PE); Muli, *Qinghai-Xizang Exped. 4521* (KUN, PE), *13802* (KUN, PE), *Y. B. Yang 7160* (CDBI), *T. S. Ying 4111* (PE), *4238* (PE), *T. T. Yu 6632* (PE), *7139* (PE), *Q. S. Zhao 6374a* (CDBI), *6474b* (CDBI), *6895* (CDBI), *7925* (CDBI); Daocheng, *K. Y. Lang et al. 2705* (KUN, PE), *Z. K. Liu 215* (CDBI), *Sichuan Veget. Exped. 678* (CDBI), *1716* (CDBI, PE), *2422* (CDBI, PE), *2630* (CDBI), 1980, *S. K. Wu s.n.* (MO); Xiancheng, *Hengduan Shan Exped. 3681* (KUN, PE), *K. Y. Lang et al. 2647* (PE), *Z. G. Liu 1440* (CDBI, PE), *Nanshuibeidiao Exped. 9655* (PE, YNU), *Qinghai-Xizang Exped. 3681* (KUN, PE), *4522* (KUN, PE), *Sichuan Veget. Exped. 948* (CDBI), *2146* (CDBI, PE), *2715* (CDBI, PE), *N. W. Yunnan Exped. 9655* (YNU); Batang, *K. Y. Lang et al. 2262* (KUN, PE), *2451* (KUN, PE), *Soulie 3638* (P), *3639* (P), *3640* (P), *3641* (P); Derong, *Qinghai-Xizang Exped. 3391* (KUN, PE).—YUNNAN: Deqen, *Hengduan Shan Exped. 2282* (KUN, PE), *3621* (KUN, PE), *N. W. Yunnan Exped. 9485* (PE, YNU), *Qinghai-Xizang Exped. 3621* (KUN, PE), *C. W. Wang 69740* (BP, PE); *T. T. Yu 9364* (PE), *9380* (PE), *9761* (PE), *10873* (PE); near Gongshan, *Ducloux 5336* (P), *Forrest 14320* (W); Zhongdian, *K. M. Feng 2476* (GH, KUN), *Hengduan Shan Exped. 996* (PE); *1482* (PE), *N. W. Yunnan Exped. 9513* (PE, YNU), *10272* (YNU), *J. S. Yang 7510* (KUN, PE), *T. T. Yu 11876* (PE), *11961* (GH, KUN, PE), *12369* (PE), *12635* (PE), *13817* (PE); Weixi, *C. W. Wang 63898* (TAI), *68130* (TAI); Lijiang, *Rock 5227* (US), *Schneider 2447* (K), *Y. Z. Zhao 21295* (PE); Lanping, *H. T. Tsai 53791* (PE); western Yunnan, *H. T. Tsai 56506* (GH).—XIZANG: Jiangda, *Qinghai-Xizang Exped. 12503* (KUN, PE); Qamdo, *B. Z. Go & W. Y. Wang 21647* (HNWP), *Qinghai-Xizang Exped. 12644* (KUN, PE), *12650* (KUN, PE), *Xizang NWPBI Exped. 2674* (HNWP); Baxoi, *Soulie 2564* (L); Markam, *Qinghai-Xizang Exped. 11994* (KUN, PE); Rawu, *Y. T. Chang & K. Y. Lang 431* (E, PE), 1984, *H. S. Kung et al. s.n.* (MO); Lotu, *Ludlow et al. 15550* (BM, E, UPS); Zogang, *H. S. Kung et al. 20* (MO, PE), *Qinghai-Xizang Exped. 12110* (KUN, PE); Gongbogyamda, *Ludlow et al. 14085* (UPS); Zayu, *P. S. Li et al. 7256* (PE); Qayu, *Ludlow & Sherriff 2066* (BM); Sog Xian, *D. D. Tao 10936* (PE); Mainling, *Ludlow et al. 5697* (UPS), *Qinghai-Xizang Exped. 74-1998* (KUN, PE); Lulong, *Ecol. Group Xizang Exped. 15548* (PE); Biru, *D. D. Dao 11281* (KUN, PE); Gyaca, *Qinghai-Xizang Exped. 750720* (KUN, PE); Lhasa, *Ludlow & Sherriff 8901* (E, P, UPS), *11038* (UPS), *Ludlow et al. 4860* (BM, UPS); Himalaya region, *Strachey & Winterbottom 2* (S); no locality, *Hugel 1153* (W).

In most of northeastern, central, and southwestern China, both subspecies of *E. angustifolium* occur in the same general areas, whereas in Xinjiang, northern Xizang (Tibet), and northwestern Sichuan, only subsp. *angustifolium* occurs. In the areas of range overlap, subsp. *angustifolium* is usually found at higher elevations than subsp. *circumvagum*. For example, on Mt. Omei, Sichuan, populations of subsp. *angustifolium* are restricted to the higher zone from 2900 to 3135 m, whereas populations of subsp. *circumvagum* are found from (550–) 1100 to 3000 m, mostly below 2500 m.

Most collections of this taxon have distinctly narrowly lanceolate leaves, but a number of specimens, especially from Xinjiang, have broader ones, more within the range of subsp. *circumvagum*. As discussed previously, however, they have pollen, pubescence, and other characters that are typical for subsp. *angustifolium*, and may be best considered as robust collections of this taxon.

4b. Epilobium angustifolium subsp. **circumvagum** Mosquin, Brittonia 18: 167. 1966. *Chamaenerion angustifolium* var. *platyphyllum* Daniels, Fl. Boulder Colo. 176. 1911. *Chamaenerion angustifolium* subsp. *cirvumvagum* (Mosquin) Moldenke, Phytologia 27: 289. 1973.—TYPE: U.S.A. Colorado: Summit Co., canyons of Green Mountain, 2130 m, 16 Jul 1906, *Daniels 268* (holotype: MSC 59114!).

Common name: Maomai Liulan (Chen et al. 1988).

Stems 30–200 cm tall, up to 2 cm in diameter, glabrous below, sparsely strigillose throughout upper stem and inflorescence. Leaves with petioles 2–7 mm long; cauline blades (6–) 9–23 cm long, (0.7–) 1.5–3.4 cm wide, herbaceous or sometimes subcoriaceous, oblong- or elliptic-lanceolate, mostly sparsely pubescent on adaxial surface with strigillose leaf midribs, especially on the abaxial surface, apex attenuate-acute, base subcuneate to attenuate, margins more or less denticulate, flat or scarcely revolute, the lateral veins usually prominent. Inflorescence a raceme 10–50 cm long, strigillose, the lower bracts 7–25 mm long, foliaceous, the upper ones linear, very reduced and subcoriaceous. Floral buds 8–18 mm long, 4–7 mm in diameter, narrowly obovoid, usually cuspidate; ovaries 0.8–2.5 cm long; pedicels 1–3 cm long; sepals 9–19 mm long, 1.5–3 mm wide, acuminate; petals 14–25 mm long, 7–15 mm wide, obovate to suborbicular, usually entire; filaments 8–17 mm long; anthers 2.5–3 mm long, 1.1–1.7 mm in diameter; pollen mixed 3- and 4-porate (rarely 5-porate), (72–) 75–95 μm in diameter; style 9–18 mm long. Capsules 5–9.5 cm long, borne on pedicels 1–3 cm long. Seeds 1–1.3 mm long, 0.3–0.4 mm wide. Chromosome numbers: n = 36, 54. Figs. 5b, f, 16.

Phenology. Collected in flower from June to September, in fruit from July to October.

Distribution (Fig. 17). In North America across the breadth of southern Canada and the adjacent United States, extending northwestward along the Mackenzie Lowlands to the Arctic and along the Pacific coast to the Pribiloff Islands; in Asia from southern Kamchatka, Sakhalin and Ussuri regions, to Honshu, Japan, and Korea, through northeastern to southwestern China, northern Burma, the southern slopes of the Himalaya, Pakistan, the Caucasus Mountains and northeastern Turkey; in China, in Heilongjiang, Jilin, Liaoning, Nei Monggol, Hebei, Shanxi, Kunyu Shan of Shandong, western Henan, Shaanxi, southern Ningxia, southeastern Gansu, eastern Qinghai, Lushan of Jiangxi, western Hubei, Sichuan, western Guizhou, northwestern Yunnan, and eastern and southern Xizang (Tibet); in mountains; (50–) 500–1700 (–2800) m in northern China, (550–) 1100–3600 (–4350) m in southwestern China.

ADDITIONAL SPECIMENS EXAMINED. **China.** HEILONGJIANG: Huling, *G. Z. Wang 441* (IFP); eastern Heilongjiang, *F. H. Chen 144* (NAS), *430* (NY), 1924, *Kozlov s.n.* (BM); Shitouhe, *K. S. Hop 16128* (PE); "Djalantum," 1938, *Skvortzov s.n.* (GH); Jixian, *Y. L. Zhang 423* (PE), *1683* (PE); "Ninguta," 1925, *Kozlov s.n.* (BM); Mudan Jiang, *Mudan Exped. 220* (PE), *352* (PE), *625* (PE), *969* (PE); Ning'am, *P. Y. Fu 3272* (IFP); Shangzhi, *Y. C. Zhu 148* (IFP, PE), *168* (IFP); Acheng, *G. Z. Wang et al. 473* (PE); Dailing, *T. N. Liou 5956* (IFP), 1934, *W. Wang s.n.* (IFP), *W. Wang et al. 753* (PE); Laoye Ling, *Licent 8519* (BM, K); Wuyuhe, *Y. C. Zhu & Balarov 39* (PE); Hailun, *N.-E. China Exped. 341* (NAS, PE); "Chaffanjon," *Marecage 1693* (P); no locality, *L. C. Zhang 57* (PE).—JILIN: Hunchuan, *P. Y. Fu 770* (PE); Helong, *Yanbian Second Group 924* (PE); Wangqing, *C. S. Wang 1391* (PE); Dunhua, *T. N. Liou 3500* (IFP, PE), *3508* (NAS); Changbai Shan, *J. J. Chien 450* (PE), *Pek. Veget. Exped. 421* (PE), *N.-E. China Teacher's Univ. 399* (KUN, NETU, PE), *W. Wang et al. 1865* (IFP, PE), *G. S. Zhou et al. 122* (PE); Fusong, *S. X. Li & S. Y. Wang 546* (PE), *T. N. Liou 1280* (IFP), *Tunghua Exped. 298* (PE), *Y. L. Zhou 1250* (PE); Jinyu, *T. N. Liou et al. 1148* (IFP, NAS); Linjiang, *S. X. Li et al. 1206* (IFP, PE); "O-muhsien," *H. W. Kung 1773* (PE), *1830* (PE); Ganhe, *J. C. Zhu 532* (PE); no locality, *T. H. Chen 430* (P, PE).—LIAONING: Shenyang to Jilin, 1886, *James s.n.* (NY).—NEI MONGGOL: Da Hinggan Ling, *P. Y. Fu et al. 3796* (IFP), *Y. C. Zhu et al. 39* (IFP); "Nan chan," *Licent 3529* (P); Rem Meng, *Z. H. Zhang 1-271* (IFP); eastern Nei Monggol, *David 2242* (P).—HEBEI: Chengde, *Nankai Univ. 17* (PE); Dongling, *Dongling Exped. 109* (KUN, PE), *1092* (PE), *H. F. Chow 40579* (PE), *Y. Liu*

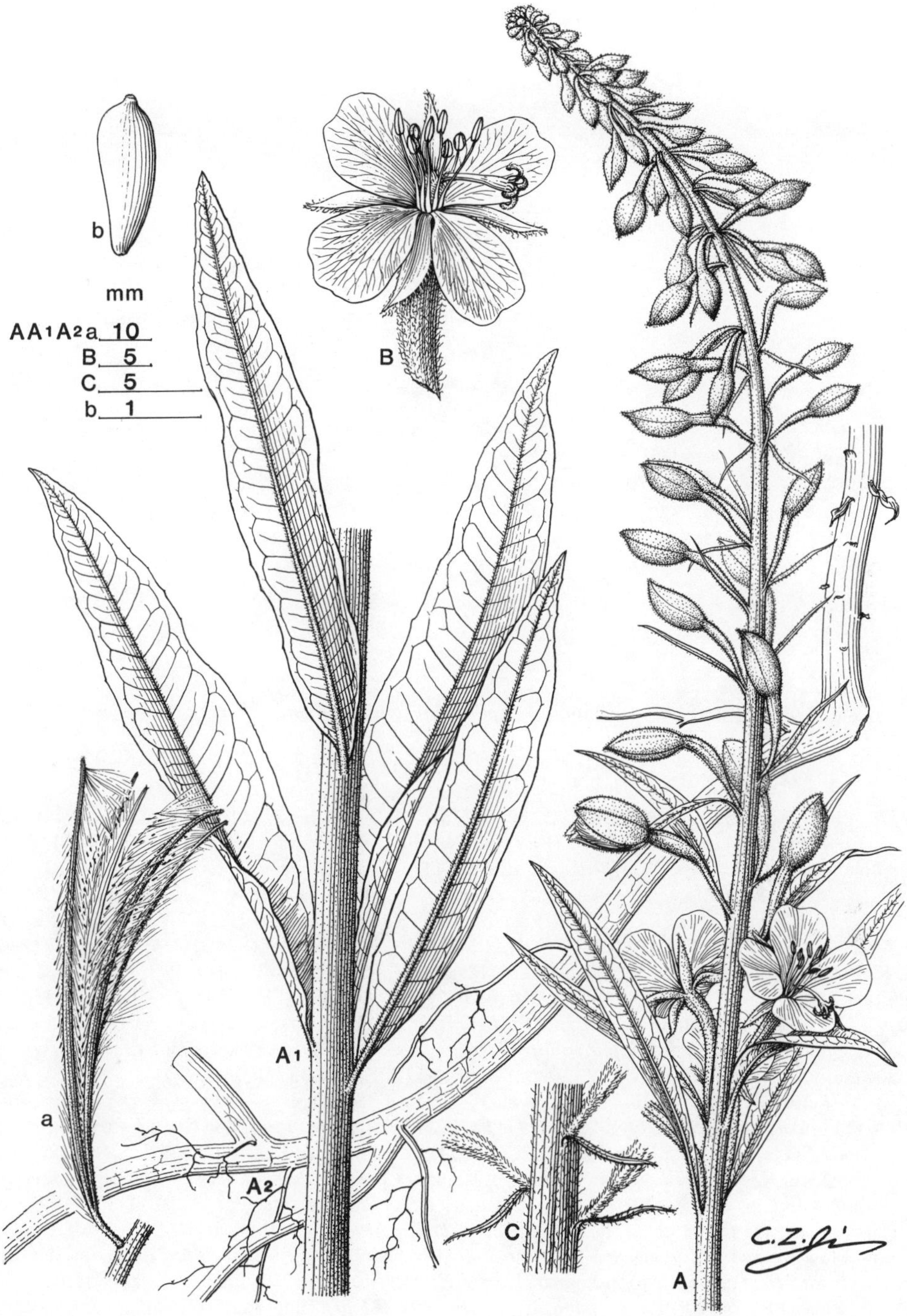

FIG. 16. *Epilobium angustifolium* subsp. *circumvagum*. A. Habit: upper stem and inflorescence. A¹. Mid-stem. A². Stem base, caudex, and root (partial). B. Flower. C. Axils from inflorescence. a. Capsule. b. Seed. (A–C based on *T. F. King 135*; a–b based on *Xian University 206*.)

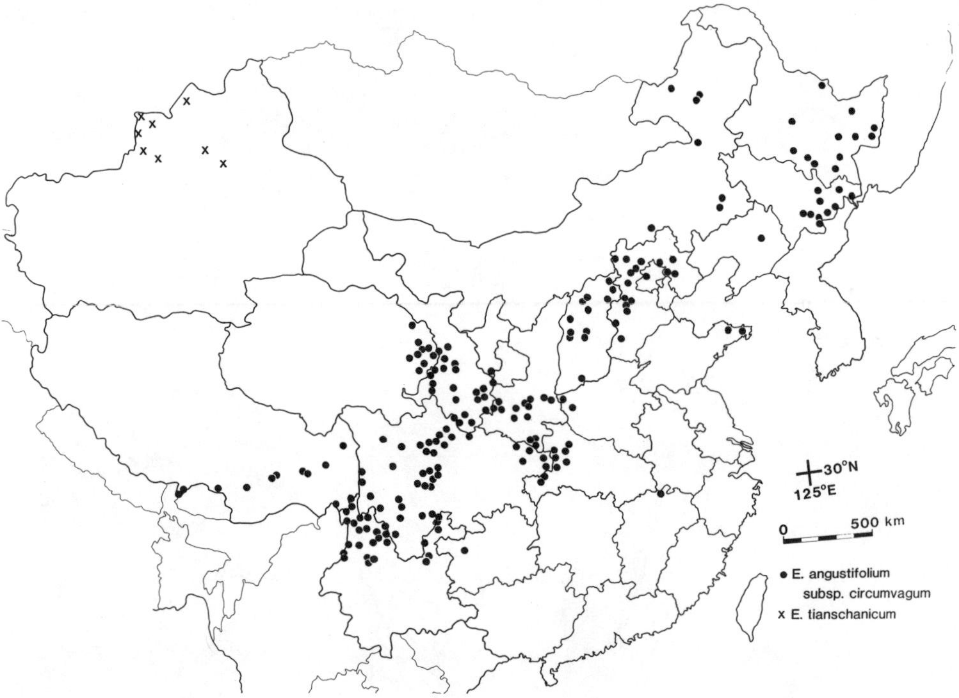

FIG. 17. Distribution in China of *Epilobium angustifolium* subsp. *circumvagum* and *E. tianschanicum*.

11520 (PE), *11682* (PE), *11813* (PE), *11832* (PE), *11860* (PE), *T. Tang 2182* (PE), *C. G. Yang 748* (PE), *1369* (PE), *1667* (PE), *K. M. Liou 413* (PE), *Wu 247* (PE); Wuling Shan, *Z. F. Fang 882* (IFP), *T. N. Liou 696* (PE), *4801* (IFP), 1933, *Nakai et al. s.n.* (TI), *W. X. Zhang et al. 53* (PE); northern Hebei, *Licent 8134* (K), *T. N. Liou 5289* (PE), *W. X. Zhang et al. 53* (PE); Beijing, s.d., *Hemeling s.n.* (BM, E, K), *Inst. Bot. Beijing 2152* (PE), *7298* (PE), *Licent 4578* (K), *D. F. Jing 247* (PE), *364* (PE), *521* (PE), *T. F. King 27* (NY, S), *521* (PE, S, UC), *Y. C. Ma 10172* (PE), *Moellendorff 484* (BM), *Schindler 177* (BM, E, K, W), *Serre 2600* (M), *C. W. Wang 60153* (PE), *W. T. Wang 343* (PE), *679* (PE), *2544* (PE), *2693* (PE), *Wu 156* (PE), *C. G. Yang 483* (PE), *256* (KUN, PE), *748* (KUN, PE), *1719* (PE); Chicheng, *Hebei Exped. 4153* (PE); Zhangbei, *Hebei Exped. 3518* (PE), *3731* (PE); Zhangjiakou, *Y. W. Tsui 2033* (PE), *2067* (PE), *2339* (PE), *2880* (PE); Zhoulu, *W. Y. Hsia 2415* (PE); Huailai, *Hebei Exped. 1723* (PE); Xiaowutai Shan, *J. C. Liu 2303* (NY), *Wu & Yang 2963* (PE), *Xiaowutai Shan Exped. 999* (PE), *1920* (PE), *2741* (PE), *3039* (PE), *10178* (PE), 1906, *Y. Yabe s.n.* (NAS), *C. Y. Yang 36292* (PE), *X. Y. Zhu 89138* (PE); Laiyuan, *Hebei Univ. Agr. 2114* (PE), *3054* (PE), *K. M. Liou 2686* (PE); Longguan, *Hebei Exped. 4282* (PE), *4286* (PE), *6548* (PE), *7670* (PE); Ling Shan, *T. F. King 135* (PE), *J. C. Liu 1469* (UC); Yi Xian, *Hebei Exped. 915* (PE); south of Beijing, *Chanet & Serre 917* (P), *2600* (PE); Nieqiu, *X. Y. Liu 793* (KUN, PE), *13103* (NAS); Pingshan, *Herb. Beijing Inst. Bot. V1057* (PE).—SHANXI: Wuzhai, *Huang He Exped. 2235* (PE); Ningwu, *C. T. Ren 6196* (UC); Jinyuan, *K. Ling 9300* (E, UC); "Teow Mow Konget Jan Yang Seu," *Licent 2849* (BM, K, P, W); "Chiao-ch'eng distr., Pa-shui-ko-shan," *Smith 7108* (UPS); western Shanxi, *Huang He Exped. 2606* (PE), *2718* (PE); Lishi, *Huang He Exped. 2425* (PE), *2492* (PE); Yuanqu, *S. Y. Bao 677* (PE); "Tung Tsa," *Couduback 5589* (UC); "Youngmongchan," *Licent 522* (BM, K, P); Wutai Shan, *Clemens 6277* (E), *6277a* (E), *Y. L. Chen 1772* (PE), *W. Y. Hsia 1393* (PE), *K. C. Kun & Y. L. Chen 837* (PE), *1081* (PE), *2467* (PE), *Licent 6603* (G), *Serre 2163* (W), *Shanxi Exped. 618* (NAS); Taigu, *Shanxi Teacher's Coll. 235* (PE); Zhongyang, *Shanxi Exped. 95* (PE); Jiaocheng, *X. Y. Liu 21449* (PE).—SHANGDONG: Yantai, Kunyu Shan, *Shangdong Exped. 3142* (PE).—HENAN: Lushi, *Henan Exped. 34596* (PE).—SHAANXI: Hua Xian, *K. S. Hao 4004* (PE), 1899, *Hugh s.n.* (BM), *T. P. Wang 15876* (KUN, PE, WUK); "Huan-ton-san," 1899, *Hugh s.n.*

(BM); "Kan-y-san," 1899, *Hugh s.n.* (BM); "Thae-hei-san," 1899, *Hugh s.n.* (BM); "Ki-fong-san," 1898–1900, *Hugh s.n.* (BM); "Miao-wang-san," *Hugh s.n.* (BM); Ankang District, *Xian Univ. 206* (WUK); Langao, *Z. P. Wei et al. 260* (WUK); Hu Xian, 1904, *Hugh s.n.* (BM), *B. Z. Go 400* (PE); Zhouzhi, *X. M. Zhang 178* (NAS); Taibai Shan, *Q. Hong & Z. S. Huang 46* (PE), *Huanghe Exped. 3135* (KUN, PE), *Y. S. Kuo et al. 538* (WUK), *Lanzhou Univ. Exped. 252* (LZU), *301* (LZU), *302* (LZU), *Y. Y. Pai 1581* (PE), *F. T. Wang et al. 591* (PE); Ningshan, *K. T. Fu 17945* (WUK), *H. W. Kung 3009* (PE); Foping, *B. Z. Go 1402* (PE); Feng Xian, *K. T. Fu 12572* (LZU, WUK), *13424* (LZU, WUK), *J. M. Liu 10617* (PE, WUK); Laoyu, *B. Z. Go 890* (PE, WUK); "Chone-pei-san," 1899, *Giraldi s.n.* (K, P, UC).—Ningxia: Jingyuan, *Jingyuan Exped. 83* (PE), *T. P. Wang 17075* (PE).—Gansu: Border between Gansu and Shaanxi, 1910, *Hosie s.n.* (K); Tianshui, *Q. F. Zhong 439* (LZU), *K. T. Fu 16055* (WUK), *W. Y. Hsia 5517* (PE), *Q. Zhao 137* (LZU); Hui Xian, *Z. X. Peng 5614* (LZU); Hui Xian to Cheng Xian, *Z. Y. Zhang 800* (WUK); Wudu, *Hopkinson 491* (P, PE); Wen Xian, *W. L. Chen & C. J. Chen 8807* (MO, PE), 1959, *Lanzhou Univ. Exped. s.n.* (LZU); Yuzhong, *C. J. Chen & X. Y. Zhu 89077* (PE), *Huang He Exped. 1716* (NAS), *W. L. Gao 102* (LZU), *Huang He Exped. 3297* (PE), *3471* (PE), *Lanzhou Univ. Exped. 183* (LZU), *211* (LZU), *Wen Xian Ecol. Pl. Exped. 211* (LZU); Xinlong Shan, *S. Y. Wang 161* (LZU); Tung Shan, *K. S. Hao 75* (PE); southern Gansu, *Z. Y. Peng 826* (LZU); "Toul Ping, near Lichen," *R. C. Ching 438* (E, US); Min Xian, *K. S. Hao 631* (PE), *Lanzhou Univ. 109* (LZU), *T. P. Wang 4662* (PE); Linxia, *Tao He Exped. 3705* (NAS, PE); Lintan, *Q. R. Wang & M. S. Wu 11069* (PE); Yuzhong, *Y. Q. He 5886* (PE), *5978* (PE), *Huanghe Exped. 3135* (PE); Lanzhou, *Y. Q. He 5154* (PE), 1964, *W. M. Go et al. s.n.* (LZU); "Founho," 1944, *C. K. Chow s.n.* (NAS); Tao River, *Rock 12954* (E, GH, K); Lianhua Shan, *Rock 12755* (GH, K, NY); "Kon-lang-kien," *Licent 4555* (K, P, W); "Lon Kia Wa Sen," *Licent 4771* (K, P); Xiahe, *K. T. Fu 1013* (PE); border between Gansu and Qinghai, *Ridley 123* (K); near "Chouni," *T. P. Wang 5368* (PE); "Lapuchan," *T. P. Wang 7337* (PE); "Tsamousche et Meischu inter Kweite et Labrang," *G. Fenzel 2525* (W).—Qinghai: Minhe, *S. W. Liu et al. 2945* (HNWP, PE), *L. H. Zhou 2506* (HNWP); Datong, *K. M. Liou 6384* (PE); Huzhu, *B. Z. Go 6643* (PE), *9547* (PE); Xining, *Hopkinson 815* (P, PE); eastern Qinghai, *H. W. Kung 554* (LZU); Huangzhong, *S. W. Liu 2084* (HNWP), *H. P. Tsui et al. 80-044* (PE); Menyuan, *K. M. Liou 6970* (PE), 1957, *Qilian Shan Exped. s.n.* (PE); Xunhua, 1980, *W. S. Zhang s.n.* (HNWP), *T. P. Wang 6008* (PE); Tongren, *S. W. Liu & D. S. Lio 1395* (HNWP), *L. H. Zhou & K. Z. Zhang 1267* (HNWP).—Jiangxi: Lushan, *Y. G. Xiong 113* (LBG).—Hubei: Fang Xian, *Wilson 1381* (E, K, NY, US, W); Chienghi, *Wilson 1336* (K); Shennongjia, *G. X. Fu & Z. S. Zhang 1030* (KUN, PE), *Shennongjia Pl. Exped. 10766* (PE), *21433* (PE), *21434* (PE), *22844* (PE), *22870* (PE), *31536* (PE), *31712* (PE), *32280* (PE); Badong, *R. H. Huang 321* (KUN); Enshi, *G. X. Fu & Z. S. Zhang 1314* (KUN, NAS, PE); western Hubei, *Henry 6031* (BM, E, GH, K, NY, US), *Henry 6031a* (G, K, P).—Sichuan: Wushan, *H. F. Zhou & H. Y. Su 109908* (PE, SZ), *109918* (PE, SZ); Wuxi, *K. L. Chu 2070* (PE), *P. Y. Li 3484* (WUK), *B. Z. Ni 547* (CDBI), *G. H. Yang 58730* (PE), *59344* (CDBI, NAS, PE), *59439* (CDBI, KUN, PE); Chengkou, *T. L. Dai 101050* (PE, SZ), *101813* (CDBI), *105622* (CDBI, KUN, NAS, PE, SZ), *105664* (CDBI, KUN, NAS, PE, SZ), *109622* (PE); Wanyuan, *Z. Y. Zhang et al. 125* (CDBI); border between Sichuan and Hubei, *G. H. Yang 59752* (PE); Kai Xian, *T. L. Dai 101810* (PE, SZ); Pingwu, *H. L. Tsiang 10801* (PE); Nanping, *W. L. Cheng & C. J. Chen 8550* (MO, PE), *8852* (MO, PE), *8855* (PE); *Q. X. Li & X. C. Zhao 3038* (PE), *Sichuan Econ. Pl. Exped. 4144* (CDBI), *H. T. Tsai 63791* (S); Songpan, *W. P. Fang 4309* (E, GH, K, P, PE, US), *Sichuan Econ. Pl. Exped. 1564* (CDBI, PE), 1914, *Weigold s.n.* (W); Aba, *W. L. Chen & C. J. Chen 7981* (MO, PE); Maowen, *Z. He & Z. L. Zhou 13246* (PE, SZ), *14241* (PE, SZ); Heishui, *X. Li 73108* (NAS, SZ), *72250* (SZ), *73350* (PE), *74061* (NAS, PE, SZ), *X. Li & J. X. Zhou 73961* (SZ, PE); Shuajingsi, *X. Li 74631* (PE, SZ), *74687* (PE, SZ), *74866* (NAS, PE, SZ); Li Xian, *S. Y. Chen et al. 5316* (SZ), *Z. He 14241* (SZ), *D. P. He 44413* (SZ), *45122* (SZ), *45262* (SZ), *45381* (SZ), *46527* (PE, SZ), *X. Li & J. X. Zhou 74217* (PE, SZ), *T. J. Zhang 53-323* (SZ); Barkam, *S. Jiang 1580* (PE), *Sichuan Forestry Exped. 195* (CDBI), *Sichuan Med. Exped. 22331* (NAS); Wenchuan, *W. P. Fang 1460* (GH, K, P, US), *1469* (E), *K. Y. Lang et al. 1272* (KUN, PE), *F. T. Wang 21766* (GH); Dajing, *X. Li 77893* (NAS, PE), *78095* (PE, SZ), *78345* (NAS), *78753* (PE); Xiaojin, *S. S. Chang & Y. X. Ren 5873* (PE, SZ), *5995* (CDBI, SZ), *6402* (CDBI, PE, SZ), *6496* (CDBI); Mt. Emei, *S. Y. Chen et al. 4158* (SZ), *Y. Chen 7133* (NAS), *7147* (NAS), *Cheo & Hsu 328* (NAS), *C. Y. Chiao & C. S. Fan 818* (GH), *Emei Bot. Exped. 104* (PE), *Y. H. Dao 51641* (SZ), *W. P. Fang 2822* (E, GH, K, NY, P, PE, US), *2941* (E, GH, PE), *12993* (BM, PE, US, W), *Z. T. Guang 6624* (PE), *X. Y. He 6091* (NAS), *T. C. Lee 3249* (PE), *3486* (US), *3879* (PE), *T. N. Liou & C. Wang 1237* (PE), *Raven et al. 26130* (MO, PE), *K. S. Shing & K. Y. Lang 34* (PE), *35* (PE), *S. C. Sun & K. Chang 1125* (GH), *T. H. Tu 613* (PE), *R. T. Wang 23439* (P, PE), *Wilson 4915* (BM, GH, P), *S. H. Wu 119* (GH, S, SZ), *J. H. Xiong et al. 31494*

(IBSC, PE), *31533* (PE), *32638* (PE), *G. H. Yang 57442* (NAS, PE, SZ), *C. W. Yao 3353* (PE), *4652* (NAS, PE), *S. Z. Yu 49923* (SZ), *X. B. Zhang 364* (NAS), *M. Z. Zheng 142* (PE); Ebian, *Sichuan Veget. Exped. 13218* (CDBI), *S. L. Sun 1060* (PE); Hongxi, *Sichuan Econ. Pl. Exped. 1551* (PE), *C. W. Yao 3066* (PE); Leibo, *T. T. Yu 3432* (PE), *3846* (PE); Meigu, *Sichuan Econ. Pl. Exped. 1993* (PE); Baoxing, *S. S. Chang & Y. X. Ren 5709* (PE, SZ), *7251* (SZ), *K. L. Chu 3597* (BM, W), *S. Y. Hu 1309* (SZ), *Sichuan Veget. Exped. 7251* (CDBI), *T. P. Soon 39065* (SZ), *T. H. Tu 4431* (PE); Tianquan, *W. K. Hu 34798* (PE), *35383* (PE, SZ), *35421* (PE, SZ), *37718* (PE, SZ); *S. Jiang 1849* (PE), *Y. B. Yang 40039* (CDBI), Luding, *Z. He 11601* (PE), *W. K. Hu & Z. He 10268* (PE, SZ), *K. C. Kuan et al. 1835* (PE), *G. Y. Xu 23135* (CDBI); Kangding, *C. Y. Chiao 1647* (F, S), *Cunningham 482* (E), *W. P. Fang 3662* (GH, K, P, PE, US), *3667* (E, GH), *36710* (PE, SZ), *36915* (PE, SZ), *Farges 314* (L), *W. G. Hu & Z. He 10680* (PE, SZ), *Henry 8945* (P), *X. L. Jian 36915* (IBG, SZ), *K. C. Kuan et al. 241* (K, PE), *H. S. Kung 6008* (MO), *6050* (PE), *Lidjestrand 17* (UPS), *C. S. Liu 849* (PE), *McLaren AC124* (E), *AC216* (PH), *Mussot 155B* (P), *Soulie 233* (G, K, L, MO, MPU, P), *Pratt 417* (BM, K); Kangding to Baoxin, *Stevens 83* (F, US); Garze, *Sichuan Med. Exped. 15420* (NAS), *T. S. Ying 9563* (PE); Dawu, *Sichuan Veget. Exped. 5881* (CDBI), *Smith 10961* (S, UPS), *Soulie 867* (G, K); Batang, *Soulie 3929* (P); Rongjing, *S. Jiang 7727* (KUN, PE); Mianning, *S. G. Wu 1764* (PE); Anning, *X. Li 78449* (NAS, PE); Jiulong, *Q. Q. Wang 21024* (CDBI); Yanyuan, *Qinghai-Xizang Exped. 12802* (KUN, PE); Muli, *T. T. Yu 6056* (PE), *14357* (BM, E, GH, PE), *Q. S. Zhan 6374* (CDBI), *Q. S. Zhao et al. 6895* (CDBI), *Qinghai-Xizang Exped. 5949* (KUN, PE); Xiangcheng, *Sichuan Veget. Exped. 948* (PE, SZ); western Sichuan, *Bock & Rosthorn 2576* (O, UPS).—Guizhou: N. W. Guizhou, s.d., *Bodinier s.n.* (P).—Yunnan: Yongshan, *H. T. Tsai 51048* (KUN, PE); Daguan, *N. E. Yunnan Exped. 307* (KUN, YNU); Qiaojia, 1914, *Maire s.n.* (E, P); "Tien-sin," 1921, *Maire s.n.* (P); "le-ma-tchouan," *Maire 243* (BM, E); "Pe long tsin," *Maire 7469* (NY, UC); Huize, *Dongchuang Exped. 63-106* (KUN); Dongchuan, Jun 1913, *Maire s.n.* (G), Aug 1913, *Maire s.n.* (E, G, P); Huaping, *S. Jiang 6220* (KUN, PE); Yongling, *Nanshui Beidiao Exped. 6101* (KUN, PE); *Schneider 1628* (E, G, GH); Ninglang, *Handel-Mazzetti 3156* (L, W); Dali, *Forrest 4457* (BM, E), *28054* (BM, E), *McLaren et al. B. 131* (E, K, P), *Chanet 1039* (K), *H. C. Wang 820* (KUN, PE), *2200* (KUN), *2325* (KUN, PE); "Ma-enl-shan," *Delavay 3847* (P); central Yunnan, 1889, *Delavay s.n.* (L, MO, P); Heqing, *R. C. Ching 23993* (KUN); Binchuan, *Chanet 917* (K); Deqen, *K. M. Feng 5563* (KUN, PE), *6250* (KUN), *6805* (KUN, PE), *Qinghai-Xizang Exped. 1974* (KUN, PE), *Nanshui Beidiao Exped. 9485* (KUN, PE), *70150* (GH, KUN, NAS), *T. T. Yu 8774* (KUN, PE), *9380* (KUN, PE), *9761* (KUN, PE); near Bai-mar-shan, *C. W. Wang 69740* (NAS, PE); "Dokeria Numah shula," *T. T. Yu 7890* (KUN); Gongshan, *Feng 9007* (PE), *Nanshui Beidiao Exped. 9153* (KUN, PE, YNU); Zhongdian, *K. M. Feng 1694* (GH, KUN, PE), *23304* (KUN), *Nanshui Beidiao Exped. 9513* (PE), *Qinghai-Xizang Exped. 996* (KUN, PE), *1482* (KUN, PE), *T. T. Yu 11876* (KUN, PE), *12635* (KUN, PE), *12639* (KUN, PE), *13817* (BM, E, KUN), 1962, *A. L. Zhang s.n.* (KUN), *Zhongdian Pl. Exped. 791* (KUN, PE), *63-2893* (KUN, PE); Lijiang, *Forrest 2759* (E, K, P), *10283* (E, K), *Rock 4782* (UC, US), *4911* (E, US), *Schneider 3447* (G, US), *T. T. Yu 15212* (KUN, PE), *Yulong Shan Exped. 817* (YNU), *Yunnan Univ. 817* (KUN); Weixi, *C. W. Wang 63898* (GH, NAS), *68130* (GH, KUN), *J. S. Yang 7008* (KUN); Lanping, *H. T. Tsai 53791* (GH), *S. K. Wu 8915* (KUN); Bijiang, *Bijiang Pl. Exped. 623* (WUK), *H. T. Tsai 58260* (GH, KUN, P, PE), *58457* (GH, KUN); N.-W. Yunnan, *H. T. Tsai 57170* (GH, KUN), *57732* (GH, NAS, PE); Pianma, *S. K. Wu 8457* (KUN); Lushui, *Nanshui Beidiao Exped. 10388* (KUN, PE); eastern Xizang, *Forrest 4457* (E); western Yunnan, 1917, *Simeon Ten s.n.* (P).—Xizang: N. E. Xizang, 1921, *Leamer s.n.* (E); Baxoi, 1905, *Manberg s.n.* (K), 1903, *A. Monbeig s.n.* (L, MO, P), *Soulie 1341* (L, MO, P), *Soulie 2564* (MO), *2964* (P); Zayu, 1984, *H. S. Kung et al. s.n.* (MO, PE), *Qinghai-Xizang Exped. 7744* (KUN, PE); Zhamu, *Y. T. Chang & K. Y. Lang 651* (PE); Bomi, *T. S. Ying & D. Y. Hong 336* (PE); Nyingzhi, *Y. T. Chang & K. Y. Lang 1135* (PE); Lulang, *Xizang Veget. Exped. 15548* (PE); Mainling, *P. S. Li et al. 5992* (PE), *Xizang Med. Exped. 3608* (PE); Qayu, *Z. C. Li & D. Chi 801* (PE), *Ludlow & Sherriff 2322* (E); Lhakang, *Ludlow & Sherriff 510* (BM); Yadong, 1904, *Bell s.n.* (P); Yadong to Pagri, *J. W. Zhang & G. X. Fu 1133* (PE); Yadong, *Xizang Exped. 1867* (HNWP, KUN, PE).

This subspecies grows further south and southeast than subsp. *angustifolium,* especially in southern Hebei, Shangdong, Shaanxi, Jiangxi, Hubei, and eastern Sichuan; in other regions, where it grows with subsp. *angustifolium,* it generally grows at lower elevations. The collection from Lushan, Jiangxi (Yujing Shan, *Y. G.*

Xiong 113, LBG), is nearly 500 km to the southeast of any other collections of this taxon, and may represent a secondary introduction into this well-collected area.

One unusual collection of subsp. *circumvagum* from Yunnan (Bijiang, 26°30′N, 99°E, at the border of Yunnan and Burma, 4000 m, *H. T. Tsai 58262,* GH) had completely glabrous styles, but agreed in other characters with most specimens.

II. Epilobium section **Epilobium.**

Epilobium section *Lysimachion* Tausch, Hort. Canal. fasc. 1. 1823.—LECTOTYPE, designated by Munz, 1965: *Epilobium hirsutum* Linnaeus.

Perennial herbs, erect, ascending or procumbent, often clumped, rarely basally woody. Stems terete, subterete, or 4-angled, simple to well-branched, pubescent to glabrous, the pubescence strigillose, villous, tomentose, canescent, or glandular, often mixed. Leaves opposite below inflorescence, very rarely throughout, or very rarely alternate or spiral throughout, sometimes scalelike below ground, the scales 3–10 mm long, coriaceous, rotund, spatulate, or elliptic; the cauline blades (0.5–) 1–12 (–23) cm long, lanceolate, ovate, elliptic to linear or rarely spatulate, membranous or rarely subcoriaceous. Inflorescence erect or sometimes nodding, simple to well-branched; bracts foliaceous, at most only somewhat smaller than cauline leaves. Flowers actinomorphic, opening in parallel to stem axis, sometimes protandrous, very rarely protogynous (not in China); floral tube formed by fused bases of sepals, petals, and filaments, 0.6–2.8 mm long, 0.8–3 mm in diameter, producing nectar from base of style, detaching from ovary prior to seed maturation; petals deeply two-cleft to emarginate; stamens in two unequal whorls of 4 each, suberect throughout anthesis; filament bases not swollen, arising near mouth of floral tube; pollen yellow, shed in tetrads; style erect throughout anthesis; stigma usually entire, cylindric to capitate, subequal to or exserted beyond anthers, sometimes 4-lobed and exserted, the lobes opening and spreading soon after flower's opening, commissural, receptive on inner surfaces. Chromosome number: n = 18.

Section *Epilobium,* as defined most clearly by Raven (1976), includes about 150 species, which occur virtually throughout the geographical and ecological range of the genus. A majority of species of *Epilobium* in any region is in this section, including 33 of the 37 species in China.

5. **Epilobium hirsutum** Linnaeus, Sp. Pl. 1: 347. 1753. *Chamaenerion hirsutum* (Linnaeus) Scopoli, Fl. Carniol. ed. 2, 1: 270. 1772.—TYPE: Europe [exact locality not specified] (holotype: LINN 486.3!).

Epilobium villosum Thunberg, Prodromus 75. 1794. *Epilobium hirsutum* var. *villosum* (Thunberg) H. Hara, J. Jap. Bot. 18: 178. 1942.—TYPE: not located. (Hara referred to var. *villosum* Haussknecht, Monogr. Epil. 55. 1884; however, the basionym was first validly published by Thunberg; also, Haussknecht's use was ambiguous, may have been informal, and even if formal was not necessarily at the varietal level [Hoch & Raven 1990].)

Epilobium tomentosum Ventenat, Descr. Pl. Nouv. Jard. Cels. t. 90. 1802. *Epilobium hirsutum* var. *tomentosum* (Ventenat) Boissier, Fl. Orient. 2: 746. 1872. *Epilobium velutinum* Nevski, Trudy Bot. Inst. Akad. Nauk SSSR, Ser. 1, Fl. Sist. Vyss. Rast. 4: 312. 1937, nom. superfl.—TYPE:

Cultivated in France at garden of J. M. Cels, possibly from Persia (holotype: not located).

Epilobium hirsutum var. *sericeum* Bentham ex C. B. Clarke in J. D. Hooker, Fl. Brit. India 2: 584. 1879. [*Epilobium sericeum* Bentham in Wallich, Numer. List 216, n. 6325. 1832, nomen nudum.]—TYPE: INDIA. Himachal Pradesh, Sirmur, *Gerard* in *Wallich 6325* (holotype: K!).

Epilobium hirsutum var. *laetum* Wallich ex C. B. Clarke in J. D. Hooker, Fl. Brit. India 2: 584. 1879. [*Epilobium laetum* Wallich, Numer. List. 216, n. 6329. 1832, nomen nudum.]—TYPE: INDIA. Uttar Pradesh, Kumaun, *Blinkworth* in *Wallich 6329B* (holotype: K!).

Epilobium nassirelmulcii Stapf, Denkschr. Kaiserl. Akad. Wiss. Wien Math.-Naturw. Kl. 51: 325. 1886.—TYPE: IRAN. In wet places in mountains around Hamadan, 26 Jul 1882, *Polak & Pichler s.n.* (lectotype, here designated: W!).

Common name: Liuyecai (Liou et al. 1977; an early reference in Chinese literature is from *Jiuhuang Bencao* [Plants for Disaster Relief Food] by Su Zhu in 1368).

Robust perennial herb, sometimes woody near base, producing long, thick, ropelike hypogeal stolons to 1 m or more in length, with scattered cataphylls and often terminating with rosette of leaves. Stems 25–120 (–250) cm tall, 3–9 mm in diameter, well-branched in upper half, densely covered with long spreading villous pubescence, with an admixture of short glandular hairs especially on inflorescence, or rarely sparsely pubescent, or rarely densely white-tomentose. Leaves mostly opposite, alternate on inflorescence, sessile and more or less clasping the stem; cauline blades 4–12 (–23) cm long, 0.3–4 (–5) cm wide, lance-elliptic to narrowly obovate or elliptic, rarely very narrowly lanceolate, villous on both surfaces, very rarely glabrescent, apex acute to acuminate, base subcuneate, margin serrulate with 20–50 teeth on each side, the lateral veins inconspicuous, 7–9 on each side of the midrib. Inflorescence erect. Flowers erect; buds 4.5–9 mm long, 2.5–5 mm in diameter, ovoid-oblong, apiculate; ovaries 2–5 cm long, grey-green to purplish, densely long-villous and short-glandular-pubescent, sometimes predominantly glandular, or rarely densely villous-tomentose and lacking glandular hairs; pedicels 0.3–1.5 cm long; floral tube 1.3–2 mm long, 2.2–3 mm in diameter, a conspicuous ring of long white hairs at the mouth within; sepals 6–12 mm long, 1–2 mm wide, oblong-linear, keeled, pubescent as on ovaries; petals 9–20 mm long, 7–15 mm wide, bright pink to dark purple, broadly obcordate, the notch 1–2 mm deep; filaments of longer stamens 5–10 mm long, those of shorter ones 3–6 mm long; anthers 1.5–2.5 mm long, 0.6–1 mm wide, oblong, cream; style 5–12 mm long, erect, white or pink, glabrous, very rarely with scattered villous pubescence; stigma deeply 4-lobed, the lobes 2–3.5 mm long, oblong, initially erect, later recurved, whitish, glabrous or very rarely scattered villous on the outer surface, exserted beyond anthers. Capsules 2.5–9 cm long, greenish brown, sometimes purplish, pubescent as on ovaries, rarely glabrescent, borne on pedicels 0.5–2 cm long. Seeds 0.8–1.2 mm long, 0.35–0.6 mm in diameter, obovoid, with very short chalazal collar, dark brown, the surface coarsely papillose; coma 7–10 mm long, tawny or dull white, detaching easily. Chromosome number: $n = 18$. Figs. 2a, 4f–h, 18.

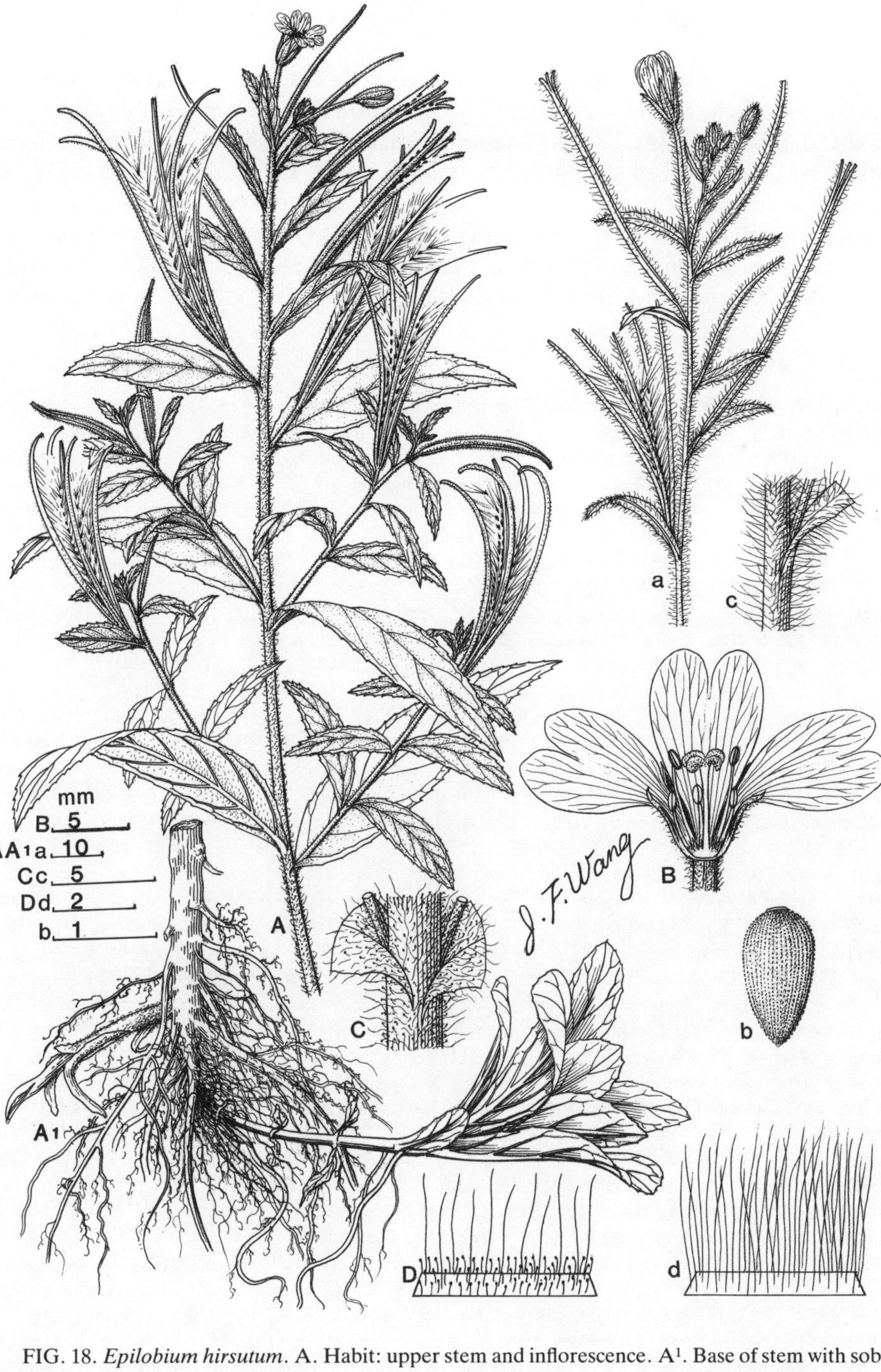

FIG. 18. *Epilobium hirsutum*. A. Habit: upper stem and inflorescence. A[1]. Base of stem with sobole terminating in leafy rosette. B. Flower (one petal and three stamens removed). C. Axils from mid-stem showing clasping leaf bases. D. Detail of hairs from mid-stem: mixture of longer villous and much shorter glandular pubescence. a. Habit: inflorescence. b. Seed. c. Axil from inflorescence. d. Detail from c: long villous pubescence. (A–D based on *C. J. Chen 83001*; a–d based on *Xinjiang Exped. 1646*.)

Phenology. Collected in flower from June to August, in fruit from July to September.

Distribution (Fig. 12). Temperate and montane Eurasia and Africa; introduced in North America; common in temperate and subtropical China, in Jilin, Liaoning, Nei Monggol, Hebei, Shanxi, Shandong, Henan, Shaanxi, southern Ningxia, Gansu, Xinjiang, Anhui, Jiangsu, Zhejiang, Jiangxi, Guangdong, Hunan, Hubei, Sichuan, Guizhou, Yunnan, and Xizang (Tibet); cultivated as an ornamental in Nanjing, Guangzhou, and some other cities; frequent and weedy in wet places near streams, ditches, marshes, gravel or sandy beds of rivers, and along roadsides; (200–) 500–2000 m in northern China, (180–) 500–2800 (–3500) m in southwestern China.

ADDITIONAL SPECIMENS EXAMINED. **China.** JILIN: Antu, *Yanbian Exped. 398* (PE).—LIAONING: Guoqian Qi, *T. N. Liou et al. 6370* (IFP); "Hengren," *C. S. Wang 4083* (IFP, PE); Lingyuang, 1959, *S. X. Li s.n.* (IFP); Xifeng, *X. P. Yu 181* (IFP), *250* (IFP); Zhangwu, *C. S. Wang et al. 3796* (PE); Jinzhou, 1910, *Y. Yabe s.n.* (PE), 1917, *Y. Yabe s.n.* (NAS); Lingyuan, *S. X. Li et al. 342* (PE); "Manchuria," s.d., *Faber s.n.* (NY).—NEI MONGGOL: Ihju Mong, *Nei Mong. Univ. Exped. 91* (NMU); Jirem Mong, *J. C. Ma & W. S. Yang 92* (NMU, PE).—HEBEI: Dongling, *T. Tang 2269* (PE); Ji Xian, *Hebei Med. Pl. Exped. 760* (PE), *Ji Xian Exped. 1984* (PE); Kuan Tso Ling, *J. C. Liu 1167* (UC); Beijing, *Beiging 45* (KUN, PE), *Bretschneider 1868* (BM, GH), 1903–1935, *Chanet & Serre s.n.* (MO), 1903–1935, *Chanet & Lazaristes s.n.* (MO, P), *J. Chang 2009* (PE), *C. J. Chen 83001* (MO, PE), *83007* (PE), *83010* (MO, PE), *H. F. Chow 40527* (PE), *41564* (PE), *David 427* (P), *D. Z. Fu & Q. Y. Xiang 83055* (PE), *W. Y. Hsia 2001* (PE), *Inst. Bot. 6748* (PE), *6796* (PE), *K. C. Kuan 7043* (PE), *J. C. Liu L. 1092* (UC), *L. 1333* (NY, UC), *K. M. Liou 497* (NY, PE), *T. N. Liou 6962* (PE), 1933, *T. N. Liou s.n.* (PE), *Y. Liu 15511* (PE), *Moellendorff 1774* (BM), *Read 730* (UC), *Serre 1430* (K), 1851, *Tatarinow s.n.* (MO), *C. W. Wang 60271* (PE), *60952* (PE), *S. T. Wang 220* (PE), *W. T. Wang 2439* (PE), *3062* (PE), *Wawra 1006* (W), Aug 1905, *Y. Yabe s.n.* (NAS, PE), Sept 1905, *Y. Yabe s.n.* (NAS), *C. G. Yang 1018* (KUN, PE); Zhoulu, *C. G. Yang 1697* (KUN, NAS, PE); Xiawutai Shan, *Smith 1168* (UPS), 1921 *Smith s.n.* (UPS); *C. W. Wang 61856* (GH); Liayuan, *Hebei Univ. Agr. 4-3725* (PE), *K. M. Liou 2848* (PE); Laishui, *C. F. Yang 361* (KUN); Zhanhuang, *Q. Y. Liu 891* (KUN, NAS, PE); Fuping, *K. M. Liou 3337* (PE); Beian He, *D. F. Jin 191* (PE); Neiqiu, *Y. Liu 13646* (KUN, NAS, PE); "Tschi-fu," *Wawra 1253* (W); no locality, *Licent 1359* (PE).—SHANXI: Xie Xian, *Shanxi Exped. 116* (NAS); Wutai Shan, *K. C. Kuan & Y. L. Chen 2542* (PE), *2549* (PE); Wutai, *W. Y. Hsia 1275* (PE); Chensi Central, *Licent 2391* (BM); Chansi Meridian, *Licent 2239* (K, P), *2269* (BM, K, P), *2323* (BM, K, P, W); Luyu Shan, *Huang He Exped. 2391* (PE); Li Shan, *Huang He Exped. 1699* (PE); Hengshan, *T. P. Wang 3947* (PE); Fenyang, *T. P. Wang 2772* (PE); Kanson, *Licent 6207* (P); "Tschili: Haitou-schan, pr.. Yati," *Licent 9880* (W); "Sin Louotch'ang," 1923, *Licent s.n.* (K); Yunch'eng, *Smith 6075* (UPS), *6080* (UPS); "Hsia-hsien," *Smith 6230* (UPS); "Neicheng," *S. Y. Bao 1529* (PE).—SHANDONG: Tai Shan, *T. Y. Chou 5328* (PE), *S. C. Cui 212* (PE), *Shandong Univ. 3* (KUN, PE), *7365* (NAS); Weihai, *Matthew 34* (HK); Lao Shan, *C. Y. Chiao 2851* (BM, F, GH, K, MICH, NY, PE, TAI, UC); Fei Xian, *T. Y. Cheo & L. Yen 181* (BM, G, GH, P, W), *298* (GH); Wenshui, *Keng 127* (GH); Pingyi, *T. Y. Chou 400* (NAS).—HENAN: Hui Xian, *K. M. Liou 5614* (PE); Jiyuan, *Henan Exped. 6084* (PE); Dengfeng, *Henan Exped. 106* (PE), *1209* (PE); Yiyang, *Henan Exped. 22054* (PE), *22095* (PE), *23549* (PE), *23974* (PE), *Z. W. Wen 51235* (NAS); Song Xian, *Henan Exped. 34878* (PE), *35038* (PE), *35186* (PE), *35285* (PE); Luoning, *C. C. Li et al. 31* (PE); Funui Shan, *Henan Forestry Dept. 453* (PE), *640* (PE); Lushi, *K. M. Liou 4752* (K, PE), *J. Q. Fu 912* (NAS); Neixiang, *K. C. Kuan & T. L. Dai 671* (NAS); Xixia, *Henan Exped. 991* (PE); Tongbai, *Henan Exped. 56069* (PE); Luanchuan, *Henan Exped. 2009* (PE), *15002* (PE), *20031* (PE); Lingbao, *Henan Exped. 14301* (PE), *14766* (PE); no locality, *Schindler 138* (BM, G, K, W).—SHAANXI: Yulin, *K. T. Fu 6928* (PE), *W. Y. Hsia 3687* (PE); Hengshan, *K. T. Fu 7113* (PE); Yan'an, *Huang He Exped. 315* (PE), *376* (PE); Nanniwan, *Huang He Exped. 968* (PE); Ganquan, *W. Y. Hsia 5792* (PE); Lochuan, *K. T. Fu 3442* (PE); Yijun, *C. L. Xiao 624* (KUN, PE); Tongchuan, *C. L. Xiao 400* (KUN, PE); Hua Shan, *K. S. Hao 3939* (PE), *4159* (PE); "Chungnanshan," *Y. Y. Pai 1078* (PE), *1172* (PE); *T. P. Wang 2174* (PE), Shang Xian, *T. P. Wang 16225* (KUN); Lantien, *T. P. Wang 15961* (KUN, PE); "Cho tse Miao," s.d., *Hugh s.n.* (BM); "Ki-fon-san," 1898, *Hugh s.n.* (BM); "Paoki, Maowansan," *Hugh 32* (BM); Miao-uansan, *Hugh s.n.* (BM); "Kusan", *Hugh 81* (BM); Ziyang, *P. Y. Li 7154* (WUK); Hu Xian, *Shaanxi-Gansu Exped. 10933* (KUN, PE); Zhouzi, *X. M. Chang 4* (NAS, WUK), *Y. Y. Pai 1372* (PE); Taibai Shan, *S. T. Sen 5779* (KUN), *T. P. Wang 1435* (PE), *J. X. Yang 1173* (PE), *1242* (WUK); Zhengba, *K. T. Fu 11658* (KUN); Hanzhong, *J. W. Wang & Z. C. Shi 48* (PE); Chenggu, *K. T. Fu 5354* (PE); Feng Xian, *Z. Y. Chang 57* (NAS), *K. T. Fu 12956* (LZU); Yueyang, *C. L. Tang 838* (NAS); Ningqiang, *P. Y. Li 132* (NMU); Mei Xian, *X. M. Chang 847* (NAS); border between N. Shaanxi and Gansu, s.d., *T. Y. Le s.n.* (PE); "Tzewuyu," *H. W. Kung 2978* (PE); no locality, *W. D. Chang 110* (PE), *Licent 2391* (K), *Limpip 2834* (WU).—NINGXIA: Longde, *T. P. Wang 13647* (HNWP).—GANSU: N. E. Gansu, *Licent 6207* (K); Tiansui, *W. H. Chang 381* (LZU), *J. L. Ding 417* (LZU), *Q. X. Ding 65* (LZU); *J. M. Liu 10030* (PE), *P. L. Liu & T. H. Jiang 67* (LZU), *Z. S. Ji 329* (LZU), *Z. A. Lu 464* (LZU), *C. S. Wang et al. 358* (LZU), *C. Y. Zhu 67* (LZU), Jinsuo Zhen, *Huang He Exped. 3442* (KUN); Hui Xian, *Y. S. Lian & J. L. Bai 820047* (PE), *Z. X. Peng 5364* (LZU); Cheng Xian, *X. Pu 80316* (LZU); Wen Xian, *Z. Y. Chang 6057* (HNWP), *6156* (HNWP), *7037* (HNWP), *7961* (HNWP), *8056* (HNWP), *8201* (HNWP), *15390* (PE), *W. L. Chen & C. J. Chen 8742* (MO, PE), *8772* (PE), *Y. Q. He 883* (PE), *1179* (PE), *Wen Xian Exped. 3052* (HNWP, LZU), *J. X. Yang & Z. X. Hu 3712* (WUK); Kang Xian, *W. Y. Hsia 6309* (PE), *Z. Y. Zhang 17181* (PE); Shimen, Wudu, *W. Y. Hsia 6390* (PE); Zhengning, *P. F. Chang 621* (LZU); Chongxin, *S. Q. Zhong & Y. Zhou 50* (PE); Chenggu, *H. J. Wang 148* (WUK); Pinglian, *Huang He Exped. 1940* (PE); "Bitin," *T. P. Wang 16843* (PE).—XINJIANG: Turpan, *H. P. Tsui et al. 80-455* (PE), *Turpan Exped. 3890* (XJBI); Toksun, *Xinjiang Exped. 7334* (XJBI, PE); Dabancheng, *C. J. Chen & Hoch 86024* (MO, PE); Manas, *Acad. Sin. Xinjiang Exped. 23-519* (PE); Urumqi, *R. C. Ching 443* (KUN, PE), 1931, *Reymond s.n.* (P, W), *T. N. Liou 2649* (HNWP, PE), *2830* (HNWP, PE), *T. N. Liou 3205* (PE), *G. L. Chu 5448* (NAS, PE); Shihezi, *N. N. Feng 265* (PE), *F. Konta & K. Y. Lang 324* (PE), *K. C. Kuan 1176* (HNWP, PE, XJBI); Toli, *A. R. Li & J. N. Zhu 10941* (KUN, PE, XJBI); Beitun, *T. Y. Chou 652047* (NAS); Huocheng, *C. W. Chang 4042* (XJBI), *Y. R. Ling 74854* (PE), *Xinjiang Exped. 1456* (HNWP), *1646* (HNWP); Tacheng, *K. C. Kuan 2820* (PE), *Xinjiang Exped. 2407* (HNWP); Wenquan, *C. J. Chen & Hoch 86312* (MO, PE); Yining, *Z. X. Peng 5054* (LZU), *Xinjiang Med. Pl. Exped. 497* (XJBI); Yili, *K. C. Kuan 3082* (PE), *Ili Exped. 21* (PE); Chabuchar, *Xinjiang Exped. 3026* (PE, WUK); Miquan, *Miquan Exped. 6500130* (HNWP); Toli, *A. R. Li & J. N. Zhu 10941* (PE); Tekes to Zhao, *Xinjiang Exped. 2732* (PE, WUK); west of Tekes, *K. C. Kuan 3639* (PE, XJBI); Gongliu, *Q. R. Wang et al. 2509* (PE, WUK), *2633* (PE); Altay, *R. C. Ching 2464* (PE); no locality, *G. L. Chang et al. 703* (LZU).—ANHUI: Meishan Reservoir, Jinzhai, *Y. L. Wu 001-1* (HSFS); Guichi, *Guichi Exped. 7184* (NAS); Huizhou, *Nakai 30291* (NAS), *30518* (NAS).—JIANGSU: Nanjing, *T. P. Chang 303* (NAS), *Z. L. Zhao 523* (NAS); Haizhou, *Chang & Cheng 924* (NAS).—ZHEJIANG: Tianmu Shan, *W. Y. Hsia 294* (PE); Jinhua, *H. Y. Ho 2959* (NAS), 1935, *H. Migo s.n.* (NAS); "Chien-teh," *K. Ling 2731* (NY).—JIANGXI: Lushan, *M. J. Wang 1185* (LBG); Jinggang Shan, *S. S. Lai et al. 4672* (LBG); no locality, *Z. B. Yang et al. 1094* (PE).—GUANGDONG: Ruyuan, *X. P. Gao 53407* (IBSC, PE); Lechang, *L. G. Chen 42977* (HK, IBSC, NAS, PE); Xinfeng, *L. Teng 7996* (IBSC, NAS, PE); Wengyuan, *X. X. Liu 25082* (HTC, IBSC, NAS, PE, TAI); Guangzhou, *W. Y. Chun 8867* (IBSC).—HUNAN: Yongshui, *Hunan Veg. Group 515* (HTC); Baojing, *L. H. Liu 9789* (HTC, IBSC); Nanyue, *Herb. Hunan Teachers Coll. 9249* (HTC); "Mangshan," *M. X. Huan 113064* (HTC, IBSC); Henshan, *Yang 6011* (PE); Xinning, *C. J. Qi & J. G. Xiao 10298* (CSFU); western Hunan, *S. G. Li 204112* (HTC); no locality, *R. H. Shan 599* (NAS).—HUBEI: Yichang, *Henry 2257* (K, P), *K. K. Tsoong 4050* (PE), *W. Y. Chun & S. S. Chien 8155* (UC); Fang Xian, *K. M. Liou 8946* (NAS, PE), *9298* (NAS), *J. Q. Xing 16962* (KUN), *17928* (KUN); Xiaozidian, *A. R. Li et al. 103* (PE); Xingshan, *Y. Chen 2011* (UC), *C. M. Hu 98* (IBSC); Badong, *L. Q. Chen et al. 1829* (PE), *M. Z. Chien 1028* (PE); "Nanto," *Wilson 1415* (K, W); western Hubei, *Wilson 2654* (K, W); "Hu Tze-see," *Silvestri 6171* (BM); "Pao-kan," *Silvestri 6174* (GH); no locality, *Henry 2753* (BM, GH, P), *7521* (GH, IBSC, K).—SICHUAN: Wu Shan, *G. H. Yang 65544* (KUN, NAS), *H. F. Zhou & H. Y Su 109589* (PE); Wuxi, *K. L. Chu 1781* (PE); Chengkou, *K. L. Chu 2102* (PE), *T. L. Dai 102259* (PE, SZ), *103354* (NAS, PE, SZ), *103409* (NAS, PE, SZ); Fengjie, *M. Y. Fang 24667* (PE), *H. F. Zhou & H. Y. Su 110893* (NAS, PE, SZ); Nanchuan, *Jinfo Shan Exped. 1887* (PE), *2310* (PE), *G. F. Li 61097* (SZ), *632976* (KUN, NAS, PE, SZ), *64097* (KUN, PE), *C. Pei 10026* (K), *Bock & Rosthorn 427* (O), *837* (O), *J. H. Xiong & Z. L. Zhou 92107* (PE, SZ); "Hsu-Tsing," *Smith 459* (UPS); Lanping, *W. L. Chen & C. J. Chen 8715* (MO, PE), *Lanping Exped. 80006* (LZU); Songpan, *Chengdu Inst. Biol. 8639* (CDBI); north of Wenchuan, *F. T. Wang 21724* (GH, PE); Lixian, *F. T. Wang 21698* (PE), *Z. He & Z. L. Zhou 13890* (PE); Mt. Emei, *T. N. Liou 10179* (PE), *C. W. Yao 4954* (PE); Tianquan, *W. P. Fang 3489* (GH, NY, PE); Luding, *W. P. Fang 3719* (GH, K, NY, P, PE), *11478* (PE), *K. C. Kuan et al. 1592* (PE); *T. P. Wang 9759* (PE); Jinchuan, *X. Li 75525* (NAS, PE, SZ), *75623* (NAS, PE, SZ), *75770*

(NAS, PE, SZ), *75778* (NAS, PE, SZ), *76704* (NAS, PE), *76813* (NAS, PE), *78251* (NAS, PE), *78402* (NAS, PE), *G. Z. Zhu & X. Li 75248* (PE, SZ), *75861* (NAS, PE, SZ), *76503* (PE, SZ); Baoxing, *K. L. Chu 3834* (BM, W), *K. C. Kuan et al. 2963* (PE), *T. P. Soong 39460* (PE, SZ), *Stevens 47* (G, GH, S); Kangding, *Liljestrand L30* (UPS), *H. S. Kung 6218* (MO, PE); Shimian, *C. C. Hsieh 41776* (PE), *42808* (PE); Megui, *Chengdu Inst. Biol. 13030* (CDBI, PE), *Liangshan Econ. Pl. Exped. 5971* (PE); Butuo, *Liangshan Econ. Pl. Exped. 5685* (PE), *5798* (PE); Yuexi, *Chengdu Inst. Biol. 14013* (CDBI, PE); Leibo, *S. Jiang 7545* (KUN, PE), *Sichuan Med. Pl. Exped. 27851* (NAS); Puge, *Liangshan Econ. Pl. Exped. 5574* (PE), *14174* (PE), *Chengdu Inst. Biol. 14174* (CDBI); Dukou, *Q. S. Zhao 5502* (CDBI); Xichang, *W. L. Chen & C. J. Chen 5587* (PE), *6138* (PE), *T. T. Yu 1251* (PE); Huidong, *N. F. Yu 60-1303* (SZ), *Q. S. Zhao & G. Hu 5845* (CDBI); Huili, *Handel-Mazzetti 5164* (S, W, WU), *T. T. Yu 1377* (GH, PE); Miyi, *Q. S. Zhao et al. 5368* (CDBI); Hanyuan, *T. P. Wang 9590* (PE), *Sichuan Med. Pl. Exped. 15963* (NAS); Daocheng, *Chengdu Inst. Biol. 2523* (CDBI, PE); Muli, *J. S. Liu 7627* (CDBI), *Qinghai-Xizang Exped. 14850* (KUN, PE), *Y. B. Yang 7359* (CDBI), *T. T. Yu 14421* (PE), *14485* (PE), *Q. S. Zhao 6099* (CDBI), *8665* (CDBI), *Q. S. Zhao et al. 8246* (CDBI); Xiancheng, *T. T. Yu 13316* (PE); no locality, *W. P. Fang 4971* (NAS), *5100* (NAS), *5377* (NAS), *9280* (NAS), *Maire 348* (BM).—GUIZHOU: Songtao, *Z. Y. Cao 757* (GZBI, PE), *N. Guizhou Bot. Exped. 1391* (GZBI, PE), *1957* (GZBI, NAS, PE); Shiqian, *Wuling Exped. 2627* (KUN, PE); Shibing, *Wuling Exped. 3417* (KUN, PE); Meitan, *N. Guizhou Bot. Exped. 2548* (PE); Guiyang, *K. Z. Yang 25* (GZAC); Weng'an, *Libo Bot. Exped. 1756* (GZBI, PE); Dushan, *Y. Tsiang 6461* (M, NAS, NY); Suiyang, *West China Acad. Sci. 3409* (PE); Pingba, *P. C. Troong 126* (PE); Dafang, *Y. Tsiang 8312* (NAS), *8942* (NAS), *Anshun Exped. 1738* (GZBI); Pingfa, *Y. Tsiang 5380* (GH, NAS, NY, PE, S, UC, W); Hezhang, *Guizhou Exped. 510* (GZBI); Ceheng, *Z. Y. Cao 555* (GZBI, PE); Xingren, *Guizhou Exped. 7537* (PE), *8428* (GZBI, NAS, PE); Xingyi, *Guizhou Exped. 6388* (GZBI, NAS, PE); Anlong, *Guizhou Exped. 5760* (PE); Shuicheng, *K. Z. Yang 141* (GZAC); no locality, *Bodinier 2452* (P), *Cavalerie 208* (MO, P), *3075* (P), *S. W. Teng 90559* (GH).—YUNNAN: Yanjin, *N. E. Yunnan Exped. 966* (KUN); Yiliang, *B. Y. Qiu 21265* (KUN), *60749* (KUN), *771063* (KUN); Zhenxiong, *H. W. Li 176* (KUN); Qiaojia, *N. E. Yunnan Exped. 1285* (KUN), *1286* (KUN); Dongchuan, Jul 1921, *E. Maire s.n.* (L, MO, P, WU), Aug 1921, *E. Maire s.n.* (WU), Jul 1921, *R. Maire s.n.* (MO); "Pau-long-se," 1911, *Maire s.n.* (MO); Luquan, *P. I. Mao 1381* (KUN), *1662* (KUN, PE); northeast Yunnan, *Ducloux 283* (P), *3162* (P), *3819* (P); Songming, *Ducloux 7536* (L, MO), *7624* (MO, P), *P. I. Mao 218* (KUN, PE), *Sino-American Bot. Exped. 1353* (A, MO, PE); Kunming, *K. M. Feng 10230* (PE), *10232* (KUN), *H. Li 556* (KUN), *565* (KUN), *801125* (KUN), *T. N. Liou 13755* (PE), *16518* (PE), *16860* (KUN, PE), *20453* (PE), *20726* (PE), *20867* (PE), 1926, *Mander s.n.* (S), *B. Y. Qiu 436* (KUN), *447* (KUN), *50233* (SHIN); *50266* (KUN, SHIN), *50818* (KUN), *54778* (KUN), *77535* (KUN), *77608* (KUN), 1978, *s.n.* (MO), 1980, *Raven et al. s.n.* (MO), *Schoch 257* (G, WU), *K. K. Tsoong 2022* (PE), 1956, *Yunnan Univ. s.n.* (YNU); Jiangchuan, *B. Y. Qiu 60259* (KUN), *60285* (KUN); Yanshan, *C. W. Wang 84031* (KUN, PE); Chengjiang, *H. Wang 41755* (GH); Mengzi, 1890, *Leduc s.n.* (P), *Hancock 316* (K, NY), *Henry 13723* (NY), *S. T. Li 45* (KUN), *205* (KUN), *C. W. Wang 81516* (KUN, PE); Huaping, *C. W. Wang 69328* (GH, NAS, PE, TAI); Jingdong, *M. K. Li 1378* (KUN, PE), *3291* (KUN); Heqing, *K. M. Feng 811* (GH, KUN); Chenyuan, *Y. Tsiang 12599* (NAS); Lanping, *Hengduan Shan Exped. 657* (PE), *H. T. Tsai 56056* (NAS, PE), *56155* (GH, KUN, NAS, PE), *56245* (GH, KUN, NAS, PE), *56248* (PE); Dali, *Delavay 3* (P), *5* (P), *1223* (P), *Delavay 2864* (MO), *2942* (P), 1884, *s.n.* (L, MO, P), *Forrest 4461* (K, P), *T. N. Liou 16505* (KUN), *Sino-American Bot. Exped. 745* (A, PE), *1560* (A, PE), *K. K. Tsoong 2988* (PE), *Y. Tsiang 11708* (NAS); Jindong, *M. K. Li 1378* (PE); Fengqing, *T. T. Yu 17586* (GH, KUN); Eryang, *R. C. Ching 23202* (KUN), *B. Y. Qiu 60931* (KUN); Tengchong, *Ducloux 4567* (P); Bijiang, *Z. G. He 586* (KUN), *590* (KUN), *598* (KUN); Weixi, *C. W. Wang 67611* (GH, PE), *67938* (KUN, PE); Yangbi, *T. N. Liou 22938* (PE); Yuangjiang, *Y. H. Li 5934* (KUN); Lijiang, *R. C. Ching 21165* (GH, PE), *R. Q. Feng 21165* (KUN), *K. M. Feng 2565* (GH, KUN), *S. T. Li 801199* (KUN), *80246* (KUN), *B. Y. Qiu 61147* (KUN), *61174* (KUN), *Qinghai-Xizang Exped. 14922* (KUN, PE), *Rock 5343* (UC), *5363* (K), *Schneider 1741* (GH), *C. W. Wang 65013* (PE), *S. W. Yu 63-134* (KUN), *Yulong Shan Exped. 769* (YNU); Zhongdian, *Hengduan Shan Exped. 2071* (KUN, PE), *2745* (PE), *3068* (PE), *T. T. Yu 13489* (GH, KUN, PE); Gongshan, *Qinghai Xizang Exped. 7957* (KUN, PE), *7958* (PE); Deqen, *Hengduan Shan Exped. 2976* (KUN, PE), *T. T. Yu 10222* (KUN, PE), *10393* (KUN, PE); Dongchuan, s.d., *Maire s.n.* (MO); "Pan-long-se," *Maire 6433* (UC); "Pan-long-te," 1912, *Maire s.n.* (G); "Tsin-long-se," *Maire 10045* (NY, UC); "Kin-tchong-chion," *Maire 7472* (NY, UC); "Cchez hiong," *Ducloux 500* (UC); "Tch'a-ho," 1912, *Maire s.n.* (P); "a-Ou-li-fai," 1912, *Maire s.n.* (G); "O'ui-fa," *Cavalerie 89* (K); "Ku-long-tchang," 1912, *Maire s.n.* (WU); "Hsin-Tien," *Y. P. Chang 739* (KUN); no locality, *H. T. Tsai 57559* (GH, KUN), *F.*

T. Wang 400 (KUN), *2225* (KUN), *T. T. Yu 13316* (GH), *Delavay 2864* (MO), 1884, *s.n.* (MO), *Ducloux 2684* (P), *Bonati 391* (UC), *478* (UC), *Esquirol 1097* (P); s.d., *Simeon s.n.* (GH).—XIZANG: Baxoi, *Soulie 1126* (L, MO), 1903, *Soulie s.n.* (P); no locality, *Hugel 1192* (W).

Epilobium hirsutum, extremely widespread in temperate Eurasia and Africa (Raven 1967a, 1968), is the most widely distributed species in China and exhibits great variability in leaf size and shape, flower size, and indumentum. Some plants from Xinjiang (including the Tian Shan) have a strikingly tomentose indumentum and resemble plants from elsewhere in Central Asia, which are sometimes segregated as *E. hirsutum* var. *tomentosum.* These plants are covered on their stems, leaves, and inflorescences with dense grey-white tomentose pubescence, usually lack glandular hairs entirely, and usually have scattered villous pubescence on the outer surface of the stigma and upper style. Other plants from Xinjiang, and most from the rest of China and elsewhere from the range of *E. hirsutum,* have the more typical pubescence pattern—mixed villous and glandular indumentum but glabrous style and stigma. In view of the extensive variability of *E. hirsutum,* we have not recognized these tomentose plants from Central Asia as a distinct taxon.

Epilobium hirsutum is one of the commonest species of the genus in Eurasia, especially the southern half, and is found in all parts of China except Heilongjiang, Taiwan, Hainan Island, the Fujian and Guangxi subtropical region, and the Qinghai-Xizang (Tibetan) Plateau.

Two collections from Yunnan (Qiaojia, *N-E Yunnan Exped. 1286,* KUN; Yanjin, *N-E Yunnan Exped. 966,* KUN) appear to be hybrids between *E. hirsutum* and probably *E. platystigmatosum.* Both specimens have smaller leaves (2–3 cm long, 0.4–0.7 cm wide) and less pubescence on the stems than is typical for *E. hirsutum,* and both have a high proportion of sterile pollen and seed.

6. Epilobium parviflorum Schreber, Spicil. Fl. Lips. 146, 155. 1771.—TYPE: GERMANY. From vicinity of Leipzig (not located).

Epilobium parviflorum var. *vestitum* Bentham ex C. B. Clarke in J. D. Hooker, Fl. Brit. India 2: 584. 1879. [*Epilobium vestitum* Bentham in Wallich, Numer. List 216, n. 6327. 1832, nomen nudum.]—TYPE: NEPAL. Sheopuri Hills, near Rivulet Rocks, N of Katmandu, 1821, *Wallich 6327* (holotype: K!; isotype: BM!).

Common name: Xiaohua Liuyecai (C. J. Wu 1979).

Robust, erect perennial herb, in autumn bearing short-stalked epigeous basal rosettes. Stems 18–100 (–160) cm tall, 3–10 mm in diameter, well-branched above, densely covered with long spreading grey-villous pubescence on the lower part and mixed villous and short glandular pubescence above, often with raised lines decurrent from the margins of petioles. Leaves alternate on upper part, opposite below, subsessile or the lower ones on petioles 1–3 mm long; cauline blades 3–12 cm long, 0.5–2.5 cm wide, narrowly lanceolate or oblong-lanceolate, more or less densely villous on both surfaces, the hairs often appressed, apex subacute, base usually rounded, the margin denticulate, often remotely so, with 15–60 teeth on each side, the lateral veins 4–8 on each side of the midrib. Inflorescence erect, often branched. Flowers erect; buds 3–5 mm long, 2–3 mm in diameter, oblong-obovoid, apiculate; ovaries 1–4 cm long, with abundant short erect glandular hairs, sometimes mixed with scattered villous hairs; pedicels 0.3–1 cm long; floral tube 1–1.9 mm long, 1.3–2.5 mm in diameter, a ring of villous pubescence within mouth of tube; sepals 2.5–6 mm long, 1–1.5 mm wide, narrowly lanceolate, keeled, covered with villous and glandular pubescence; petals 4–8.5 mm long, 3–4.5 mm wide, pink to bright rose-purple or rarely white, broadly obovate, the notch 1–3.5 mm deep; filaments of longer stamens 2.6–6 mm long, those of shorter ones 1.2–3.5 mm long; anthers 0.5–1.3 mm long, 0.35–0.6 mm in diameter, oblong; style 2.6–6 mm long, erect, white to pink, glabrous; stigma deeply 4-lobed, the lobes 1–1.8 mm long, oblong, initially erect, later recurved, whitish, subequal to anthers or barely exserted beyond them at anthesis. Capsules 3–7 cm long, with short erect glandular hairs, sometimes mixed with long eglandular ones, rarely glabrescent, borne on pedicels 0.5–1.8 cm long. Seeds 0.8–1.1 mm long, 0.4–0.5 mm in diameter, obovoid, with short, inconspicuous chalazal collar, brown, the surface coarsely papillose; coma 5–9 mm long, dingy or tawny, detaching easily. Chromosome number: n = 18. Fig. 19.

Phenology: Collected in flower from June to September, in fruit from July to October.

Distribution (Fig. 20). Eurasia, from Japan, north-central China, and the Himalaya to the Caucasas and Europe; also in northern Africa, and introduced in North America and New Zealand; in China, mainly in the Huang He (Yellow River) Basin (including Nei Monggol, Hebei, Shanxi, Shangdong, Henan, Shaanxi, and Gansu), in the central Chang Jiang (Yangtze River) Valley (including Hunan, Hubei, Sichuan, Guizhou, and northeastern Yunnan), and also in the Tian Shan region of Xinjiang; usually in disturbed wet places near streams, bogs, and rivers, and on open waste slopes and meadows in the mountains; (350–) 500–1800 (–2500) m.

Additional Specimens Examined. **China.** Nei Monggol: Ihju Meng, *Nei Mong. Exped. 227* (NMU).—Hebei: Beijing, *Beijing Exped. 3062* (PE), 1882, *Carles s.n.* (E), *Chanet & Serre 177* (MO, P), *674* (MO, P), *679* (MO, P), *Herb. Beijing Inst. Bot. 1660* (PE), *Hosticka s.n.* (PR), *K. M. Liou 505* (PE), *524* (PE), *J. C. Liu 1166* (UC), *P. Liu 2044* (UC), 1961, *X. Y. Liu 547* (PE); Yi Xian, *Hebei Univ. Agr. 556* (PE); Laiyuan, *K. M. Liou 2848* (PE), *2923* (PE); Fuping, *Hebei Exped. 19* (PE); Neiqui, *Y. Liu 13170* (NAS, NMU, PE); She Xian, *Hebei Exped. 78* (PE).—Shanxi: Wutai, 1959, *K. C. Kuan & Y. L. Chen s.n.* (PE); Taiyuan, *T. P. Wang 2696* (PE); Zhongyang, *Zhongyang Exped. 428* (PE); Fengyang, *T. P. Wang 2778* (K, PE); Lingchuan, *K. M. Liou 7830* (PE); Jincheng, *S. Y. Bao 418* (PE), *1485* (PE); *1490* (PE), *1628* (PE); Xia Xian, *Q. H. Go 706* (PE); Ruicheng, *S. Y. Bao 491* (PE).—Shangdong: Yidu, *Shangdong Exped. 5261* (NAS); Qufu, s.d., *Panvel s.n.* (P).—Henan: Lin Xian to Hui Xian, *Henan Exped. 4218* (PE), *4250* (PE); Lushi, *K. M. Liou 4797* (PE), *Henan Exped. 34337* (PE); Yiyang, *Henan Exped. 23549* (PE); Song Xian, *Henan Exped. 1740* (PE); Xixia, *Henan Exped. 935* (KUN, PE), *991* (PE), *1305* (PE), *1327* (PE); Launchuan, *Henan Exped. 15002* (PE).—Shaanxi: Hua Shan, *K. S. Hao 4210* (PE), *K. T. Fu 6825* (PE); Shang Xian, *T. P. Wang 16225* (PE); Lantien, *T. P. Wang 15949* (KUN, PE); Foping, *T. N. Liou & P. C. Tsoong 3205* (KUN, PE), *X. M. Chang 497* (NAS); Pingli, *P. Y. Li 9121* (WUK); Yulin, *K. T. Fu 6995* (PR); Mt. Taibai, *Z. X. Peng & S. Z. Chang 1501* (LZU), *J. X. Yang 1242* (HTC, PE); Ziyang, *P. Y. Li 4734* (PE), *4810* (PE), *4881* (PE), *6343* (PE); Yang Xian, *K. T. Fu 5148* (KUN), *T. N. Liou & P. C. Tsoong 3449* (PE); Chenggu, *J. X. Yang 2138* (PE); Baoji, *Hugh 32* (BM); 1898, *Hugh s.n.* (BM), *K. T. Fu 13809* (LZU); "Mt. Ki-fon-san," 1898, *Hugh s.n.* (BM); Baocheng, *K. T. Fu 5519* (KUN, PE); Liuba, *K. T. Fu 13627* (LZU).—Gansu: Huating, *T. P. Wang 17023* (PE), *19977* (PE); Tianshui, *W. S. Shao 2* (LZU), *L. K. Deng 244* (LZU), *G. L. Chang & X. Pu 79104* (LZU), *G. M. Xu 83-225* (PE); Cheng Xian, *J. Z. Sung 82* (LZU); Kang Xian, *Kang Xian Exped. 387* (LZU); Wen Xian, *Z. Y. Chang 6138* (HNWP), *7433* (HNWP), *13454* (WUK), *W. L. Chen & C. J. Chen 8802* (MO, PE), *J. X. Yang & Z. X. Hu 3404* (PE); Xigu, *T. P. Wang 14598* (PE); no locality, *T. T. Li 331* (PE).—Xinjiang: Urumqi, *T. N. Liou 3207* (PE), *Xinjiang Exped.*

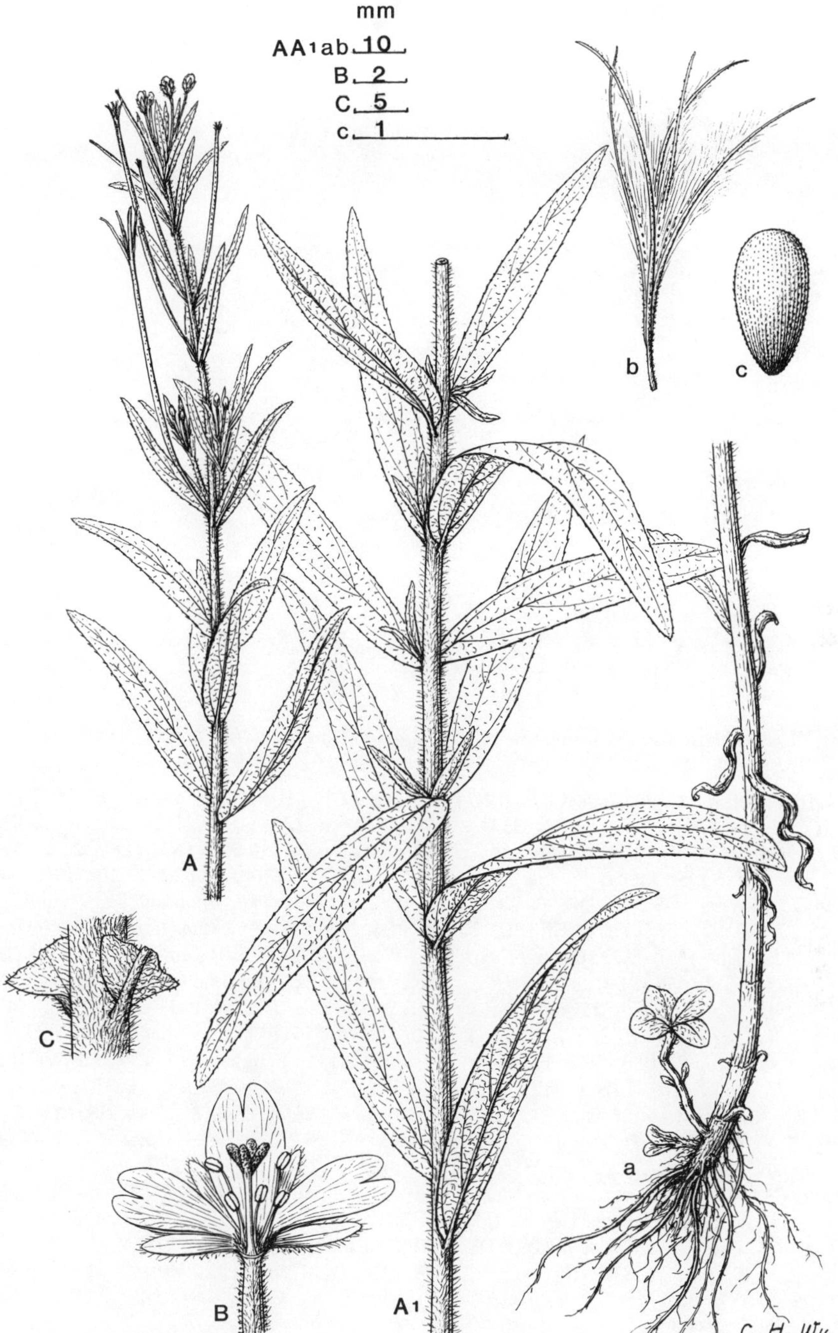

FIG. 19. *Epilobium parviflorum*. A. Habit: upper stem and inflorescence. A¹. Mid-stem. B. Flower (one petal and three stamens removed). C. Axils from mid-stem. a. Habit: base of stem with short-stalked leafy rosettes. b. Capsule. c. Seed. (A–C based on *J. X. Yang 1242*; a–c based on *H. Z. Li 658*.)

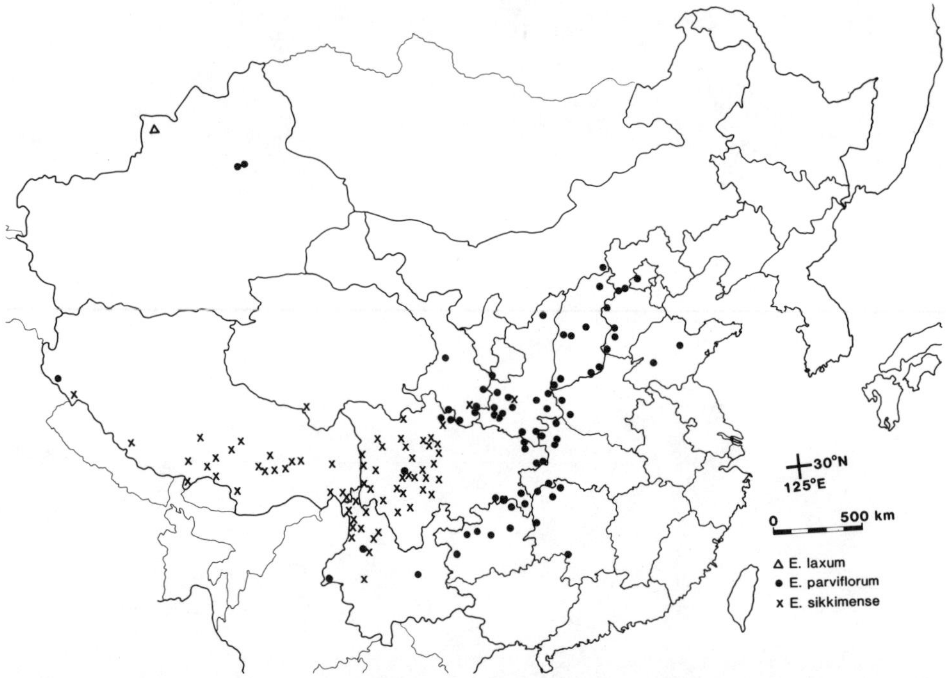

FIG. 20. Distribution in China of *Epilobium laxum*, *E. parviflorum*, and *E. sikkimense*.

4736 (XJBI); Tianchi, *C. J. Chen & Hoch 86-052* (MO, PE).—HUNAN: Longshan, *B. M. Yang 2117* (HTC); Fenghuang, *Henan Exped. 935* (HTC), *Wuling Exped. 1224* (IBSC, PE); Lingling to Xinning, *Handel-Mazzetti 11263* (S, W, WU); Sangzhi, *Hunan-Guizhou Exped. 3439* (KUN), *3562* (KUN), *3581* (KUN), *3647* (KUN); Shimeng, *Foping Shan Exped. A416* (PE); Cili, *X. Y. Xi 84210* (PE).—HEBEI: Jun Xian, *J. Q. Xing 16492* (KUN); Fangshan, *K. M. Liou 9298* (PE); Shennongjia, *Shennongjia Pl. Exped. 21273* (PE), *21670* (PE), *30978* (PE), *31478* (PE), *1980 Sino-American Exped. 1160* (GH), *1161* (KUN), *1832* (KUN, PE), *1835* (GH); Zhuxi, *T. P. Wang 11792* (PE); Badong, *Wilson 1415* (E, NY, W); Hefeng, *H. Z. Li 658* (PE); Jianshi, *H. C. Zhou 1102* (E, NY, PE); no locality, *Henry 2950* (GH, MO), *4843* (GH, MO), *Silvestri 1578* (K).—SICHUAN: Wushan, *H. F. Zhou & H. Y. Su 25962* (PE, SZ), *109608* (PE, SZ), *110071* (PE, SZ); Chengkou, *T. L. Dai 101656* (PE), *P. Y. Li 5450* (PE), *6745* (PE); Wanyuan, *P. Y. Li 5596* (PE), *5878* (PE), *5954* (PE), *6742* (PE); Fengjie, *Z. Y. Chang 25756* (PE), *H. F. Zhou & H. Y. Su 110586* (NAS, PE, SZ); Qian Jiang, *Z. C. Zhao 88-1306* (CDBI, PE), *88-1426* (CDBI, PE); Nanchuan, *Jinfo Shan Exped. 2130* (PE), *J. H. Xuing & Z. L. Zhou 93081* (KUN, PE); Youyang, *S. X. Tan 230* (PE); Zhaohua, *Y. Q. He 1691* (PE); Nanping, *W. L. Chen & C. J. Chen 8582* (PE); Songpan, *Z. He & Z. L. Zhou 13386* (PE); Luding, *K. C. Kuan et al. 1592* (PE); no locality, *T. Tang 21698* (NAS).—GUIZHOU: Shigian Xian, *Sino-Amer. Guizhou Bot. Exped. 2402* (A, PE), *2404* (A, PE); Songtao, *N. Guizhou Bot. Exped. 1574* (GZBI); Fanjing Shan, *Hunan-Guizhou Exped. 2611* (KUN); "San-Chouen," *Cavalerie 3982* (GH, K, MO); Dafang, *Y. Tsiang 8904* (NAS, NY, PE, UC, W); Xifeng, *S. W. Teng 90559* (GH); Pingba, *Anshun Exped. 1738* (PE).—YUNNAN: Kunming, *T. N. Liou 19499* (PE); Eryuan, *R. C. Ching 33202* (PE); Longchuan, *G. D. Dao 13524* (KUN).—XIZANG: "Batli", *Height 5839* (P).

Epilobium parviflorum resembles *E. hirsutum* in having relatively large flowers, a 4-lobed stigma, and villous pubescence, but it differs by having short but distinct petioles (the base not clasping), somewhat smaller flowers, a stigma with erect lobes surrounded by the anthers at anthesis (i.e., it is at least partly self-pollinating), and in perennating by rosettes rather than thick, ropelike stolons.

Epilobium parviflorum is moderately variable in overall stature, mix of pubescence types, and size of its leaves and flowers. Small-flowered specimens of *E. hirsutum* may sometimes be mistaken for *E. parviflorum;* otherwise, it is very distinct from all other species.

7. **Epilobium blinii** H. Léveillé, Repert. Spec. Nov. Regni Veg. 7: 338. 1909.—TYPE: CHINA. Yunnan, boggy areas, Jun 1906, *Maire 393* (holotype: not located; isotype: UC!).

Epilobium forrestii Diels, Notes Roy. Bot. Gard. Edinburgh 5: 254. 1912.—TYPE: CHINA. Yunnan, stony, boggy situations in side valleys on the eastern flank of the Dali Range, 25°40′N, 2400–3000 m, Jul–Aug 1906, *Forrest 4460* (holotype: E!; isotypes: BM! P!).

Common name: Changzhu Liuyecai.

Ascending or erect perennial herb, forming stalked leafy basal rosettes. Stems 10–45 cm tall, 1–4 mm in diameter, simple or with a few branches, covered with appressed-strigillose pubescence or rarely subglabrous, with raised strigillose lines decurrent from margins of the petioles. Leaves alternate on inflorescence, opposite below, the basal ones 1.5–4.5 cm long, 0.6–1.2 cm wide, obovate to oblanceolate, with petioles 1–5 mm long, the upper ones subsessile; cauline blades 1–3 cm long, 0.4–0.9 cm wide, narrowly elliptic or lance-elliptic, sparsely strigillose on both sides, especially on veins, apex obtuse, base subcuneate or rarely rounded, margin remotely denticulate with 3–7 weak teeth on each side, rarely subentire, the lateral veins inconspicuous, 3–4 on each side of midrib. Inflorescence suberect or nodding. Flowers erect; buds 3–5 mm long, 2.5–3 mm in diameter, oblong-ovoid, acute or short-apiculate; ovaries 1.5–2.5 cm long, densely strigillose, sometimes mixed with glandular hairs; pedicels 1.2–3.5 cm long; floral tube 1.2–2.5 mm long, 1.8–3 mm in diameter, with a ring of scattered hairs at the mouth within; sepals 5–7.5 mm long, 1–2.2 mm wide, narrowly lanceolate, acute, densely strigillose with some glandular hairs; petals 10–15 mm long, 5.5–10 mm wide, rose or purple, obovate, the notch 1.5–2.5 mm deep; filaments of longer stamens 3.5–7 mm long, those of shorter ones 1.5–4 mm long; anthers 1.5–2.5 mm long, 1–1.4 mm in diameter, oblong; style 8–16 mm long, erect, with scattered spreading hairs on the lower portion; stigma deeply to shallowly 4-lobed, the lobes oblong or triangular, spreading to recurved, 1–2 mm long and conspicuously broad, with scattered spreading hairs on the papillose upper surface, exserted beyond anthers at anthesis. Capsules 3–5.5 cm long, strigillose, often densely so, sometimes mixed with glandular hairs, borne on pedicels 1.5–3.5 cm long. Seeds 1.2–1.5 mm long, 0.4–0.5 mm in diameter, narrowly obovoid, distinctly narrowed at micropylar end, with a short chalazal collar, brown, the surface minutely papillose; coma 7–10 mm long, tawny, detaching easily. Chromosome number: $n = 18$. Figs. 5g–i, 6b, 7d, 21.

Phenology. Collected in flower from April to July (rarely to September), in fruit from May to August (rarely to October).

Distribution (Fig. 22). Endemic to southwestern China in southwestern Sichuan and northern Yunnan; sporadic in sphagnum bogs and other permanently wet areas; 1500–2700 (–3300) m, most commonly below 2000 m.

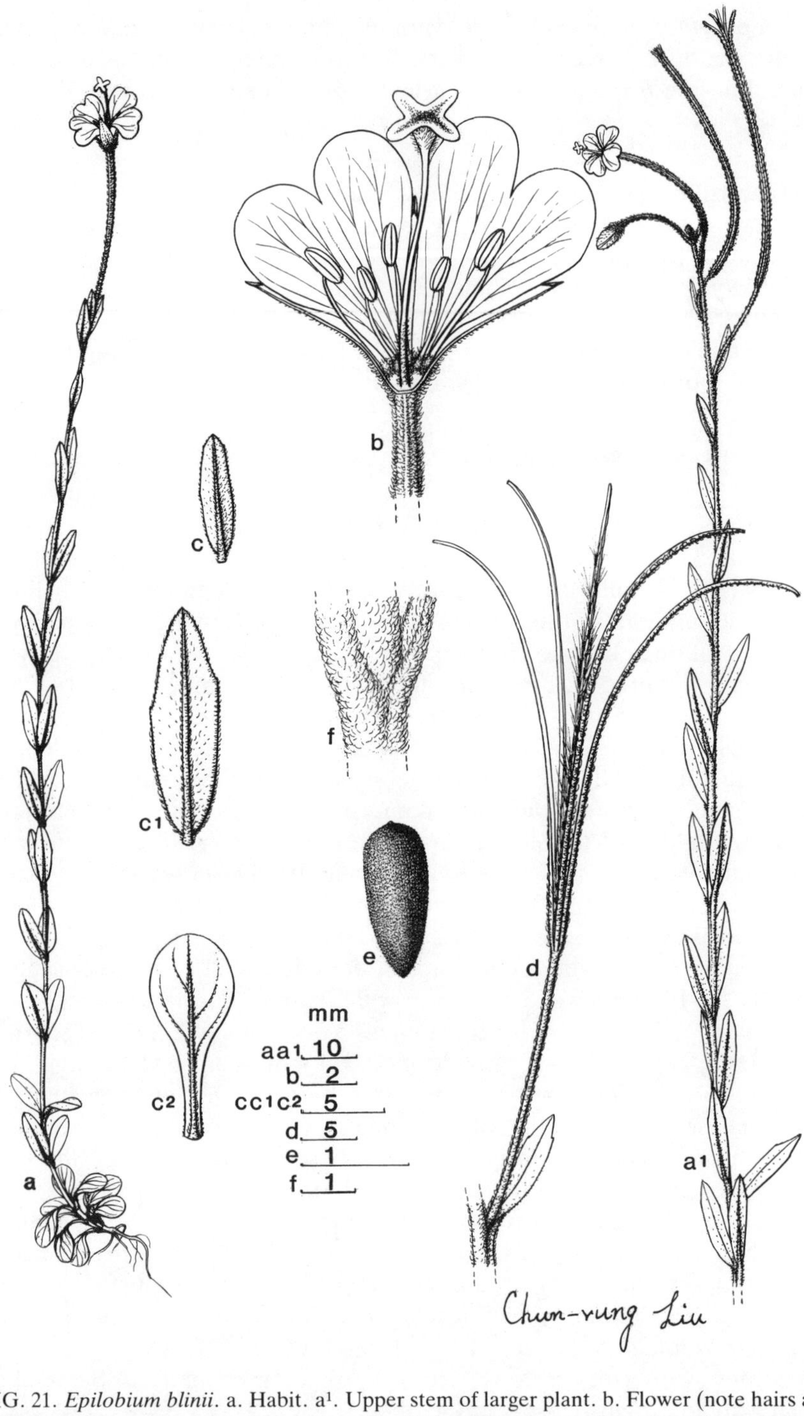

FIG. 21. *Epilobium blinii*. a. Habit. a^1. Upper stem of larger plant. b. Flower (note hairs at mouth of floral tube and near base of style). c. Upper leaf. c^1. Mid-cauline leaf. c^2. Basal leaf. d. Capsule. e. Seed. f. Axils from mid-stem. (Based on *K. M. Feng 22707*.)

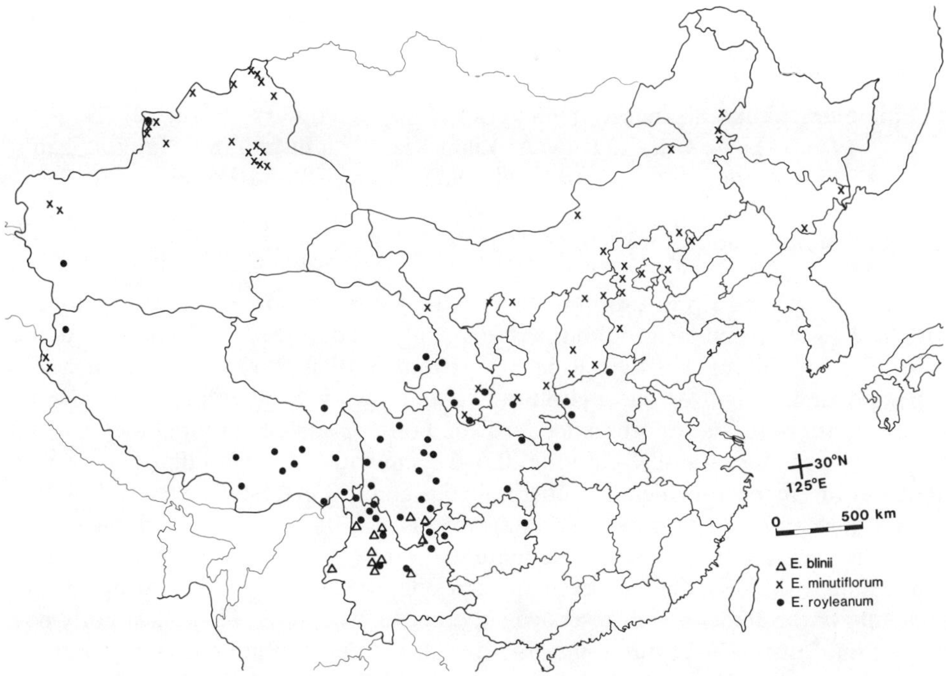

FIG. 22. Distribution in China of *Epilobium blinii*, *E. minutiflorum*, and *E. royleanum*.

ADDITIONAL SPECIMENS EXAMINED. **China.** SICHUAN: Jinyang, *Sichuan Exped. 13477* (CDBI, PE); Xichang, *Sichuan Exped. 5001* (CDBI).—YUNNAN: Qiajia, *Ducloux 4157* (P); Kunming, *P. Y. Chin 45* (A, KUN), *Ducloux 2283* (P), *Maire 1549* (BM, E), *1549a* (BM, E), *2216* (E, K), *2216a* (E, K), *Schoch 149* (G, K, WU), *C. W. Wang 62969* (GH, KUN, NAS, PE), *F. T. Wang F34* (YNU), *1765* (KUN), 1964, *Yunnan Univ. s.n.* (YUN); Mt. "Lo Shiueh," *McLaren V64* (E, MICH, UC); Binchuan, *Ducloux 7079* (UC), *7573* (P); Dali, *R. C. Ching 22707* (KUN, PE), *Delavay 1503* (P), *2006* (P), *K. M. Feng 22707* (PE), *Forrest 4460* (BM, E, P), *7146* (BM, E, K), *Handel-Mazzetti 6211* (S, W, WU), *Sino-American 1984 Bot. Exped. 1019* (KUN, MO, PE), *Sino-British Exped. to Cangshan 0474* (A); *C. W. Wang A328* (KUN), *63412* (GH, KUN, NAS, PE), *H. C. Wang 892* (PE), *2200* (KUN), *S. Chow 1282* (KUN); "Yn Chan Pou," 1887, *Delavay s.n.* (L, P); Eryuan to Heching, *Delavay 932* (P); Tengchong, *Forrest 7508* (E); Ninglang, *Schneider 1658* (E, G, GH); Lijiang, *Rock 4266* (BM, E, NY, P, UC); Weixi, *K. M. Feng 4777* (KUN, PE); no locality, *Yunnan Exped. 170* (PE).

Epilobium blinii is a highly distinctive Chinese endemic with large flowers, an exserted 4-lobed stigma and pubescent style, and long-pedicellate capsules. Among the 33 collections of this species for which collection dates are available, 24 were collected in Yunnan before 1940; three in the 1940s; and only three since 1950, two from Cang Shan, Dali, in 1964 and 1984 and one from Kunming. In addition, two recent collections (1976 and 1978) of this species were made from a small area in Xichang District, southwestern Sichuan. Collecting activity in this region has been moderate, especially in interesting areas like sphagnum bogs; therefore, we suggest that the decline in collections of this species reflects a real decline in viable populations. The probable reason is increased disturbance of wetland areas, especially by drainage of bogs and wet meadows for agricultural purposes.

According to D. Boufford (pers. comm.), who collected *E. blinii* on the Sino-

American expedition in 1984, this species flowers and fruits much earlier than other species of *Epilobium* in the same region.

8. Epilobium nankotaizanense Yamamoto, Suppl. Icon. Pl. Form. 2: 29, pl. 2. 1926.—TYPE: CHINA. Taiwan, Yilan Xian, Nanhutashan ("Nankotaizan"), 3350–3650 m, Jun (Jul on label) 1922, *S. Sasaki s.n.* (holotype: TI!).

Common name: Nanhu Liuyecai (Raven 1977).

Loosely matted alpine perennial herb with wiry roots spreading 10 cm or more, with fleshy soboles below ground that creep and root at nodes, and that have dense, minute scales. Stems 3–18 cm long, 0.5–1 mm in diameter, usually branched in upper portion, strigillose throughout and sometimes with glandular hairs. Leaves opposite, alternate on inflorescence, crowded on upper stems, on petioles 1–3 mm long; cauline blades 0.8–2.1 cm long, 0.5–1.2 cm wide, broadly elliptic to obovate or ovate, rarely to suborbicular, subcoriaceous and rather fleshy, sparsely strigillose on margins and veins, otherwise subglabrous, often with scattered bacilliform cystoliths, apex obtuse, base attenuate or rarely subtruncate, margin weakly denticulate with 3–7 indistinct teeth on each side, the lateral veins indistinct, 3–5 on each side of the midrib. Inflorescence suberect to ascending. Flowers slightly nodding in bud; buds 10–12 mm long, 5–7 mm in diameter, oblong-ovoid; ovaries 1–1.8 cm long, strigillose with an admixture of glandular hairs; pedicels 0.3–0.8 cm long; floral tube 1.2–2.8 mm long, 1.6–2.6 mm in diameter, with a ring of spreading hairs at the mouth within; sepals 11–17 mm long, 2–3 mm wide, very narrowly lanceolate, subacuminate; petals 16–33 mm long, 8–17 mm wide, rose-purple, narrowly obovate to obovate-oblong, abruptly clawed at base, the notch 3–10 mm deep; filaments of longer stamens 7.5–14 mm long, those of shorter ones 4.5–9 mm long; anthers 1.8–3 mm long, 0.5–0.6 mm wide, oblong-lanceolate; style 13–23 mm long, erect, with a few spreading hairs near the base; stigma 1.5–2.2 mm long, 1.2–2 mm in diameter, subglobose with four shallow but evident lobes, the lobes 0.6–1.2 mm long, triangular-ovate, obtuse, spreading, exserted well beyond anthers at anthesis. Capsules 2–4.5 cm long, 3–4 mm in diameter, sparsely strigillose and glandular-pubescent, borne on pedicels 0.4–0.7 cm long. Seeds 1.6–1.8 mm long, 0.5–0.6 mm in diameter, narrowly obovoid, with a short chalazal collar, brown, the surface finely reticulate; coma 6–9 mm long, tawny, persistent. Chromosome number: n = 18. Figs. 6c, 23.

Phenology. Collected in flower from July to August, in fruit from August to September.

Distribution (Fig. 15). Endemic to Taiwan; local on open scree slopes in the high mountains; 2600–3750 m.

ADDITIONAL SPECIMENS EXAMINED. **China.** TAIWAN: Yilan Xian, Nanhutashan ("Nankotaizan"), *Hidetaro 1337* (PE), *C. C. Hsu 5983* (TAI, TI), *6031* (TAI, TI), *C. S. Kuoh 3009* (TAI), *3011* (TAI), 1940, *Masamune & Fukuyama s.n.* (TAI), 1931, *Masamune & Mori s.n.* (TAI), *Ohwi 3982* (P, TNS, UPS), 1938, *Okamoto s.n.* (KYO), *C.-I Peng 5743* (HAST, MO), *8345* (HAST, MO), 1919, *Pringle s.n.* (TAIF), 1921, *Sasaki s.n.* (TAI), 1934, *Sasioka s.n.* (TAI), 1937, *Simada-Hidetaro s.n.* (TAI), 1940, *Sinozawa s.n.* (TNS), 1937, *Suzuki s.n.* (TAI), 1937, *Suzuki-Tokio et al. s.n.* (TAI), *Tamura & Koyama 23677* (KYO, S, TNS), *Tamura & Shimizu 20742* (E, S, SHIN), *20803* (SHIN), *Yamazaki et al. 321* (BM, TI); Hualian Xian, "Kuarenkotyo," 1936, *Suzuki-Tokio & Fukuyama s.n.* (TAI); "Chuochen-

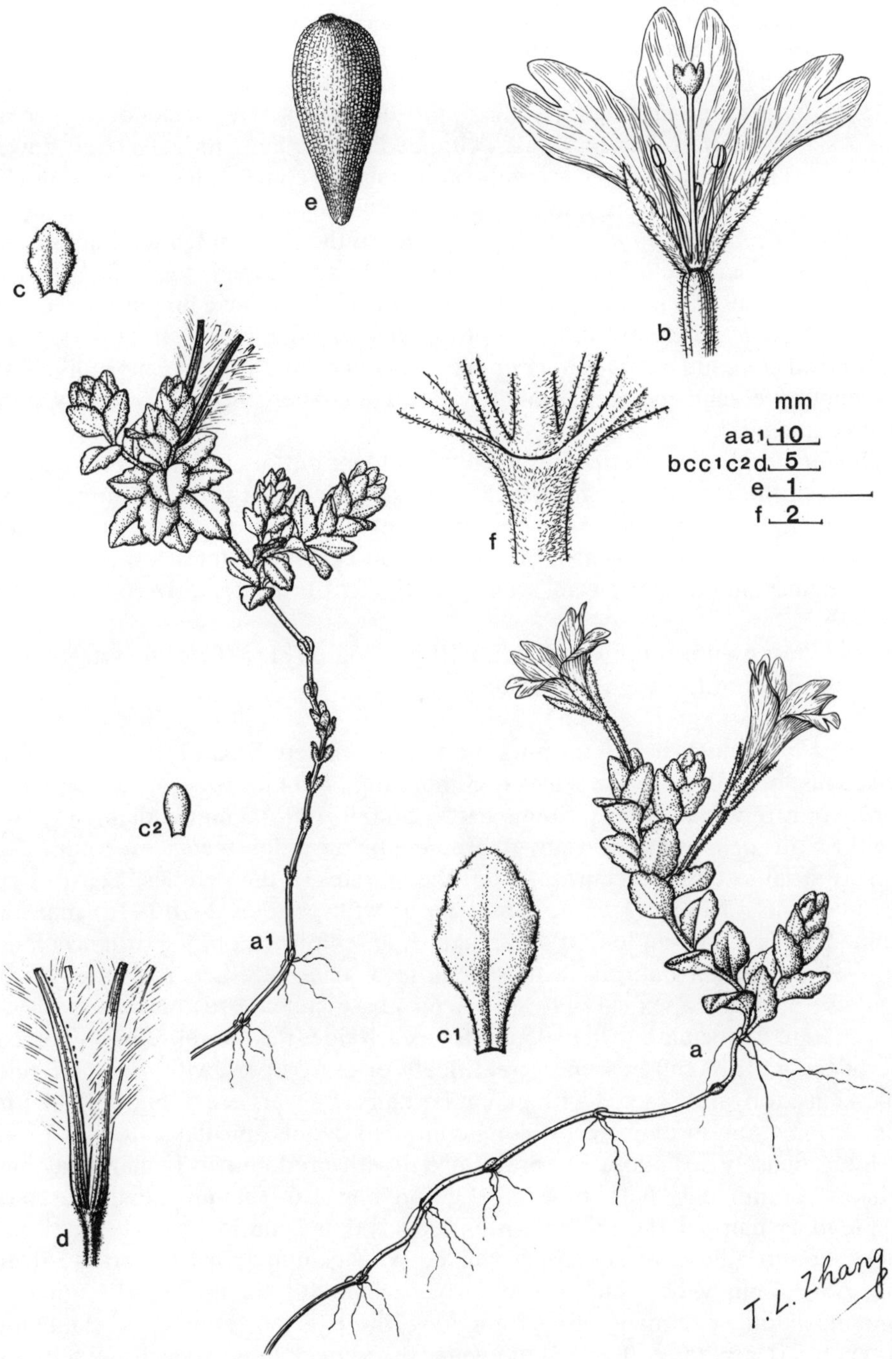

FIG. 23. *Epilobium nankotaizanense*. a. Habit: plant with long creeping rhizome; note minute basal cataphylls. a[1]. Plant showing portion of rhizome as positioned in scree. b. Flower (one petal and three stamens removed). c. Upper leaf. c[1]. Mid-cauline leaf. c[2]. Cataphyll from rhizome. d. Capsule. e. Seed showing reticulate surface. f. Axils from mid-stem. (Based on *Yamazaki et al. 321*.)

shan," 1929, *Shikano s.n.* (TAI); Taizhong Xian, Hsuehshan ("Mt. Silvia"), 1932, *Sasaki s.n.* (TAI); "Mt. Maborasu," 1931, *Shikano s.n.* (TAI); no locality, collector, or date, *081590* (TAI), *081591* (TAI).

Epilobium nankotaizanense is one of the most distinctive species of the genus in East Asia by virtue of its habit as a clumped scree plant, its very large flowers, globose and shallowly 4-lobed stigma, and persistent coma. It has no obvious close relatives in East Asia. There are striking morphological similarities between this species and *E. obcordatum,* a beautiful species of the western United States, which also occurs on scree and has very large flowers and a 4-lobed stigma (Raven 1976; Seavey & Raven 1978). Some other species of *Epilobium* have amphi-Pacific distributions (e.g., *E. ciliatum,* and possibly *E. fauriei* H. Léveillé of Japan), but *E. nankotaizanense* and *E. obcordatum* are so disjunct that it seems more likely that they may have achieved their similarity through convergent evolution in similarly extreme habitats.

Epilobium nankotaizanense is locally abundant but of extremely restricted distribution in the mountains of northeast-central Taiwan. It is one of four endemic species of *Epilobium* in Taiwan; like the other three it grows in a very restricted habitat, but unlike them it has retained certain character states, especially the 4-lobed stigmas and large flowers, which are thought to be primitive (Raven 1976).

9. Epilobium roseum Schreber, Spicil. Fl. Lips. 155. 1771.—TYPE: GERMANY. From vicinity of Leipzig (not located).

Erect perennial herb, forming fleshy, elongated basal turions that leave coriaceous, broadly obovate scales 6–8 mm long, 4–6 mm wide, or forming loose soboles or rarely thin stolons. Stems 10–50 cm tall, 1.5–4.5 mm in diameter, erect, strigillose throughout upper stem, glabrescent below inflorescence except for 2 or 4 sparsely strigillose lines decurrent from the margins of the petioles. Leaves opposite, alternate on inflorescence, subsessile or with petioles 2–10 (–15) mm long; cauline blades 2.2–6 cm long, 0.8–2.5 cm wide, elliptic to oblong or lanceolate to narrowly ovate, subglabrous with strigillose abaxial veins on upper leaves and strigillose margins, apex subobtuse to acute, base cuneate to rounded or subcordate, margin denticulate with 9–45 teeth on each side, the lateral veins 4–7 on each side of the midrib. Inflorescence erect, leafy or more open, with dense strigillose pubescence, or rarely mixed with glandular hairs. Flowers erect; buds 3.5–4.5 mm long, 1.6–2.4 mm in diameter, oblong-elliptic to ovoid, apiculate; ovaries 1.4–2.8 cm long, densely strigillose or rarely mixed with predominantly glandular hairs; pedicels 3–8 mm long; floral tube 0.8–1.2 mm long, 1.6–1.8 mm in diameter, a ring of spreading hairs at the mouth within; sepals 3.5–5 mm long, 1–1.2 mm wide, lanceolate, strigillose or sometimes mixed with glandular hairs; petals 5–8 mm long, 2.5–3.6 mm wide, pink or rose purple, obcordate, the notch 1–1.6 mm deep; filaments of longer stamens 2.5–3.2 mm long, those of shorter ones 1.6–2 mm long; anthers 0.6–1 mm long, 0.3–0.5 mm wide, ovoid or oblong; style 1.5–2 mm long, glabrous; stigma clavate or broadly clavate to subcapitate, 1.1–2 mm high, 0.7–1.2 mm in diameter, surrounded by at least the longer stamens at anthesis. Capsules 3–6 cm long, pubescent as on ovaries, borne on pedicels 0.6–1.5 cm long. Seeds 1–1.2 mm long, 0.45–0.55 mm in diameter, obovoid, with very short chalazal collar, dark brown, the surface papillose; coma 4–6.5 mm long, white, detaching easily. Chromosome number: n = 18.

The collections of *E. roseum* from China, all from the Tian Shan and Altay Mountains of Xinjiang, constitute a small sample of this variable species at the easternmost edge of its distribution. Among these collections we have recognized two taxa that are well-delimited in China, and between which there is little intergradation. However, in other portions of its range in western and central Asia there are many intermediate collections among the three subspecies of *E. roseum* circumscribed by Raven (1962b, 1964; Chamberlain & Raven 1972). Until we can complete a thorough review of this highly variable species throughout its range, we are treating the Chinese collections as subspecies in accordance with Raven's earlier treatment, even though the two taxa are quite distinct as they occur in China.

This is the first report of *E. roseum* in China, although it was previously reported from the Tian Shan region of U.S.S.R. by Pavlov (1934) and Steinberg (1949) as *E. nervosum* Boiss. & Buhse (=*E. roseum* subsp. *subsessile*).

9a. Epilobium roseum subsp. **roseum.**

Common name: Changbing Liuyecai.

Variable herb, forming fleshy, elongated basal turions with coriaceous broadly obovate scales 6–8 mm long, 4–6 mm wide. Leaves on petioles 3–10 (–15) mm long; cauline blades 2.2–5 cm long, 0.8–1.8 cm wide, elliptic to oblong or lanceolate, apex subobtuse, base cuneate to broadly so, margins weakly denticulate with 9–25 teeth on each side, the lateral veins 4–6 on each side of midrib. Inflorescence open, with very reduced bracts, strigillose, sometimes with glandular hairs. Stigma broadly clavate to subcapitate, 1.1–1.2 mm high, 0.7–1 mm in diameter. Capsules 4–6 cm long, sparsely strigillose. Chromosome number: n = 18. Fig. 24.

Phenology. Collected in flower from July to September, in fruit from August to September.

Distribution (Fig. 25). Europe and southwestern U.S.S.R. through Turkey, Iran, and the Caucasus region of the U.S.S.R., possibly Afghanistan, to the Tian Shan and Altay regions of Soviet Central Asia and China; in China only in the Tian Shan-Altay regions of Xinjiang; in damp areas near streams, roads, and marshes; 1850–2200 m.

ADDITIONAL SPECIMENS EXAMINED. **China.** XINJIANG: Yiwu, *Q. R. Wang et al. 4610* (PE); Xiaodonggou, NE of Altay, *Chen & Hoch 86071* (MO, PE), *86084* (MO, PE); NW of Altay, *Chen & Hoch 86144* (MO, PE); Fuhai, *G. L. Zhu et al. 5771a* (PE); Habahe, *Y. R. Ling 74-1149* (PE); Tacheng, *Y. R. Ling 74-1311* (PE); Wenquan, *Xinjiang Exped. 3502* (PE, XJBI); Huhai, Wenquan, N. Tian Shan, 1956, *Kaproon s.n.* (PE), *4* (PE), *12* (PE), *96* (PE); Yining, *W. Y. Wu et al. 93* (PE); Zhasu, *Xinjiang Exped. 2922* (PE); Xinertai, *Y. R. Ling 74-214* (PE); Shawan, *Xinjiang For. Exped. 78* (PE); Baihaba (at boundary station), *Xinjiang Exped. 2105* (HNWP); Ningjia He, Tianshan, *R. C. Ching 3457* (PE).

This taxon has been treated previously as a species of Europe and Asia Minor (e.g., Steinberg 1949). Although Raven (1962b) reported its range to Caucasia and Iran, there are no previous reports of *E. roseum* sensu stricto from areas east of the

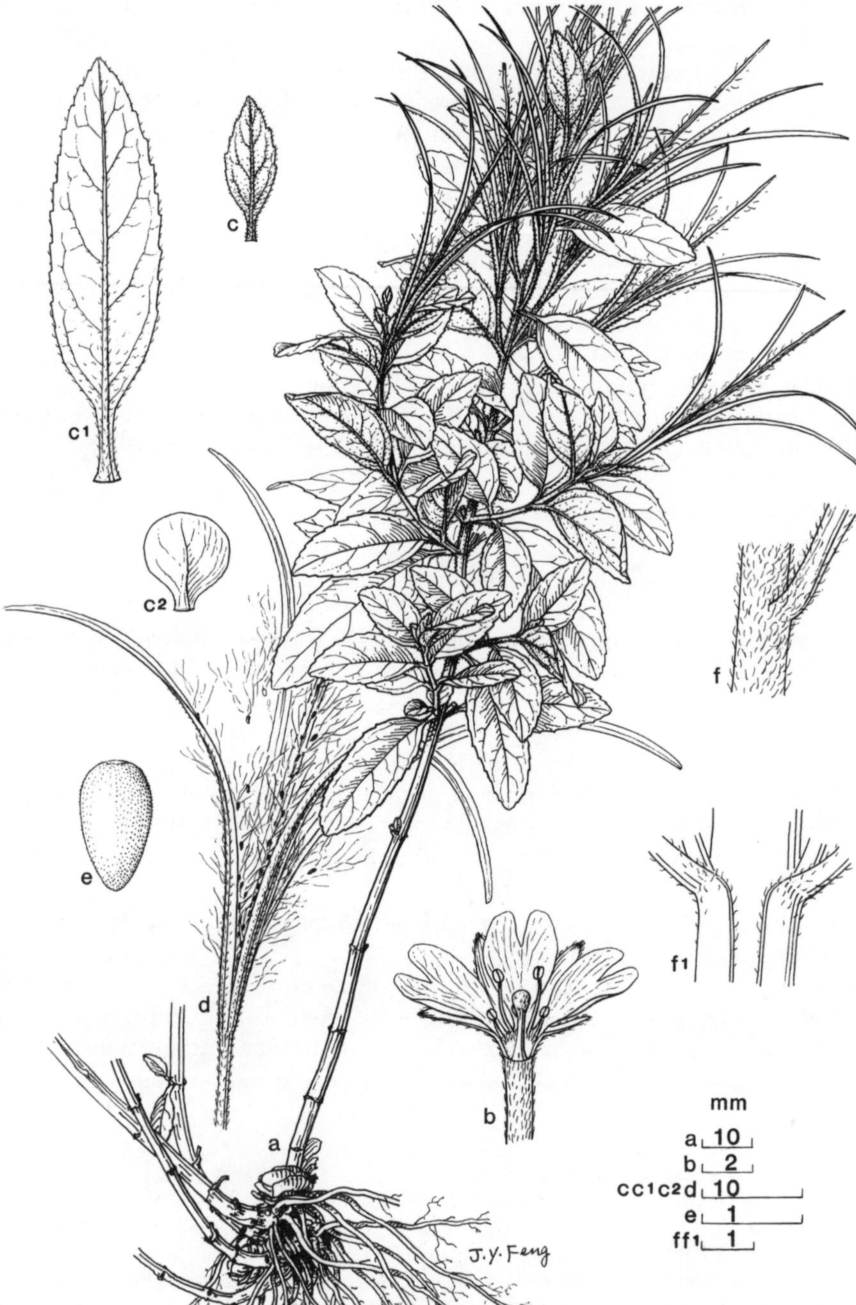

FIG. 24. *Epilobium roseum* subsp. *roseum*. a. Habit; note turion and scales at base of stem. b. Flower (one petal and three stamens removed). c. Upper leaf. c[1]. Mid-cauline leaf. c[2]. Scalelike leaf from basal turion. d. Capsule. e. Seed. f. Axil from inflorescence. f[1]. Axils from mid-stem. (Based on *Kaproon s.n.* in 1956.)

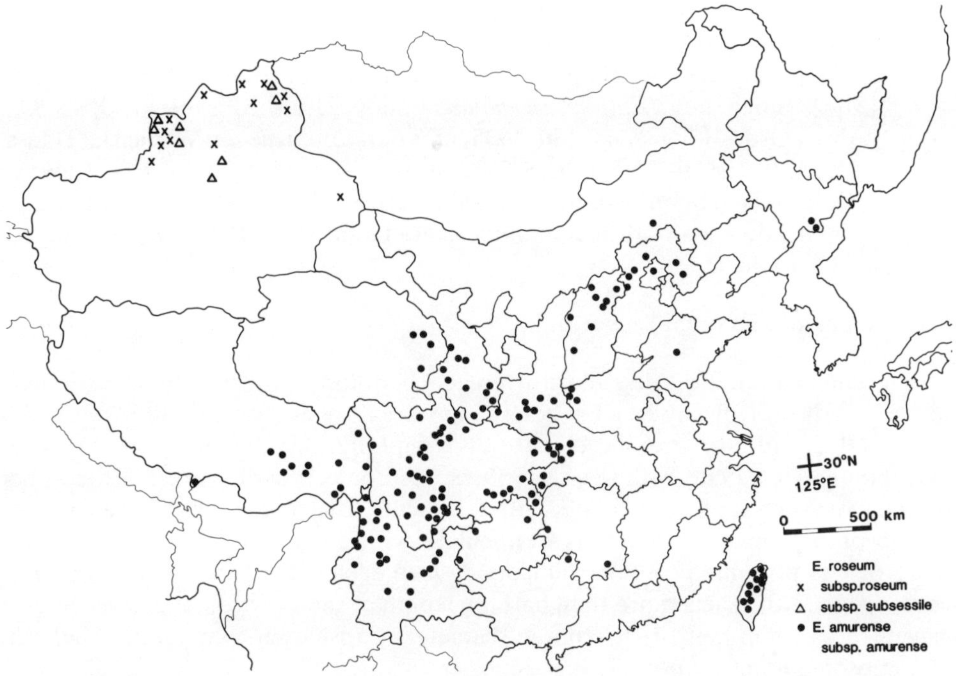

FIG. 25. Distribution in China of *Epilobium amurense* subsp. *amurense*, *E. roseum* subsp. *roseum*, and *E. roseum* subsp. *subsessile*.

Caucasus. Yet, our Chinese collections, and other recent collections kindly sent to us from the Tian Shan region of the Soviet Union by A. K. Skvortsov, are clearly *E. roseum* sensu stricto. They differ from other taxa in the Tian Shan region by their combination of fleshy basal turions, elliptic and petiolate leaves, and subcapitate stigmas. Some collections differ from typical representatives of *E. roseum* from Europe in having subcapitate stigmas rather than the more common clavate ones, or in having glandular hairs mixed with short recurved eglandular ones on the inflorescence. Other collections (e.g., Wenquan, *Xinjiang Exped. 3502,* PE, XJBI; Baihaba, *Xinjiang Exped. 2105,* HNWP), however, have clavate stigmas and eglandular pubescence, and are indistinguishable from European collections. It seems best at this time to recognize these collections from Xinjiang within the context of the established treatment of this species, and call attention to them for further study.

9b. Epilobium roseum subsp. **subsessile** (Boissier) Raven, Notes Roy. Bot. Gard. Edinburgh 24: 194. 1962. *Epilobium smyrnaeum* Boissier & Balansa in Boissier, Diagn. ser. II, 2: 52. 1856. *Epilobium roseum* β. *subsessile* Boissier, Fl. Orient. 2: 749. 1872.—TYPE: TURKEY. Izmir Prov., near Izmir (Smyrna), "Yaila de Bozdagh (Tmolus occidental)," Jul 1854, *Balansa s.n.* (lectotype, designated by Raven 1962b: G-005537!).

Epilobium nervosum Boissier & Buhse, Nouv. Mém. Soc. Imp. Naturalistes Moscou 12: 88. 1860.—TYPE: IRAN. Nemadabad by Tabriz, edges of ditches, flowering and fruiting, 11 Jul 1847, *Boissier & Buhse 673* (holotype: LE!; isotype: G!).

Epilobium almaatense Steinberg in Shishkin & Bobrov, Fl. URSS 15: 590. 1949, nom. nov. *Epilobium laetum* Popov, Bjull. Moskovsk. Obsc. Isp. Prir., Otd. Biol. n.s. 44: 130. 1935, non *Epilobium laetum* Wallich ex Haussknecht, 1884.—TYPE: U.S.S.R. Prov. Kazakhstan, Alma-Atinskaya Oblast, Tian Shan region, in subalpine areas, 2700 m, along mossy banks at the headwaters of the Almatinka river, 12 Jul 1933, *Popov s.n.* (holotype: LE, not located).

Common name: Duomai Liuyecai.

Perennial herb, forming filiform epigeous stolons with widely spaced small leaves, or shorter, fleshy basal soboles. Leaves subsessile, or the lower ones with petioles 0–2 mm long; cauline blades 3.4–6 cm long, 0.9–2.5 cm wide, the lower ones oblong-lanceolate with apex subobtuse and base broadly cuneate, the upper ones narrowly ovate to lanceolate with apex acute to acuminate and base rounded or subcordate, margins remotely denticulate with 23–45 teeth, the lateral veins more or less prominent on abaxial side, 5–7 on each side of the midrib. Inflorescence leafy, with bracts more than half the length of the ovaries, strigillose. Stigma clavate, 1.5–2 mm high, 1–1.2 mm in diameter. Capsules 3–6 cm long, strigillose. Chromosome number: n = 18. Fig. 26.

Phenology. Collected in flower from June to August, in fruit from July to September.

Distribution (Fig. 25). From southeastern Europe and southeastern U.S.S.R. through Turkey, Caucasia, Iran, Afghanistan, and Soviet Central Asia to northwestern China; in China only in the Tian Shan and Altay regions of Xinjiang; in wet places along streams and ditches in mountains; 1500–2100 m.

ADDITIONAL SPECIMENS EXAMINED. **China.** XINJIANG: Xiaodonggou, NE of Altay, *C. J. Chen & Hoch 86070* (MO, PE), *86083* (MO, PE); NW of Altay, *C. J. Chen & Hoch 86143* (MO, PE); Manas, *Xinjiang Exped. 321* (XJBI); Daliu, N. Tian Shan, *K. C. Kuan 1382* (PE, XJBI); Bole, *Xinjiang Exped. 3757* (PE, XJBI); Wenquan, *C. J. Chen & Hoch 86331* (MO, PE), *Xinjiang Exped. 3740* (XJBI); Gongliu, *Q. R. Wang 2664* (PE); Huocheng (Suiding), Ili, *T. Y. Chou et al. 650903* (KUN, NAS, PE), *650904* (PE); without definite locality, Tian Shan, *Xinjiang Exped. 15506* (XJBI, YNU).

Although noted to be in the Tian Shan region in the *Flora URSS* (Steinberg 1949, as *E. nervosum*), *E. roseum* subsp. *subsessile* has never been reported from China; these collections from Xinjiang are the first records. This taxon is distinguished from subsp. *roseum* by its subsessile leaves, clavate stigma (in this region), and sparsely pubescent stems with eglandular inflorescence. Plants from Xinjiang resemble typical specimens of subsp. *subsessile* from farther west, although only our own recent collections include intact turions.

10. Epilobium tianschanicum Pavlov, Ucen. Zap. Moskovsk. Gosud. Univ. 2: 327. 1934.—TYPE: U.S.S.R. Kirgiz S.S.R., prov. Syr-Dar'ya, Talasskiy Alatau mountains, "Russian Turkestan," banks of Topschak-ssu River, 19 Jul 1931, *Pavlov 622* (holotype: MW, not located; isotype: LE!).

Common name: Tianshan Liuyecai.

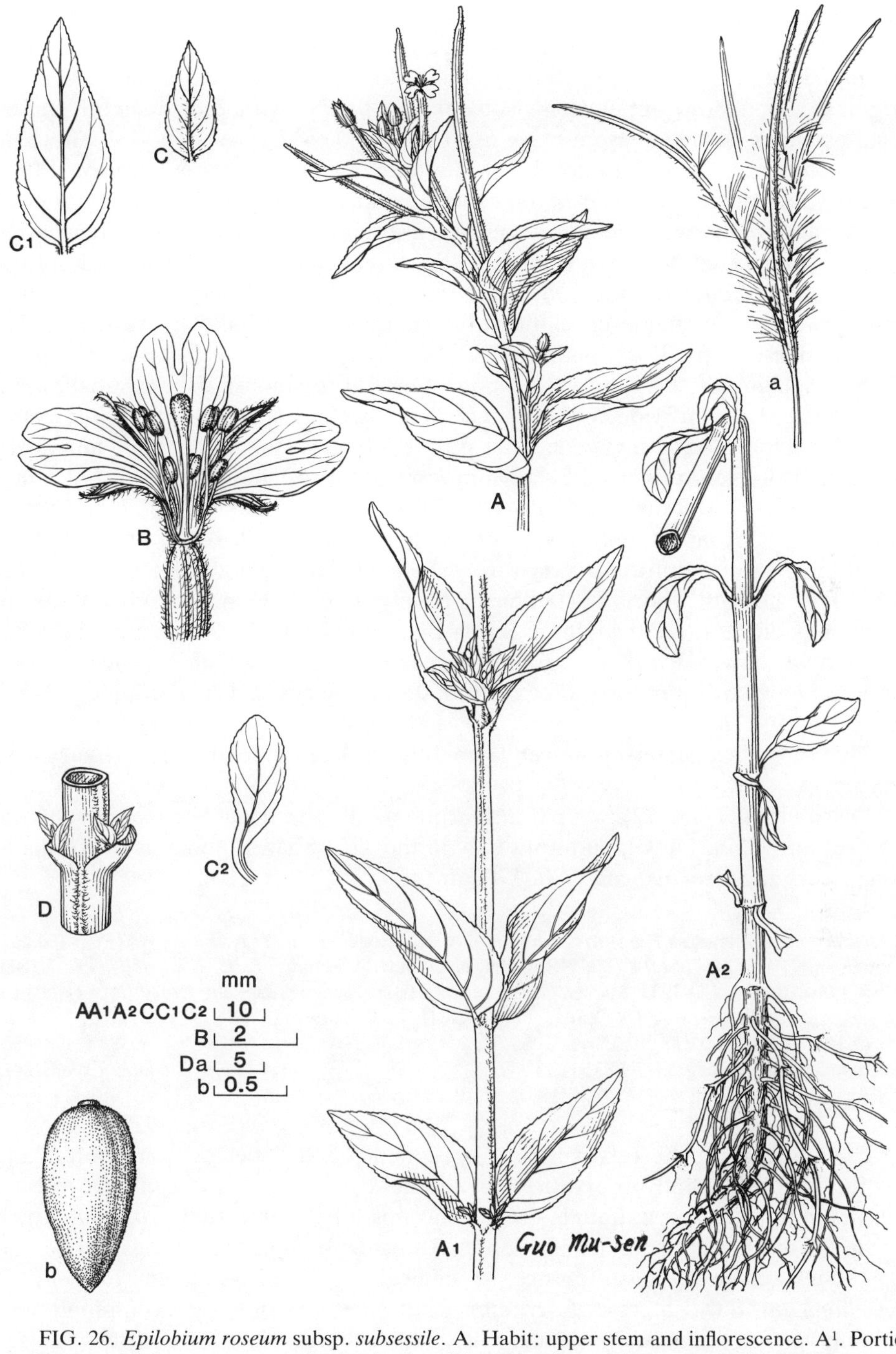

FIG. 26. *Epilobium roseum* subsp. *subsessile*. A. Habit: upper stem and inflorescence. A[1]. Portion of mid-stem. A[2]. Lower stem, caudex, and root, with stolons (note minute cataphylls). B. Flower (one petal and one stamen removed). C. Upper leaf. C[1]. Mid-cauline leaf. C[2]. Basal leaf. D. Axils from mid-stem. a. Capsule. b. Seed. (A–D based on *T. Y. Chou 650903*; a–b based on *Xinjiang Exped. 3757*.)

Clumped perennial herbs, forming fleshy basal soboles or leafy rosettes that leave brown scales or leaves loosely arranged on the rhizome. Stems 30–50 cm tall, 3–4 mm in diameter, simple, often basally decumbent, often purplish or dark, strigillose throughout upper stem, subglabrous below, with 2 or 4 slightly raised strigillose lines decurrent from margins of the petioles. Leaves opposite, alternate on the inflorescence, with petioles 2–4 mm long; cauline blades 3–5 cm long, 0.9–1.4 cm wide, narrowly ovate or lanceolate, sparsely strigillose on veins of both surfaces and on margins, apex acute, base subrounded or broadly cuneate, margin serrulate with 14–25 teeth on each side, the lateral veins 4–5 on each side of the midrib. Inflorescence slightly nodding, strigillose. Flowers erect; buds 3.5–4.2 mm long, 1.8–2.4 mm in diameter, elliptic-oblong, apiculate, strigillose; ovaries 1.5–2.5 cm long, densely strigillose; pedicels 0.7–2 cm long; floral tube 1–1.2 mm long, 2–2.2 mm in diameter, a ring of villous pubescence at the mouth within; sepals 4–5.5 mm long, 1–1.2 mm wide, oblong-lanceolate, keeled, sparsely strigillose; petals 5.5–6.5 mm long, 2.5–3 mm wide, rose-purple, obovate, the notch 0.7–1 mm deep; filaments of longer stamens 2.5–3.5 mm long, those of shorter ones 1.5–2.5 mm long; anthers 0.7–0.8 mm long, 0.4–0.5 mm in diameter, oblong; style 3–3.8 mm long, erect, glabrous; stigma clavate or rarely subcapitate, 1.3–1.6 mm high, 0.8–1.1 mm in diameter, entire or very rarely shallowly indented at the apex, subequal with longer or both whorls of stamens. Capsules 4–6 cm long, sparsely strigillose, borne on pedicels 1.5–2.9 cm long. Seeds 1–1.3 mm long, 0.38–0.46 mm in diameter, narrowly obovoid, with a very short chalazal collar, brown, the surface coarsely papillose; coma 6–7 mm long, dingy white, detaching easily. Chromosome number unknown. Fig. 27.

Phenology. Collected in flower from July to August, in fruit from August to September.

Distribution (Fig. 17). Apparently endemic to the Tian Shan region of the U.S.S.R and China; in China found only in the Tian Shan Mountains in Xinjiang; along rivers and streams; ca. 1000–1700 m.

ADDITIONAL SPECIMENS EXAMINED. **China.** XINJIANG: Bogda Shan, *T. N. Liou 3394* (PE); Fukang, Xiaotianchi, *T. Y. Chou 651198* (HNWP, KUN, NAS, XJBI); Fukang, *K. C. Kuan 4374* (PE, XJBI), *Xinjiang Inst. Biol. 141* (XJBI); Manas, *G. J. Liu 66* (XJBI); Tacheng, *Xinjiang Exped. 2478* (HNWP); Gongliu, *C. W. Chang et al. 3474* (XJBI), *3482* (XJBI), *Y. R. Ling 74370* (PE); Takes, *G. J. Liu et al. 1091* (XJBI), *Y. W. Xiu 177* (PE); Wenquan, *C. J. Chen & Hoch 86311* (MO, PE), *86322* (MO, PE); Huocheng, *C. W. Chang 3347* (XJBI), *Y. R. Ling 74370* (PE), *74855* (PE), *Xinjiang Exped. 1441* (HNWP); Ili, *T. Y. Chou et al. 650904* (NAS, HNWP).

Examination of the relatively few specimens of this species from both China and the U.S.S.R., and observations in the field reveal *E. tianschanicum* to be a well-delimited species, although similar and possibly related to *E. roseum,* which reaches its eastern limits in the Tian Shan-Altay region. The two species can be told apart consistently by a difference in pedicel length (1.5–2.9 cm long in *E. tianschanicum,* 0.6–1.5 cm in *E. roseum*), but otherwise only by a combination of leaf and vestiture characters. More collections from throughout Central Asia are needed to define more precisely the limits of these taxa.

Raven (1962a, 1964) placed *E. tianschanicum* in the synonomy of *E. cylindricum* of the Himalayas and southcentral China, but did not have the opportunity to examine the type or other collections from Soviet Central Asia. Indeed, *E.*

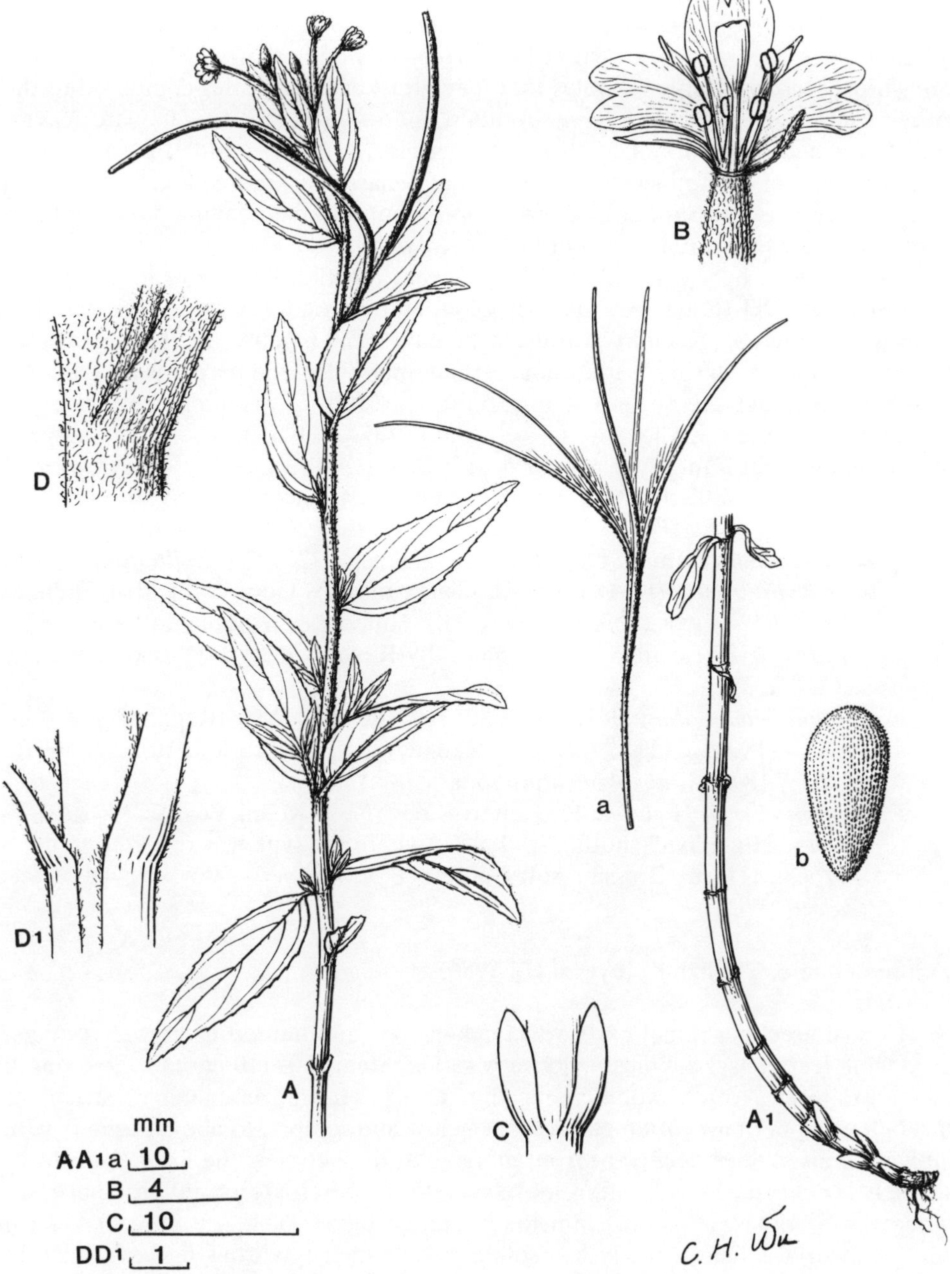

FIG. 27. *Epilobium tianschanicum*. A. Habit: upper stem and inflorescence. A[1]. Base of stem with turion scales. B. Flower (one petal and three stamens removed). C. Basal scales, following nodal elongation of turion. D. Axil from inflorescence. D[1]. Axils from mid-stem. a. Capsule. b. Seed. (A–D[1] based on *G. J. Liu 66*; a–b based on *K. C. Kuan 4374*.)

tianschanicum shares a number of features with *E. cylindricum,* including a generally similar habit, long pedicels, coarsely papillose seeds, and a vestiture pattern of subglabrous stems with 4 decurrent lines from the petioles. *Epilobium tianschanicum,* however, has fleshier soboles that leave thicker basal scales compared to the more rosuliferous *E. cylindricum;* broader, more lanceolate to subovate leaves, with fewer and less-marked teeth on the margins, a more rounded base, a more abruptly acute apex, and shorter petioles; and a narrower, more clavate stigma. In some aspects it is also similar to *E. royleanum,* but differs in having pubescent leaf base lines, a feature notably absent in *E. royleanum.*

An apparent hybrid involving *E. tianschanicum* (Xinjiang, Fukang, *Xinjiang Inst. Biol. 141,* XJBI) has 26% fertile pollen and markedly reduced seed set. The plant is extremely large and vigorous, with leafy basal rosettes and leaves to 7 cm long, lanceolate, with sharp, irregular serrulate margins. One parent is probably *E. tianschanicum,* but the second is uncertain—possibly *E. minutiflorum.* Another collection from the same locality (*K. C. Kuan 21374,* XJBI) has rosettes and petioles 3–5 mm long; it might be a hybrid with *E. roseum* (it lacks enough flowers to get an estimate of pollen fertility), but it seems best placed here for the present.

11. Epilobium cylindricum D. Don, Prodr. Fl. Nepal. 222. 1825. *Epilobium roseum* var. *cylindricum* (D. Don) C. B. Clarke in J. D. Hooker, Fl. Brit. India 2: 585. 1879.—TYPE: NEPAL. Sheopuri Hill, N of Katmandu, Aug 1821, *Wallich 6328* (lectotype, designated by Raven 1962a: BM!; isolectotypes: E! G! K! W!).

Epilobium beauverdianum H. Léveillé, Repert. Spec. Nov. Regni Veg. 8: 138. 1910.—TYPE: CHINA. Sichuan, Kangding, "Ta-Tsien-lou, principaute de Kiala," 1893, *Soulie 449* (holotype: G!).

Epilobium christii H. Léveillé, Repert. Spec. Nov. Regni Veg. 9: 19. 1910.—TYPE: "Himalaya" (not located at G, where the type was reported to be—"type in Herb. Boissier sub nomine *E. cylindrici*"—nor elsewhere; cf. Raven 1962a).

Common name: Yuanzhu Liuyecai (Li 1986).

Robust erect perennial herb with thick caudex and numerous fibrous rootlets, producing leafy basal soboles or loose rosettes. Stems 10–110 cm tall, 3–9 mm in diameter, terete, often woody near the base, well-branched above, strigillose throughout upper part, often glabrescent below and rarely also above, usually with indistinct raised lines decurrent from margins of the petioles, the lines glabrous or sparsely strigillose. Leaves often loosely opposite, alternate on inflorescence, attenuate to petioles 3–7 (–10) mm long; cauline blades 3–12 cm long, 0.4–2 cm wide, subcoriaceous, narrowly lanceolate to sublinear, glabrous on both surfaces except for sparsely pubescent veins and margins, apex acute, base cuneate, margins densely serrulate with (20–) 30–50 teeth on each side, the lateral veins 4–5 on each side of the midrib. Inflorescence erect, densely strigillose, very rarely with a few glandular hairs. Flowers suberect; buds 2.5–4.5 mm long, 2–2.5 mm in diameter, ovoid, long-mucronate; ovaries 1.2–3.5 cm long, strigillose and usually eglandular; pedicels 0.5–1.5 cm long; floral tube 1–1.5 mm long, 1.3–2 mm in diameter, subglabrous within; sepals 3–5 mm long, 1–1.3 mm wide, lanceolate, keeled; petals

3.6–7 mm long, 1.8–4 mm wide, pink or rose-purple, rarely white, obcordate, the notch 0.8–1 mm deep; filaments of longer stamens 2.2–4 mm long, those of shorter ones 1–2.4 mm long; anthers 0.5–0.7 mm long, 0.3–0.4 mm in diameter, oblong-ovoid; style 2–4 mm long, white, erect, glabrous; stigma capitate or broadly clavate, 0.8–2.2 mm high, 0.6–1.8 mm in diameter, white, surrounded by anthers at anthesis. Capsules 4–8.5 cm long, strigillose or sparsely strigillose, borne on pedicels (0.5–) 1–2.5 cm long. Seeds 0.8–1 mm long, 0.32–0.45 mm in diameter, narrowly obovoid, with inconspicuous chalazal collar, brown, the surface papillose; coma 5–8 mm long, dull white, detaching easily. Chromosome number: n = 18. Figs. 7a, 28.

Phenology. Collected in flower from June to September, in fruit from July to September (rarely December).

Distribution (Fig. 29A). From northern Afghanistan and Pakistan throughout the Himalayan region in northern India, Nepal, Sikkim, and Bhutan to southwestern China, and possibly northwards into the Tian Shan region of the U.S.S.R.; in China distributed in southeastern Gansu, western Hubei, Sichuan, eastern Guizhou, northern Yunnan, and southeastern Xizang; in swamps, wet places along lakeshores and riverbeds, and often along roads in disturbed places in the mountains; (400–) 1300–3200 m.

ADDITIONAL SPECIMENS EXAMINED. **China.** GANSU: Kangning, *Wudu Pl. Exped. 571* (LZU); Wenxian, *Z. Y. Chang 7491* (HNWP), *W. L. Chen & C. J. Chen 8813* (MO, PE), *Wenxian Exped. 1293* (LZU).—HUBEI: Shennongjia, *H. J. Li 1243* (PE), *Shennongjia Pl. Exped. 21629* (PE), *31240* (PE), *31514* (PE), *1980 Sino-Amer. Bot. Exped. 1831* (GH, KUN, MO, PE); Xingshan, *H. J. Li 995* (PE); Western Hubei, *Henry 6429* (BM, GH, K), *Wilson 2296* (W); Hefeng, *Y. M. Wang 5240* (PE, HIB).—SICHUAN: Chengkou, *T. L. Dai 104773* (CDBI, NAS, PE, SZ), *104800* (CDBI, KUN, NAS, PE, SZ), *104860* (NAS, PE, SZ), *105906* (NAS, PE, SZ); s.d., *Farges s.n.* (P), *Y. L. Qiao 1519* (KUN); Wuxi, *K. H. Yang 58632* (PE); Nanchuan, *Bock & Rosthorn 332* (O); Pingwu, *W. P. Fang 4190* (E, GH, NY, P, PE); Nanping, *Q. X. Li & X. C. Zhao 2741* (PE); Barkam, *Sichuan Med. Pl. Exped. 22143* (NAS); Jinchuan, *G. Z. Zhu & X. Li 75287* (PE), *X. Li 78056* (NAS, PE); Guan Xian, *Z. Y. Li 87-1711* (PE); Ebian, *Z. W. Yao 2979* (NAS, PE); Baoxin, *Sichuan Econ. Pl. Exped. 274* (PE); Tianquan, *Sichuan Econ. Pl. Exped. 662* (CDBI, PE), *T. H. Tu 4616* (PE); Kangding, *S. H. Kung 6218* (PE), *Sichuan Med. Pl. Exped. 15005* (NAS), *Soulie 449 p. p.* (K, P); Qianning, 1974, *s.c. 5680* (CDBI, PE); Dawu, *Sichuan Med. Pl. Exped. 15531* (NAS); Serxu to Dainkog, *W. L. Chen & C. J. Chen 7066* (MO, PE); Leibo, *T. T. Yu 3683* (PE), Shimian, 1955, *C. C. Hsieh s.n.* (PE); Jiulong, *W. L. Chen & C. J. Chen 6418* (MO, PE); Daocheng, *Daocheng Pl. Exped. 2424* (CDBI); Ganluo, *Sichuan Econ. Pl. Exped. 4175* (KUN, PE); Yuexi, 1960, *Sichuan Med. Pl. Exped. 4175* (KUN, PE); Yuexi, 1960, *Sichuan Med. Pl. Exped. s.n.* (NAS); Mianning, *W. L. Chen & C. J. Chen 6184* (PE); Yajiang, *Z. P. Huang 637* (PE); Yanguan, *Quinghai-Xizang Exped. 12661* (KUN, PE); Muli, *W. L. Chen & C. J. Chen 5662* (MO, PE), *6017* (MO, PE), *6018* (MO, PE), *J. S. Liu 7627* (CDBI), *Quinghai-Xizang Exped. 14989* (PE), *Q. S. Zhao et al. 8272* (CDBI), *8276* (CDBI); Xiangcheng, *Qinghai-Xizang Exped. 3710* (KUN, PE); western Sichuan, *S. Y. Hu 1164* (KUN), *S. Jiang 6159* (KUN, PE).—GUIZHOU: Yinjiang, *N. Guizhou Bot. Exped. 786* (GZBI), *Y. Tsiang 7759* (NAS, NY, PE, W); Loushan Guan, *Guizhou Inst. Chin. Med. 5* (GZMI); Bijie, *K. Z. Yang 176* (GZBI); Longfouda Shan, *Guizhou Inst. Bot. 149* (GZBI).—YUNNAN: Daguan, *N. E. Yunnan Exped. 248* (KUN, PE, YNU); Zhengxiong, *P. H. Yu 1148* (KUN); Dongchuan, 1911, *Maire s.n.* (G, MO); Luquen, *B. X. Sun 64142* (YNU); Yongsheng, *Qinghai-Xizang Exped. 597* (KUN, PE); Diancang Shan, *K. K. Tsoong 4597* (PE, UC); Dali, *Schneider 3074* (GH); Eryuan, *H. C. Wang 2357* (KUN, PE); Lijiang, *Z. S. Diau 14851* (PE), *S. T. Li 170* (KUN), *Forest 10328* (BM, E, K, UC), *Rock 10619* (E, S), *Qinghai-Xizang Exped. 741* (KUN, PE); *14969* (KUN, PE), *14989* (KUN, PE); Zhongdian, *Hengduan Shan Exped. 1582* (PE); Weixi, *H. T. Tsai 57623a* (KUN, NAS), *59535* (GH, KUN, NAS, PE), *63041* (GH, KUN, NAS, PE), *C. W. Wang 64320* (GH, KUN, NAS, PE), *67909* (GH, KUN, PE), *68072* (KUN): Deqen, *K. M. Feng 23962* (KUN, PE), *23991* (KUN, PE), *Hengduan Shan Exped. 2977* (PE), *Qinghai-Xizang Exped. 2549* (PE), *2976* (KUN, PE), *T. T. Yu 10305* (BM, E, PE),

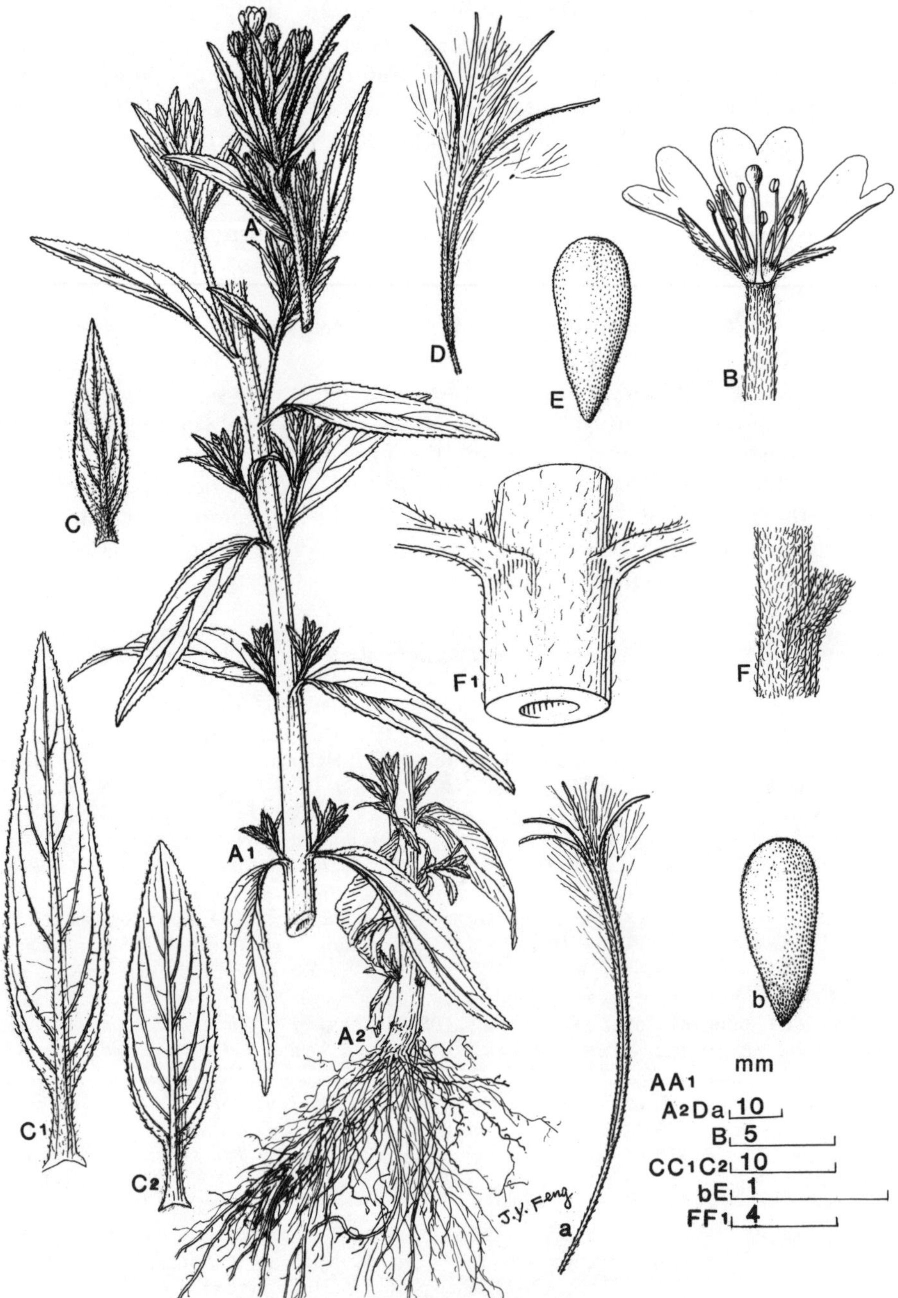

FIG. 28. *Epilobium cylindricum*. A. Inflorescence. A[1]. Portion of mid-stem. A[2]. Base of stem. B. Flower (one petal and two stamens removed). C. Upper leaf (bract). C[1]. Mid-cauline leaf. C[2]. Basal leaf. D. Capsule. E. Seed. F. Axil from inflorescence. F[1]. Axils from mid-stem. a. Capsule. b. Seed. (A–F[1] based on *G. Z. Zhu & X. Li 75287*; a–b based on *N. E. Yunnan Exped. 248*.)

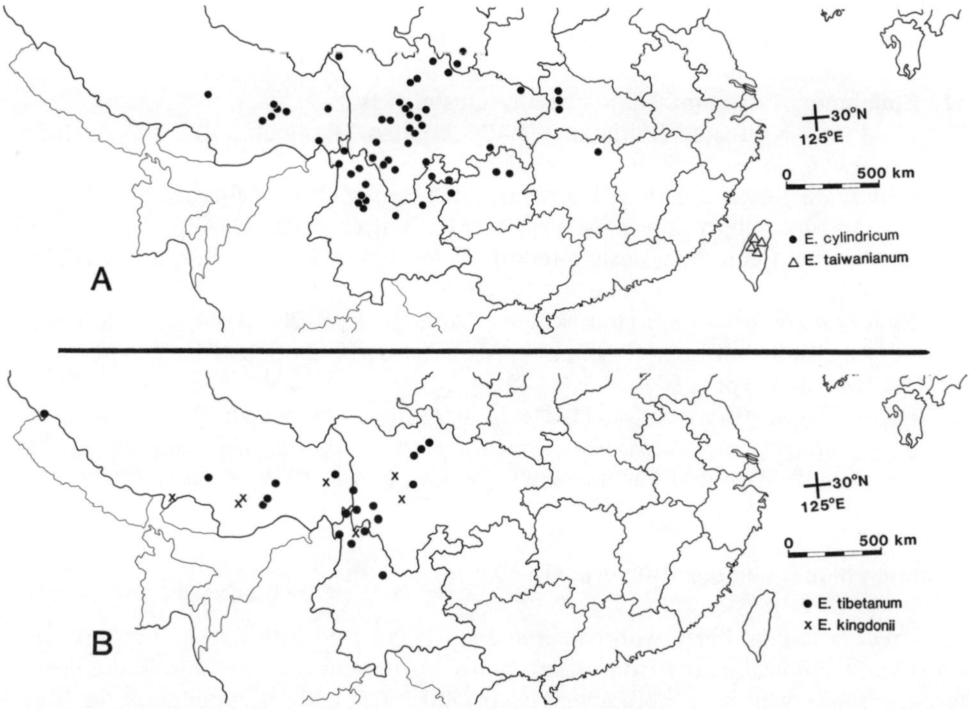

FIG. 29. Distribution in China of: A. *Epilobium cylindricum* and *E. taiwanianum*; B. *Epilobium kingdonii* and *E. tibetanum*.

19588 (E, GH, KUN, PE), *Soulie 1127* (P), *1256* (P); Gongshan, *D. M. Feng 7130* (KUN, PE), *8563* (KUN, PE), *Qinghai-Xizang Exped. 9184* (KUN, PE), *9831* (KUN, PE); "Tchen-fong-chan," 1894, *Delavay s.n.* (G, P); no locality, *H. T. Tsai 57150* (GH, PE), *57623* (GH, PE), *Y. Tsiang 11522* (KUN, NAS, PE), *K. K. Tsoong 4597* (PE, UC), *S. W. Union Univ. 11719* (KUN).—XIZANG: Zhag'yab, *Qinghai-Xizang Exped. 12996* (PE); Bomi, *Y. T. Chang & K. Y. Lang 854* (PE), *C. Y. Wu 5772* (KUN): Zayu, 1984, *H. S. Kung et al. s.n.* (MO), *Z. C. Ni et al. 237* (PE), *Xizang Exped. 860* (NWP), *1193* (NWP); Kongbo, *Ludlow et al. 14248* (BM); Dongjug, *Kingdon-Ward 12103* (BM); Nyingchi, *P. C. Kuo & W. Y. Wang 21938* (PE), *Ludlow et al. 6803* (BM), *Ludlow & Sherriff 8851* (BM); Mainling, *Qinghai-Xizang Exped. 750788* (KUN, PE); Lhasa, *Richardson 193* (BM); Nyalam, *Xizang Med. Pl. Exped. 1334* (PE).

Epilobium cylindricum is very common, especially in the western Himalaya; with *E. tibetanum, E. sinense,* and *E. platystigmatosum,* it constitutes a complex of morphologically similar species that is one of the most characteristic and common elements of *Epilobium* in the Sino-Himalayan region. Although these four species are well delimited, they share a similar overall aspect; they are mostly well branched and have narrow, denticulate leaves. The species most similar to *E. cylindricum,* one from which it can be distinguished only with difficulty in the absence of mature seeds, is *E. tibetanum.* Another characteristic Sino-Himalayan species probably related to this group is *E. royleanum,* but it is clearly distinguished morphologically by its broader leaves, among other features.

Epilobium cylindricum, and to a lesser degree other species in this complex, often grow in disturbed damp places and are among the most weedy taxa of

Epilobium in the region. In this regard, they are the ecological equivalents to *E. ciliatum* in East Asia, North America, and elsewhere.

12. Epilobium tibetanum Haussknecht, Oesterr. Bot. Zeitschr. 29: 53. 1879.—TYPE: KASHMIR. Baltistan, Thalle La near Bagmaharal, 30 Aug 1856, *Schlagintweit 5909* (holotype: JE!).

Epilobium pseudobscurum Haussknecht, Oesterr. Bot. Zeitschr. 29: 53. 1879 ("pseudo-obscurum").—TYPE: INDIA. Punjab, Spiti, 31 Aug 1847, *Thomson s.n.* (lectotype, designated by Raven 1962a: K!; isolectotypes: CGE! P! W!).

Epilobium leiospermum Haussknecht, Monogr. Epil. 206. 1884.—TYPE: KASHMIR. Baltistan, near Tashing, NW of Astor, 10–24 Sep 1856, *Schlagintweit s.n.* (holotype: JE!).

Epilobium nuristanicum K. H. Rechinger, Biol. Skr. 10(3): 61. 1958.—TYPE: AFGHANISTAN. Nuristan, wet meadow above Pashki in Parum Valley, between Shtive and Pashki, 2300–2500 m, 2 Aug 1935, *Kerstan 1382* (holotype: W!; isotype: HAL!).

Common name: Guangzi Liuyecai (Li 1986).

Erect perennial herb, with rhizome densely covered with fibrous rootlets, and short leafy soboles arising from base. Stems 13–100 cm tall, 2–7 mm in diameter, terete, usually well branched, sparsely strigillose throughout upper stem, glabrescent below, with indistinct raised lines decurrent from margins of petioles, sometimes incomplete through internodes. Leaves opposite, alternate on the inflorescence, on petioles 2–5 mm long; cauline blades 1.2–6.5 cm long, 0.5–1.5 cm wide, lanceolate or sometimes narrowly ovate, glabrous except for scattered hairs on veins and margin, subcoriaceous, apex acute or acuminate, base cuneate or rarely subrounded, margin serrulate with 15–35 teeth on each side, the lateral veins 4–5 on each side of the midvein. Inflorescence erect. Flowers erect; buds 3–5 mm long, 2–3 mm in diameter, ovoid, mucronate; ovaries 1.4–2.3 cm long, densely strigillose and eglandular; pedicels 0.4–1.2 cm long; floral tube 1–1.3 mm long, 2–2.5 mm in diameter, usually glabrous within; sepals 3.5–5 mm long, 1.3–1.7 mm wide, oblong-lanceolate, keeled; petals 5–8 mm long, 3–3.5 mm wide, pink to rose-purple, rarely white, obovate, the notch 0.7–1.2 mm deep; filaments of longer stamens 2.8–3.8 mm long, those of shorter ones 1–2 mm long; anthers 0.7–0.9 mm long, oblong; style 2.5–3.8 mm long, erect, glabrous; stigma capitate to broadly clavate, 1–1.8 mm high, 1–1.7 mm in diameter, white, surrounded by anthers at anthesis. Capsules 4.2–8.8 cm long, sparsely strigillose, borne on pedicels 0.8–2.5 cm long. Seeds 1.1–1.3 mm long, 0.45–0.6 mm in diameter, obovoid or obpyriform, rounded with inconspicuous chalazal collar, brown, the surface reticulate; coma 7–9 mm long, dull white, easily deciduous. Chromosome number: n = 18. Figs. 6f, g, 8g–i, 30.

Phenology. Collected in flower from July to September, in fruit from August to October.

Distribution (Fig. 29B). From northern Afghanistan and northern Pakistan through Kashmir, northern India (Punjab and Kumaun), Nepal, Sikkim, and Bhutan to western China; in China occurring in western Sichuan, northwestern Yun-

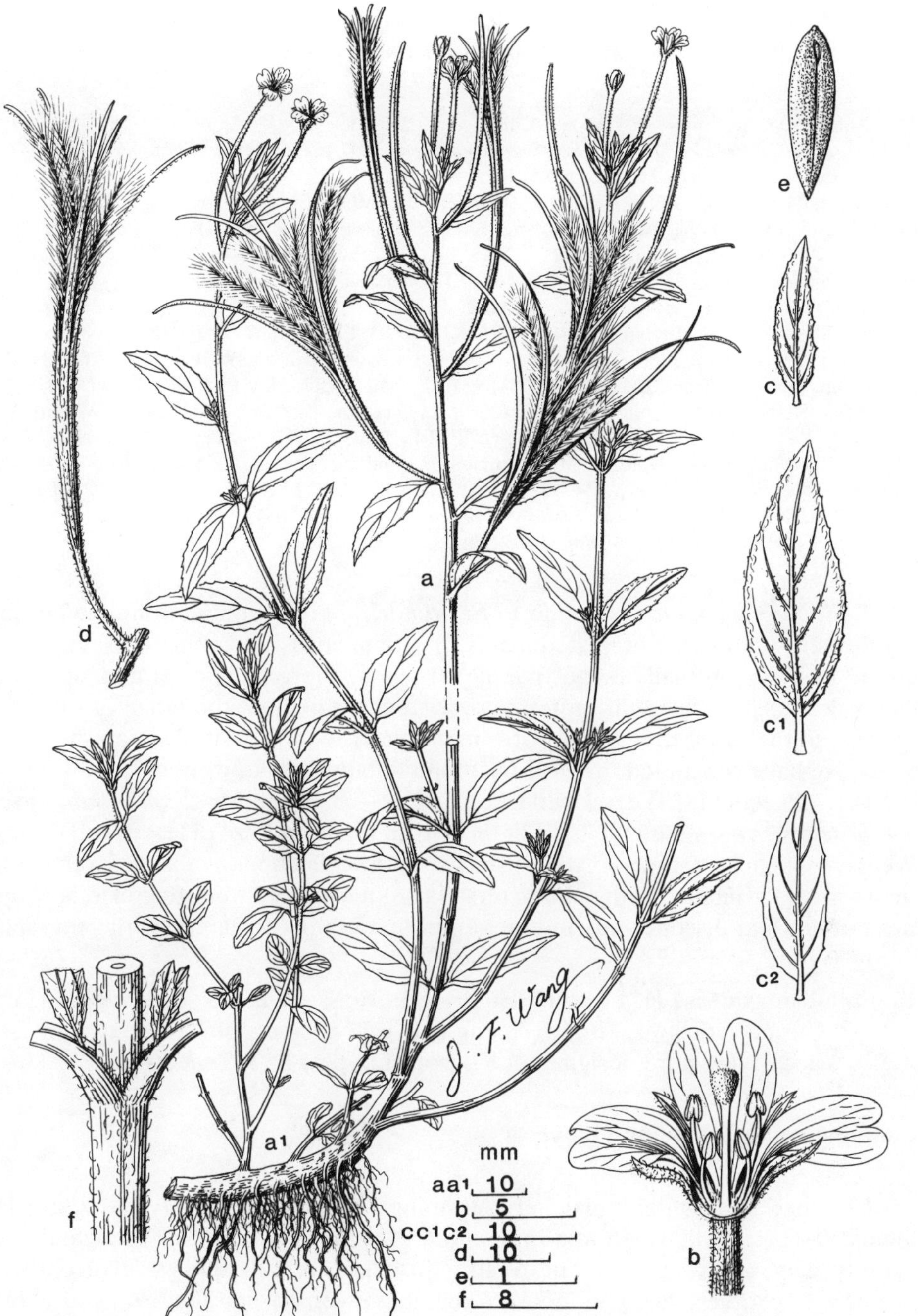

FIG. 30. *Epilobium tibetanum*. a. Inflorescence. a[1]. Habit (except main stem). b. Flower (one petal and three stamens removed). c. Leaf from inflorescence (bract). c[1]. Mid-cauline leaf. c[2]. Basal leaf. d. Capsule. e. Seed. f. Mid-cauline axils. (Based on *Xizang Exped. 4065*)

nan, and southwestern and southeastern Xizang (Tibet); scattered but locally common in moist places by ditches, streams, or rivers in the mountains; 2350–4500 m.

ADDITIONAL SPECIMENS EXAMINED. **China.** SICHUAN: Heishui, *X. Li et al. 73786* (NAS, PE); Barkam, *X. Li 71899* (PE, SZ); Xiaojin, *S. S. Chang et al. 6451* (CDBI, PE); "Chonghua," *X. Li 76813* (NAS, PE); Jinchuan, *X. Li 78512* (NAS, PE); Kangding, *K. Y. Lang et al. 805* (KUN, PE), *Smith 10817* (S, UPS); Jiulong, *W. L. Chen & C. J. Chen 6392* (MO, PE), *6420* (MO, PE); Xichang, *W. L. Chen & C. J. Chen 6127* (MO, PE); Batang, *Qinghai-Xizang Exped. 5186* (KUN, PE); Muli, *W. L. Chen & C. J. Chen 6023* (MO, PE), *6035* (MO, PE); Daocheng, *Daocheng Pl. Exped. 2524* (CDBI, KUN), *Qinghai-Xizang Exped. 4322* (KUN, PE), *5498* (KUN, PE); Xiangcheng, *Qinghai-Xizang Exped. 3124* (KUN, PE), *3682* (KUN, PE), *4821* (KUN, PE); Derong, *Qinghai-Xizang Exped. 3369* (KUN, PE).—YUNNAN: Huaping, *S. Jiang 6221* (KUN, PE); Lijiang, *S. T. Li 170* (KUN); Weixi, *H. T. Tsai 59868* (KUN, NAS, PE), *C. W. Wang 68072* (PE); Gongshan, *C. W. Wang 66762* (KUN, PE); Zhongdian, *Qinghai-Xizang Exped. 3078* (KUN, PE), *3089* (KUN, PE), *C. Y. Wu 4129* (KUN), *J. S. Yang 8495* (KUN); Deqen, *Qinghai-Xizang Exped. 2549* (KUN, PE), *2977* (KUN, PE); Western Yunnan, *T. T. Yu 10163* (GH, PE); no locality, *H. T. Tsai 59865* (KUN).—XIZANG: Zhag'yab, *Qinghai-Xizang Exped. 12996* (KUN, PE); Mt. Nanjiabawa, Medog. *P. X. Li & S. Z. Cheng 1017* (PE); Mainling, *Qinghai-Xizang Exped. 74-4637* (PE); Nyingchi, *P. S. Li & S. Z. Cheng 6290* (PE), *J. X. Yang 2328* (HNWP, KUN); Lhasa, *Ludlow & Sherriff 9040* (BM, GH); Burang, *Xizang Exped. 4065* (HNWP, PE, XJBI); "Hasora," *Wallich 6863* (P).

Epilobium tibetanum is similar to *E. cylindricum* in overall morphology, including the terete and well-branched stems, narrow leaves, and general aspect; but it differs in having reticulate (not papillose) seeds, shorter petioles, and somewhat larger flowers. The seed difference is consistent throughout the range of these two species, so that even though they are similar and are difficult to distinguish without seeds, we have concluded that it is best to maintain them as distinct species.

Haussknecht (1884) described three species—*E. tibetanum, E. pseudobscurum,* and *E. leiospermum*—on the basis of the stigma shape and size of leaves and flowers. We now recognize only one. Analysis of the many collections available for this study shows great variability in the characters Haussknecht used to distinguish these species but no clear discontinuity in that variation nor any correlation with geography.

13. Epilobium sinense H. Léveillé, Bull. Herb. Boiss., sér. 2, 7: 590. 1907.—TYPE: CHINA. Guizhou, Pingfa, "Yang-Kia-Tchong," 29 Jul 1905, *Cavalerie 2438* (lectotype, here designated: P!; isolectotypes: E! P-3 sheets!).

Common name: Zhonghua Liuyecai.

Clumped robust perennial herb, forming short, densely leafy basal soboles. Stems 10–50 cm tall, 1.5–5 mm in diameter, terete, simple or few-branched, very densely leafy, glabrescent except for strigillose hairs on raised lines decurrent from margins of petioles. Leaves opposite near base, mostly spirally arranged above, densely crowded throughout, on petioles 2–11 mm long; cauline blades 1.2–7 cm long, 0.3–1 cm wide, narrowly spatulate to sometimes oblong or lance-linear, or rarely narrowly obovate, glabrous except for sparse pubescence on adaxial midvein and margin, subcoriaceous, apex obtuse, base narrowly cuneate, margins weakly and remotely denticulate with 3–12 teeth on each side, the midrib conspicuously pale, prominent on abaxial side, the lateral veins 4–5 on each side of the midrib.

Inflorescence erect. Flowers erect; buds 4–4.5 mm long, 2–2.5 mm in diameter, elliptic-oblong, acute, subglabrous; ovaries 1.2–2.5 cm long, sparsely strigillose and eglandular; pedicels 0.7–2 cm long; floral tube 1–1.2 mm long, 1.5–2.2 mm in diameter, with a sparse ring of erect hairs within; sepals 4.5–6.5 mm long, 1–1.2 mm wide, oblong-lanceolate, short-mucronate; petals 5.5–8 mm long, 3–4.5 mm wide, white, pink, or sometimes rose-purple, obovate, the notch 1–1.7 mm deep; filaments of longer stamens 3–4 mm long, those of shorter ones 1.5–2 mm long; anthers 0.7–1 mm long, 0.3–0.45 mm in diameter, oblong-elliptic; style 3.5–4 mm long, glabrous; stigma capitate or sometimes broadly clavate, 0.8–1.7 mm high, 0.7–1.7 mm in diameter, at least the longer stamens shedding pollen directly onto stigma at anthesis. Capsules 2.5–5.5 cm long, glabrescent or sparsely strigillose, borne on pedicels 1.3–4 cm long. Seeds 1.2–1.3 mm long, 0.5–0.6 mm in diameter, oblong-obovoid, rounded or sometimes subacuminate with a very short chalazal collar, brown, the surface finely papillose; coma 6–8 mm long, reddish, detaching readily. Chromosome number: $n = 18$. Fig. 31.

Phenology. Collected in flower from June to August (rarely to September), in fruit from August to October (rarely to December).

Distribution (Fig. 32). Endemic to China, occurring in Henan, Shaanxi, Gansu, Hunan, Hubei, Sichuan, Quizhou, and northern Yunnan; in wet places along streams and rivers, occasionally in other exposed, rocky places; 550–2400 m.

ADDITIONAL SPECIMENS EXAMINED. **China.** HENEN: Luanchuan, *Henan Exped. 20726* (KUN, PE); Lushin, *K. M. Liou 4921* (PE).—SHAANXI: Hua Shan, *K. S. Hao 4157* (PE); Mts. Qing Ling, *Qing Ling Exped. 68* (PE); Foping, *T. N. Liou & P. C. Tsoong 3212* (KUN, PE); Liuba, *K. T. Fu 6228* (KUN, PE), *Z. X. Fu & Y. H. Kuo 783* (WUK); Nanzheng, *T. P. Wang 20768* (WUK); Hanzhong, *J. W. Wang & Z. C. Shi 133* (PE); "Uo-mi-san," 1899, *Hugh s.n.* (BM).—GANSU: Hui Xian, *Z. X. Peng 5759* (LZU), *Xiaolong Shan Exped. 744* (LZU); Kang Xian, *Z. Y. Zhang 16440* (PE), *16918* (PE), *17081* (PE); Wen Xian, *S. Pu 80207* (LZU), *Z. Y. Zhang 8323* (HNWP), *14757* (PE); Piandonggou, *Lanzhou Univ. Exped. 236* (LZU).—HUNAN: Yongsui, *Hunan Exped. 596* (IBSC, PE); Sangzhi, *Beijing Bot. Exped. 3628* (PE), *X. G. Li 204319* (HTC); W. Hunan, *B. M. Yang 2080* (HTC).—HUBEI: Fang Xian, *Z. Y. Liu 267* (PE); Shennongjia, *Shennongjia Pl. Exped. 10899* (PE), *11736* (PE), *21569* (PE), *21796* (PE), *22746* (PE), *31061* (PE), *31551* (PE), *K. S. Chou & H. M. Dong 76003* (BM, E, MO, NCU, PE), *1980 Sino-Amer. Exped. 24* (GH, KUN, PE), *599* (GH, KUN, PE); Xingshan, *W. Y. Chun 5338* (UC); W. Hubei, *Henry 6453* (GH, K), *W. C. Cheng & C. T. Hwa 712* (GH, K, UC); Zigui, *T. P. Wang 11824* (PE); Badong, *T. P. Wang 11633* (KUN, PE): Hefeng, *H. J. Li 6732* (KUN, PE), *J. J. Li 9115* (PE); Enshi, *L. Y. Dai & C. H. Qian 621* (PE); Lichuan, *W. C. Cheng & C. T. Hwa 712* (PE), *L. Y. Dai 1250* (PE), *L. Y. Dai & C. H. Qian 821* (KUN, PE); Laifeng, *H. J. Li 7086* (KUN, PE), *7470* (PE).—SICHUAN: Chengkou, *T. L. Dai 103116* (PE), *103669* (NAS, PE), *104277* (NAS, PE); *104597* (NAS, PE, SZ), *104773* (PE); Wanyuan, *P. Y. Li 5552* (PE), *5890* (PE), *5986* (PE); Wushan, *H. F. Zhou & H. Y. Su 110006* (PE, SZ), *G. H. Yang 59929* (CDBI, NAS); Wuxi, *G. H. Yang 65091* (CDBI, NAS, PE); Shizhu, *D. F. Fu & Y. L. Cao 1114* (CDBI, PE), *W. H. Wang 1722* (CDBI), *Y. Chen 2919* (CDBI); Wulong, *W. H. Wang 3681* (CDBI); Nanchuan, *C. H. Hsiung 91865* (PE, SZ), *Jingfo Shan Exped. 1897* (PE); Emei, *Henry 287* (NY); Hongya, *W. P. Fang 8589* (PE), *Z. W. Yao 2498* (NAS, PE); Hanyuan, *T. P. Wang 8638* (PE); "Tsekou," *Soulie 1128* (P).—GUIZHOU: Pingfa, *Cavalerie 1784* (P); Xifeng, *S. W. Teng 90435* (GH); Bijie, *P. H. Yu 360* (KUN, PE); Nayong, *Bijie Exped. 578* (GZBI, KUN, PE); Shuicheng, *Shuicheng Exped. 525* (GZBI); Xingyi, *Guizhou Exped. 7185* (GZBI, PE); no locality, *Cavalerie 4390* (P).—YUNNAN: Yiliang, *B. Y. Qiu 711221* (CDBI, KUN), *771225* (CDBI, KUN); Songming, *Songming Pl. Exped. 51331* (KUN); Kunming, *Cavalerie 2438* (E, UPS), *Z. H. Yang 101230* (KUN), *B. Y. Qiu 77586* (CDBI, KUN); Jiangchuan, *B. Y. Qiu 60427* (CDBI, KUN), *Ducloux 2685* (P); on the way from Heqing to Dali, *Schneider 3074* (G); W. Yunnan, *Delavay 931* (P); Deqen, *Deqen Pl. Exped. 23962* (PE); Baoshan, *J. Chen 910* (KUN); no locality, *Ducloux 603* (P), *3820* (P), *Maire 394* (UC), *477* (UC), *1994* (E), *2217* (E), *2217a* (E), *2218* (E), *Simeon Ten 1185* (C), *Uni Univ. 20218* (KUN).

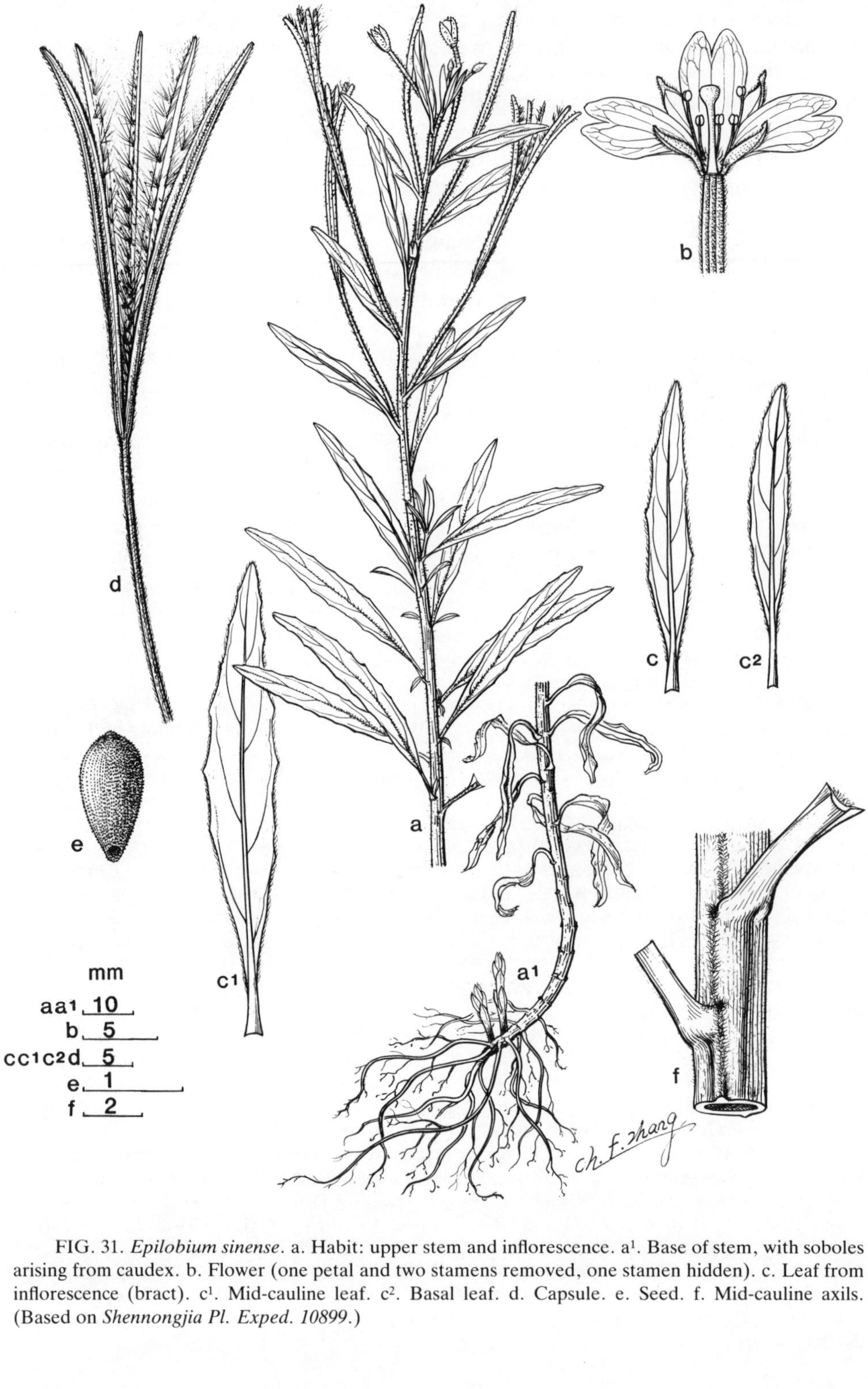

FIG. 31. *Epilobium sinense*. a. Habit: upper stem and inflorescence. a[1]. Base of stem, with soboles arising from caudex. b. Flower (one petal and two stamens removed, one stamen hidden). c. Leaf from inflorescence (bract). c[1]. Mid-cauline leaf. c[2]. Basal leaf. d. Capsule. e. Seed. f. Mid-cauline axils. (Based on *Shennongjia Pl. Exped. 10899*.)

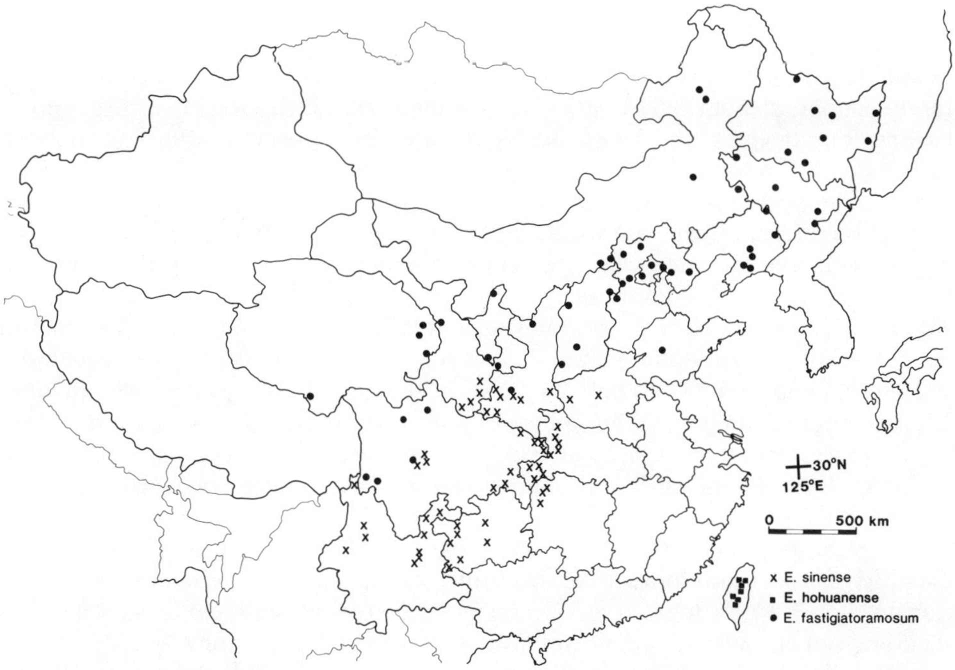

FIG. 32. Distribution in China of *Epilobium fastigiatoramosum*, *E. hohuanense*, and *E. sinense*.

This very distinctive species, frequent in western and central China, occurs in the same areas as the similar *Epilobium cylindricum* and *E. platystigmatosum,* but it is readily distinguished from them by its crowded, spirally arranged, often spatulate leaves with conspicuous pale midribs, and stems with distinct, raised, strigillose lines.

14. Epilobium platystigmatosum C. B. Robinson, Philipp. J. Sci. 3: 210. 1908.—TYPE: THE PHILIPPINES. Luzon, Province of Benguet, at Pauai, on damp soil at 2040 m, 8 Nov 1905, *Merrill 4744* (lectotype, here designated: K!). The holotype specimen at the herbarium of the Bureau of Science in Manila was destroyed in 1945; we select the only known isotype as the lectotype.

Epilobium formosanum Masamune, Trans. Nat. Hist. Soc. Taiwan 29: 62. 1939.—TYPE: CHINA. Taiwan, Taihokusyu, in grassy, rocky places between Yappitu and Dozan, 14 Apr 1933, *Suzuki-Tokio 8709* (lectotype, here designated: TAI!).

Epilobium cephalostigma var. *linearifolium* Hisauti, J. Jap. Bot. 14: 143. 1938.—TYPE: JAPAN. Hondo, Prov. Shinano, Todai-gawa, Kami-Inagun, Todai River, 13 Aug 1937, *Hisauti s.n.* (holotype: TI!).

Epilobium sohayakiense Koidzumi, Acta Phytotax. Geobot. 8: 61. 1939.—TYPE: JAPAN. Sikoku: prov. Tosa, Okunanogawa, Oct 1891, *Koidzumi s.n.* (lectotype, here designated: KYO!).

Common name: Guezhu Liuyecai (Raven 1977).

Loosely clumped erect perennial herb, bearing loose, leafy soboles from non-scaly underground caudex. Stems (9–) 15–70 cm tall, 2–6 mm in diameter, terete, usually very branched throughout, the primary branchlets 2 or 3 times furcate, very rarely simple, slender, densely leafy, sometimes reddish, strigillose throughout, lacking raised lines through internodes. Leaves opposite, alternate on the inflorescence, on petioles 1–4 mm long; cauline blades 1–4.5 cm long, 0.15–0.5 cm wide, sublinear to narrowly lanceolate, glabrous except for sparse pubescence on veins and margin, the abaxial surfaces purplish at least on older leaves, apex acute or obtuse, base attenuate to narrowly cuneate, margins weakly denticulate with 3–8 teeth on each side, the lateral veins 4–5 on each side of the midrib. Inflorescence slightly nodding before anthesis. Flowers erect; buds 2–3.2 mm long, 1.2–2 mm in diameter, elliptic or oblong-elliptic, apiculate; ovaries 1–1.2 cm long, densely strigillose, delicate; pedicels 0.4–1.1 cm long, slender; floral tube 0.6–0.8 mm long, 1.5–1.8 mm in diameter, with a sparse ring of spreading hairs at the mouth within; sepals 2.5–3.2 mm long, 1–1.2 mm wide, oblong-lanceolate, mucronate; petals 3–5 mm long, 1.8–2.5 mm wide, white, pink, or rarely rose-purple, obovate, the notch 0.4–0.6 mm deep; filaments of longer stamens 1.6–2 mm long, those of shorter ones 0.6–1.2 mm long; anthers 0.4–0.6 mm long, 0.25–0.35 mm in diameter, broadly ovoid; style 1.8–2 mm long, erect, glabrous; stigma capitate or sometimes broadly clavate, 0.7–1.4 mm high, 0.5–1.3 mm in diameter, at least the longer stamens shedding pollen onto stigma at anthesis. Capsules 2.3–5 cm long, glabrescent or sparsely strigillose, borne on pedicels 0.8–2.2 cm long. Seeds 0.8–0.9 mm long, 0.3–0.4 mm thick, oblong-obovoid, with very short chalazal collar, brown, the surface coarsely papillose; coma 6–8 mm long, dingy white, easily detaching. Chromosome number: $n = 18$. Fig. 33.

Phenology. Collected in flower from August to October, in fruit from September to November.

Distribution (Fig. 34). Japan and eastern China to the northern Philippines; in China occuring in Hebei, Henan, Shaanxi, Gansu, Qinghai, Taiwan, Guangxi, Hubei, Sichuan, and Yunnan; on wet, often disturbed grassy slopes and along streams and rivers in mountains; (400–) 1000–2000 (–3500) m.

Additional Specimens Examined. **China.** Hebei: Zanhuang, *X. Y. Liu 1004* (PE), *J. W. Wang 271* (PE); Neiqiu, *Y. Liu 13323* (KUN, NMU, PE), *13581* (PE); She Xian, *Hebei Pl. Exped. 79* (PE); Zhengding, *Hebei Exped. V538* (KUN).—Henan: Song Xian, *Henan Exped. 2197* (NAS, PE); no locality, *Hugh 51* (BM).—Shaanxi: Shangxien, *T. P. Wang 16027* (PE); Lueyang, *K. T. Fu 5779* (PE); Hanzhong, *J. W. Wang & Z. C. Shi 374* (PE); Nanzheng, *K. T. Fu 10953* (KUN); "Chui-Kio-Tsuen," 1898, *Giraldi s.n.* (K).—Gansu: Tianshui, *Q. X. Ding 203* (LZU), *S. Z. Zhang 289* (LZU); Wushan, *Wushang Exped. 1031* (LZU); Kang Xian, *W. Y. Hsia 6329* (PE), *Z. Y. Zhang 16838* (PE); Cheng Xian, *T. P. Wang 15018* (PE); Wen Xian, *W. L. Chen & C. J. Chen 8803* (MO, PE); Xigu, *T. P. Wang 14878* (PE).—Qinghai: Xunhua, *B. Z. Kuo & Y. C. Yang 9728* (HNWP).—Taiwan: Taibei Xian, *T. C. Huang 4226* (TAI), 1933, *Suzuki Tokio s.n.* (TAI); Yilan Xian, *C. C. Hsu 5801* (TAI, TI), *M. T. Kao 3160* (TAI), *Shimizu & C. C. Chuang 20454* (S, SHIN), 1928, *S. Suzuki s.n.* (TAI), 1925, *S. Suzuki s.n.* (TAI), *Tamura & Koyama 23616* (S); Nanhutashan, *Yamazaki et al. 1118* (TI), 1969, *T. Yamazaki et al. s.n.* (TAI); Xinzhu Xian, *H. J. Chang 2008* (MO, TAI), *C. S. Kuoh 3222* (TAI), *C. F. Hsieh 1257* (TAI); Taihei Shan ("Taipingshan"), *Suzuki 237* (GH); 1925, *Suzuki s.n.* (NAS), *Bartlett 6049* (MICH, NY, UC); Miaoli Xian, Kuanwu, *C.-I Peng 8412* (HAST, MO); Hualian Xian, *C. C. Hsu & R. Hsu 3767* (TAI), *C. C. Chung & M. T. Kao 4454* (TAI), *Keng & Kao 2619* (L), *M. T. Kao 7767* (PH, TAI), *C.-I Peng 5821* (HAST), *Steenis 20685* (L), 1933, *Suzuki-Tokio s.n.* (TAI), *Tamura et al. 21688* (SHIN), *Keng & Kao 2634* (L); Taizhong Xian, *M. T. Kao 5007* (MO); Nantou Xian, *C. S. Kuoh 3664* (TAI), *M. T. Kao 7246* (TAI), *C. C. Hsu 6110* (TAI), *C.-I Peng 5336* (HAST, MO), 1938, *Akiyama-Sijes s.n.*

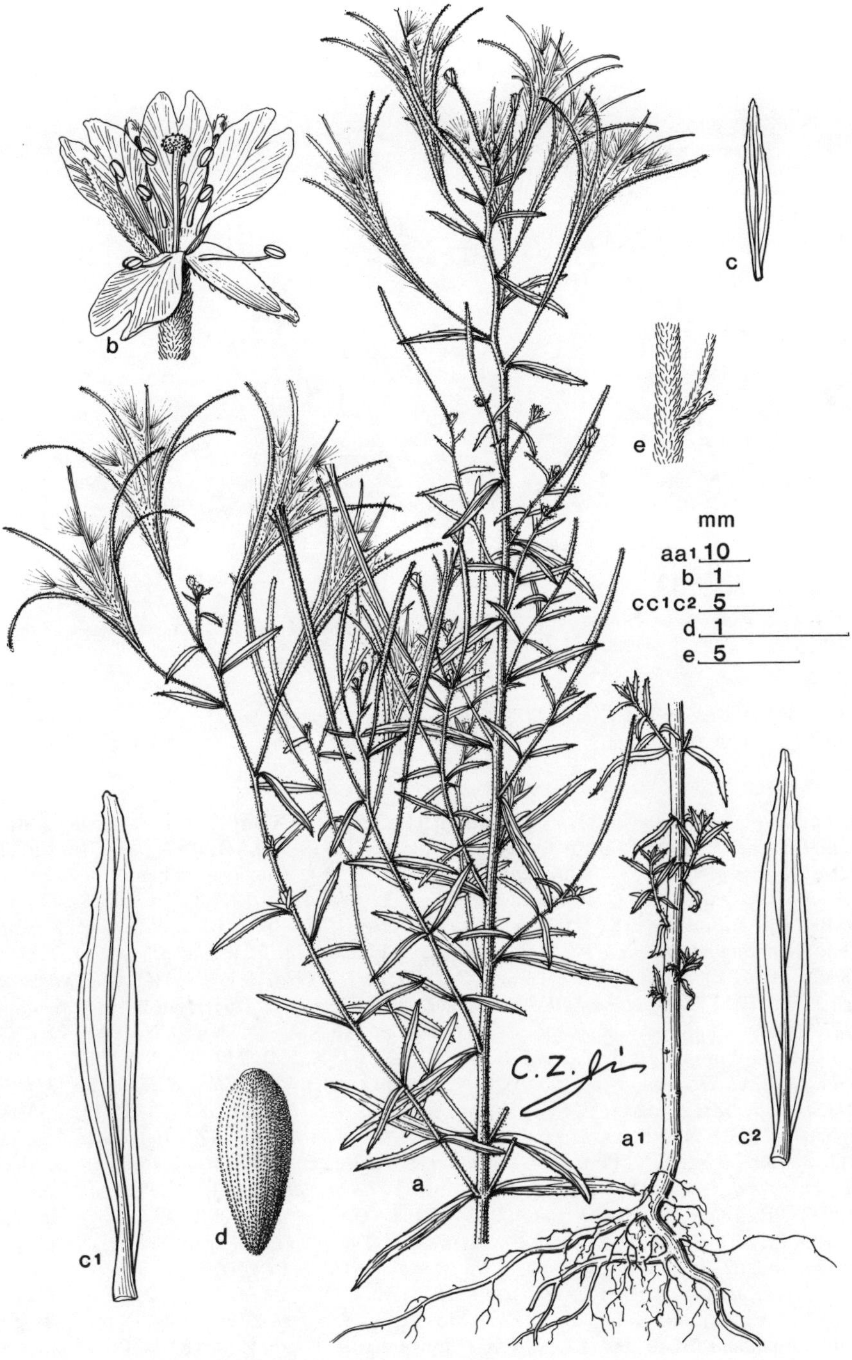

FIG. 33. *Epilobium platystigmatosum*. a. Habit: upper portion of plant, including inflorescence. a[1]. Base of stem. b. Flower. c. Leaf from inflorescence (bract). c[1]. Mid-cauline leaf. c[2]. Basal leaf. d. Seed. e. Axil from inflorescence. (Based on *K. T. Fu 5779*.)

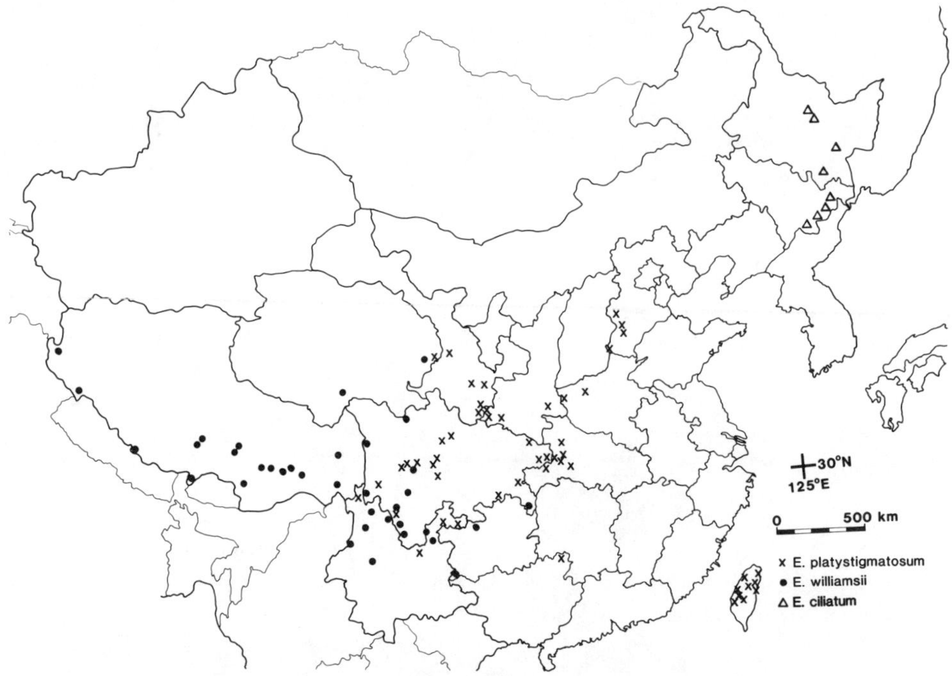

FIG. 34. Distribution in China of *Epilobium ciliatum* subsp. *ciliatum*, *E. platystigmatosum*, and *E. williamsii*.

(TAI); Yushan ("Mt. Nitaka"), 1908, Nakahara s.n. (PENN); Jiayi Xian, 1957, *T. I. Chuang & Kao s.n.* (TAI), *C.-I Peng 8925* (HAST, MO); Taidong Xian, *C. C. Hsu 3289* (TAI); 1967, *C. C. Hsu s.n.* (TAI); 1932, *Hosokawa s.n.* (TAI); Bunkito, *Faurie 560* (BM, G, P); Zhongyang Shan Mai, *M. Kato 418* (UC).—GUANGXI: Ziyuan, *X. Q. Wang 78079* (NAS); "Tzu Yuen District," *Z. S. Chung 83513* (GH).—HUBEI: Badong, *Wilson 2585* (K, NY, W); Western Hubei, *Henry 1037* (BM), *6808* (GH, K), *Silvestri 6169* (E); Shennongjia, *Shennongjia Pl. Exped. 32235* (PE), *32825* (PE), *1980 Sino-Amer. Exped. 319* (GH, KUN, MO, PE), *380* (GH), *722* (GH, KUN, MO, PE), *1013* (GH, KUN, MO, PE); Xingshan, *Q. N. Chen 2051* (PE); Yichang, *Henry 1043* (K); Zigui, *T. P. Wang 12097* (PE); Jianshi, *L. Y. Dai & C. H. Qian 1219* (PE).—SICHUAN: Chengkou, *T. L. Dai 103867* (NAS), *106539* (KUN, NAS, PE), *107393* (NAS, PE, SZ), Wushan, *H. F. Zhou & H. Y. Su 109736* (PE, SZ), *110148* (NAS, PE, SZ), *110348* (NAS, PE, SZ), *G. H. Yang 59692* (CDBI, NAS, PE); Fengjie, *H. F. Zhou & H. Y. Su 111107* (PE, SZ), *111143* (PE, SZ), *111528* (PE, SZ); Shizhu, *F. D. Fu & Y. L. Cao 1133* (CDBI, PE), *Shizhu Pl. Exped. 3384* (CDBI); Nanchuan, *C. H. Hsing & Z. L. Zhou 93862* (PE, SZ), *Jinfo Shan Exped. 2311* (PE), *G. F. Li 64321* (KUN, PE), *K. C. Kuan et al. 2229* (PE), *F. T. Wang 10066* (PE); Maowen, *Maowen Pl. Exped. 8886* (CDBI, PE); Wenchuan, *K. Y. Lang et al. 1254* (KUN, PE); Mt. Emei, *W. P. Fang 15077* (PE, SZ), *C. H. Hsing et al. 32532* (PE), *K. C. Kuan et al. 2510* (PE), *Raven et al. 26143* (KUN, MO, PE); Tianquan, *K. C. Kuan et al. 1907* (PE), *Tianguan Bot. Exped. 782* (KUN); Luding, *K. C. Kuan et al. 1587* (PE); Kangding, *H. S. Kung et al. 44* (MO, PE), *H. L. Tsiang 37783* (PE, SZ); Baoxing, *K. C. Kuan 2974*X (PE), *T. H. Tu 4910* (PE); Muli, *Q. S. Zhao et al. 8272* (CDBI); Daocheng, *Z. G. Liu 577* (CDBI); no locality, *Z. X. Zhao 818* (PE).—YUNNAN: Zhengxiong, *P. H. Yu 1148* (PE); Yilian, *N.-E. Yunnan Exped. 583* (KUN, YNU); Songming, *B. Y. Qiu 55180* (KUN, PE); Deqen, *K. M. Feng 23959* (KUN, PE); "Mi-Monatague," *Maire 3289* (UC).

This species, widely distributed in eastern Asia, is morphologically similar to a distinctive group of species in south and east Asia that includes at least *E. cylindricum, E. tibetanum,* and *E. sinense.* It is distinguished from *E. cylindricum* and *E.*

tibetanum by its more clumped and branched habit, slender stems, and smaller, narrower leaves with less toothed margin, and from *E. sinense* by its more branched and slender stems covered with strigillose pubescence throughout without raised lines, narrower leaves, and smaller seeds.

Epilobium platystigmatosum also resembles the Taiwan endemic *E. hohuanense* in its clumped habit and rather narrow leaves. The two species differ consistently in morphology, however, and their altitudinal distributions do not overlap: *E. platystigmatosum* grows at lower elevations (mostly 1100–2500 m) than does *E. hohuanense* (2650–3600 m).

Most collections of *E. platystigmatosum* from Taiwan have a broad clavate stigma. In this, they differ from collections from mainland China or Japan, where most specimens have a capitate stigma. The collections from Taiwan otherwise do not differ notably from those gathered elsewhere.

15. Epilobium hohuanense Ying, Quart. J. Chin. Forest. 8: 121. 1975.—TYPE: CHINA. Taiwan, Hualian Xian–Nantou Xian, Hohuanshan, *S. S. Ying 1571* (holotype: NTUF!, photo and fragment of holotype: MO!).

Common name: Hehaun Liuyecai (Ying 1975).

Caespitose or clumped perennial herb, spreading by numerous thin, fleshy soboles that leave small scattered scales at base of stem. Stems 5–20 (–30) cm tall, 1–2 mm in diameter, terete, simple except for branching from caudex or rarely branched above, ascending or sometimes decumbent, strigillose throughout, sometimes with broad indefinite lines of hairs decurrent from bases of petioles, often purplish, densely leafy on upper half. Leaves opposite below the inflorescence, the lower ones on petioles 1–3 mm long, the upper ones subsessile; cauline blades (0.5–) 1–2 cm long, 0.15–0.7 cm wide, elliptic or oblong to lanceolate, often narrowly so, often spathulate in lower pairs, subglabrous or with scattered hairs on margins and abaxial midrib, apex subacute in upper leaves to obtuse in the lower, base attenuate to narrowly cuneate, margins denticulate with 4–10 teeth on each side, the lateral veins inconspicuous, 3–6 on each side of midrib, especially older leaves often more or less purplish. Inflorescence erect. Flowers suberect; buds 2–4 mm long, 1–2 mm in diameter, elliptic; ovaries 0.8–1.5 cm long, moderately to densely strigillose; pedicels 0.4–0.8 cm long; floral tube 0.6–1.1 mm long, 1.2–1.6 mm in diameter, with sparse ring of hairs at mouth within or subglabrous; sepals 2–3.5 mm long, 0.7–1 mm wide, oblong, acute, strigillose, green or purplish; petals 3.5–6.5 mm long, 2.3–3.2 mm wide, white, often turning pink or rose, obovate, the notch 0.5–1.2 mm deep; filaments of longer stamens 1.8–3.4 mm long, those of shorter ones 1–1.8 mm long; anthers 0.5–0.9 mm long, 0.3–0.5 mm wide, broadly rectangular, cream; style 2.5–3.5 mm long, glabrous; stigma subcapitate to broadly clavate, 0.6–1 mm long, 0.45–0.8 mm wide, the longer stamens shedding pollen onto stigma. Capsules 2.6–5.5 cm long, glabrescent to sparsely strigillose, borne on pedicels 0.9–2.2 cm long. Seeds 1–1.3 mm long, 0.36–0.46 mm in diameter, oblong to narrowly obovoid, the chalazal neck very short and inconspicuous, light brown, the surface papillose; coma 5–9 mm long, white, rather persistent. Chromosome number unknown. Fig. 35.

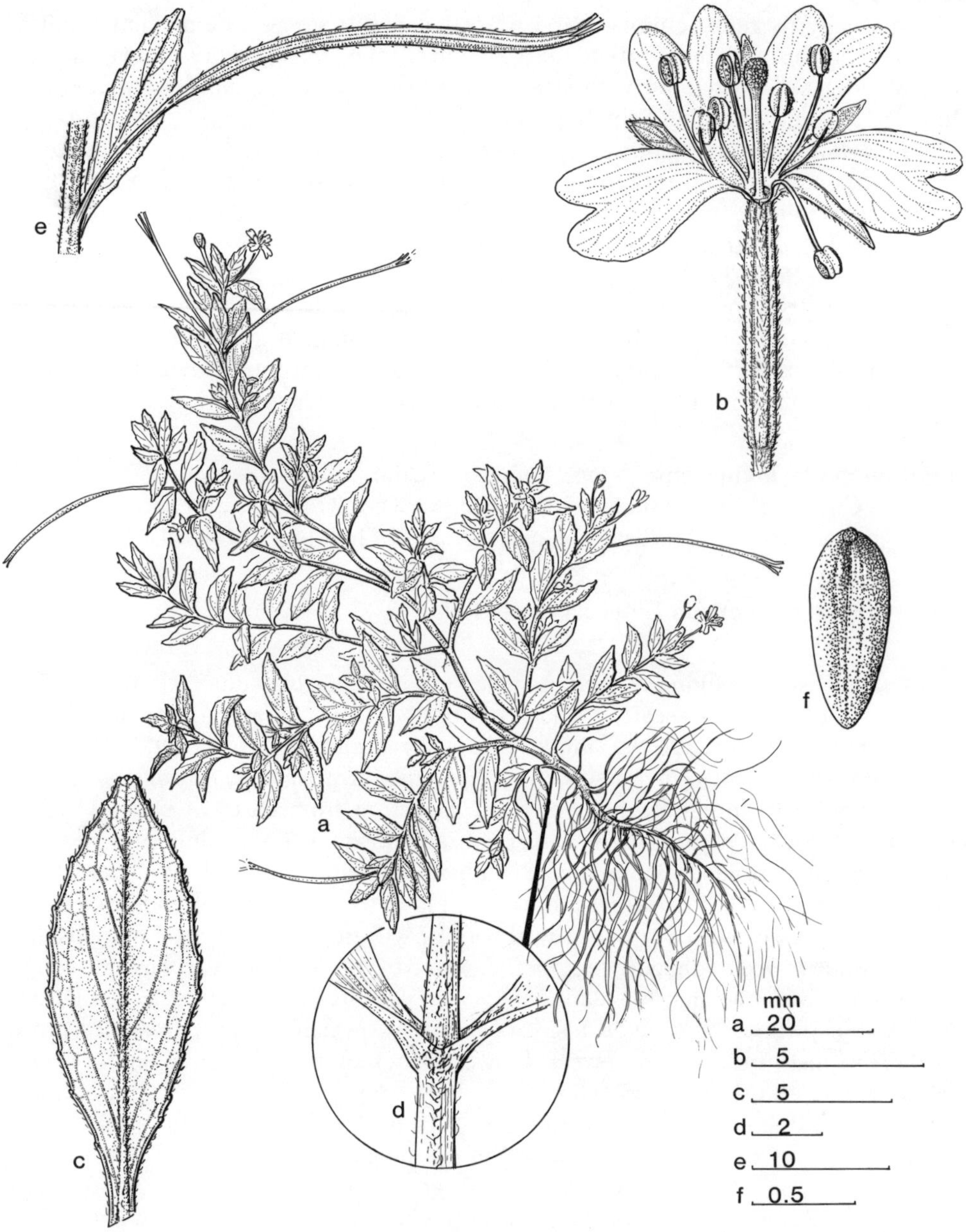

FIG. 35. *Epilobium hohuanense*. a. Habit. b. Flower. c. Mid-cauline leaf. d. Axils from lower stem. e. Capsule and subtending bract. f. Seed (adaxial side). (Based on *Hoch 3303*.)

Phenology. Collected in flower from late July to September, in fruit from August to early November.

Distribution (Fig. 32). Endemic to China, only on Taiwan; on loose scree or gravel in open moist places or rarely in shaded places in the mountains; 2650–3600 m.

ADDITIONAL SPECIMENS EXAMINED. **China.** TAIWAN. Yilan Xian, Nanhutashan *C. C. Hsu 5998* (KYO, S, TAI, TI), *Ohwi 4158* (TNS), *C.-I Peng 5796* (HAST, MO), 1937, *Suzuki-Tokio et al.* (TAI), 1969, *Yamazaki et al. s.n.* (TI); Taipingshan, 1968, *Namba et al. s.n.* (TI); Chung-yang-chien, *Tamura & Koyama 23678* (TI, TNS); Xinzhu Xian, Mt. Pintien, *Shimizu 20377* (SHIN, TI); Taizhong Xian, Hsuehshan, *Hoch 3382* (MO), 1936, *K. Mori s.n.* (TI), *C.-I Peng 7984* (HAST, MO), *8380* (HAST, MO), ("Mt. Tugitaka"), 1930, *S. Onuma s.n.* (TAI), ("Mt. Silvia"), 1936, *Mori s.n.* (TAI), ("Mt. Nenggao"), *Bartlett 6214* (GH, MICH, UC, US); Hualian Xian, Konankuan, *C. C. Chuang & M. T. Kao 4257* (PH, TAI); Hohuanshan, *M. T. Kao 7802* (TAI, TI); Nantou Xian, Hohuanshan, *Hoch 3303* (MO), *3306* (MO), *3310* (MO), *C.-I Peng 8297* (HAST, MO), *8300a–j* (HAST, MO), *8301* (HAST, MO), *8302* (HAST, MO), *8330* (HAST, MO); Kuankao, *C.-I Peng 8184–8188* (HAST, MO); Yushan, *Hoch 3288* (MO), *Price 962* (K), 1924, *Suzuki s.n.* (TAI), 1969, *Yamazaki & Yamazaki s.n.* (TI); Nengkao Shan, *Tamura & Koyama 23295* (TNS), 1939, *Moritani s.n.* (TNS), ("Nokozan"), 1919, *Matuda s.n.* (TI); Chilaichushan, 1929, *Suzuki s.n.* (TAI); Jiayi Xian, Yushan, *Koyama 23875* (SHIN, TI), 1924, *Suzuki s.n.* (TAI); Alishan, 1918, *Suzuki s.n.* (TAI); Tatachia, *C.-I Peng 9009* (HAST, MO); Gaoxiong Xian, Patungkuan, *C.-I Peng 8624* (HAST, MO), *8628–8629* (HAST, MO); Tienchih, *T. C. Huang & K. C. Yang 1845* (A, TAI).

This species, like *E. nankotaizanense* and *E. taiwanianum,* is endemic to the high mountains of Taiwan, a restricted and isolated region, where it is locally common on exposed, gravelly ridges and scree areas. Ying described *E. hohuanense* in 1975, but the journal in which it appeared is rather obscure, and Raven's (1977) treatment of *Epilobium* for the *Flora of Taiwan* was already in press; therefore, *E. hohuanense* was not included in that treatment. *Epilobium hohuanense* is similar to *E. platystigmatosum* in habit, general pattern of vestiture, and other features, but there are no unambiguously shared derived characters between them.

Some specimens of *E. hohuanense* (e.g., Hualian Xian, Hohuanshan, *Kao 7802,* TAI, TI; *Peng 8297,* HAST, MO; *Hoch 3310,* MO, PE) have extremely small leaves (5–9 mm long, 1.5–2.5 mm wide) but otherwise share the distinctive clumped habit and other features of this species. Some small, lax specimens (Yilan Xian, Taipingshan, 1968, *Namba et al. s.n.,* TI; Taizhong Xian, Hsuehshan, 1930, *Onuma s.n.,* TAI) resemble *E. fauriei* H. Léveillé of Japan, but that species perennates by characteristic "gemmae" in the leaf axils and has very long pedicels; it is not known to occur outside of Japan. One collection from a cool, very shaded habitat (Taizhong Xian, Hsuehshan, *Hoch 3382*) has notably large leaves crowded on the stems, but no other differences.

Although a number of sheets contain mixed collections of *E. hohuanense* and other species (Yilan Xian, 1937, *Suzuki et al. s.n.,* TAI; Taizhong Xian, 1936, *Mori s.n.,* TAI), and some collections are rather intermediate, only a single apparent hybrid involving this species has been detected. That collection (Jiayi Xian, Yushan, *Koyama 23875,* SHIN, TI) comprises a number of large (25–35 cm tall) robust, well-branched stems with leaves similar to those of *E. hohuanense* and bases (clustered scales, not clearly caespitose) more like those of *E. taiwanianum* or *E. brevifolium* subsp. *trichoneurum.* The capsules contain a few ripe seeds and many tiny shriveled ones, and the pollen is 36% fertile.

16. Epilobium taiwanianum C. J. Chen, Hoch & Raven, sp. nov.—TYPE: CHINA. Taiwan, Taizhong Xian, Hsuehshan (Snow Mt.), subalpine cirque, *Rhododendron-Juniperus* and alpine grassland community, on moist gravel by rock outcrops near summit, 3600–3700 m, 8 Aug 1986, *Hoch & C.-I Peng 3406* (holotype: MO!; isotypes: HAST! PE!).

Common name: Taiwan Liuyecai.

Herba perennis, turionibus globosis carnosis 1–5 cm hypogaeis et squamis brunneis praedita. Caulis erectus teres, undique strigillosus. Folia ovata vel lanceolata, subsessilia vel inferiora petiolata (1–2 cm longa). Petala 4–6.5 mm longa, roseo-purpurea; stigma capitatum vel late clavatum. Capsulae confertae, 3–5 cm longae; pedicelli minus quam 1 cm longi. Semina 1.1–1.5 mm longa, 0.45–0.6 mm crassa, vix papillosa.

Perennial suberect herb, often clumped, each stem from a separate clustered turion with a rhizome 1–6 cm long, bearing fleshy basal turions that leave brownish coriaceous scales on rhizome, the scales 3–10 mm long, 2.5–3 mm wide, narrowly obovate to spathulate. Stems 7–25 cm tall, 1–2 cm in diameter, simple or sparsely branched, strigillose throughout, especially on upper half, lacking any raised decurrent lines. Leaves opposite below, alternate on inflorescence, with petioles 1–2 mm long near base, the others subsessile; cauline blades 1–2.5 cm long, 0.5–1 cm wide, ovate to lanceolate or lance-elliptic, spatulate in basal pairs, subglabrous except for strigillose margins and veins, apex acute to obtuse below, base cuneate to broadly so, margins denticulate with 3–9 teeth on each side, the lateral veins indistinct, 3–4 on each side of midrib. Inflorescence nodding before anthesis, rather open, with bracts much shorter than the ovaries they subtend. Flowers erect; buds 3–4 mm long, 2–2.5 mm in diameter, ovoid, apiculate, strigillose; ovaries 1–2.4 cm long, strigillose, rarely with a few glandular hairs; pedicels 4–7 mm long; floral tube 0.6–1 mm long, 1.2–1.8 mm in diameter, with scattered hairs at the mouth within; sepals 3–5 mm long, 0.8–1.2 mm wide, oblong-lanceolate, apiculate, keeled, strigillose; petals 4–6.5 mm long, 2–4 mm wide, rose-purple, narrowly obovate, the notch 1–1.5 mm deep; filaments of longer ones 2.5–3 mm long, those of shorter ones 2–2.2 mm long; anthers 0.7–0.9 mm long, 0.6–0.7 mm wide, subrectangular; style 2.5–3.5 mm long, erect, whitish, glabrous; stigma capitate or broadly clavate, 1–1.6 mm long, 0.9–1.3 mm wide, surrounded by anthers at anthesis. Capsules 2.5–5 cm long, 1–2 mm in diameter, strigillose, borne on pedicels 0.5–1 cm long. Seeds 1.1–1.5 mm long, 0.45–0.6 mm in diameter, narrowly obovoid, with very short chalazal collar, light brown, the surface scarcely papillose; coma 5–7 mm long, white, detaching readily. Chromosome number unknown. Fig. 36.

Phenology. Collected in flower from July to September, in fruit from August to October (rarely to December).

Distribution (Fig. 29A). Endemic to China, only in Taiwan; in high mountains on subalpine scree or gravelly slopes among rocks where moisture is sufficient, rarely in shaded places; 3000–3900 m.

ADDITIONAL SPECIMENS EXAMINED. **China.** TAIWAN: Xinzhu Xian, Tapachienshan, *C. S. Kuoh 3262* (TAI); Yilan Xian, Nanhutashan, 1937, *Suzuki-Tokio et al. s.n.* (TAI), *Ohwi 4025* (K), *Tamura & Shimizu 20709* (SHIN), *Tamura et al. 20954* (SHIN); Syukoran, 1935, *Suzuki-Tokio s.n.* (TAI); Miaoli Xian, Tapachienshan, *C.-I Peng 8527* (HAST, MO), *8528* (HAST, MO); Taizhong Xian, Hsuehshan, *M. Tamura 21186* (SHIN), *C. Peng 7979* (HAST, MO), *8379* (HAST, MO), ("Mt. Tugitaka"), 1930, *S. Onuma s.n.* (TAI); Nantou Xian, Hohuanshan, *Hoch 3311* (MO).

Epilobium taiwanianum, an endemic to the high mountains of Taiwan, is distinguished from other species on the island by the combination of fleshy turions leaving scales at the stem base and strigillose pubescence all over the stem, lacking raised lines. This species appears to be closely related to *E. royleanum;* the specimens from Taiwan are quite similar, for example, to the type specimen of *E. himalayense*

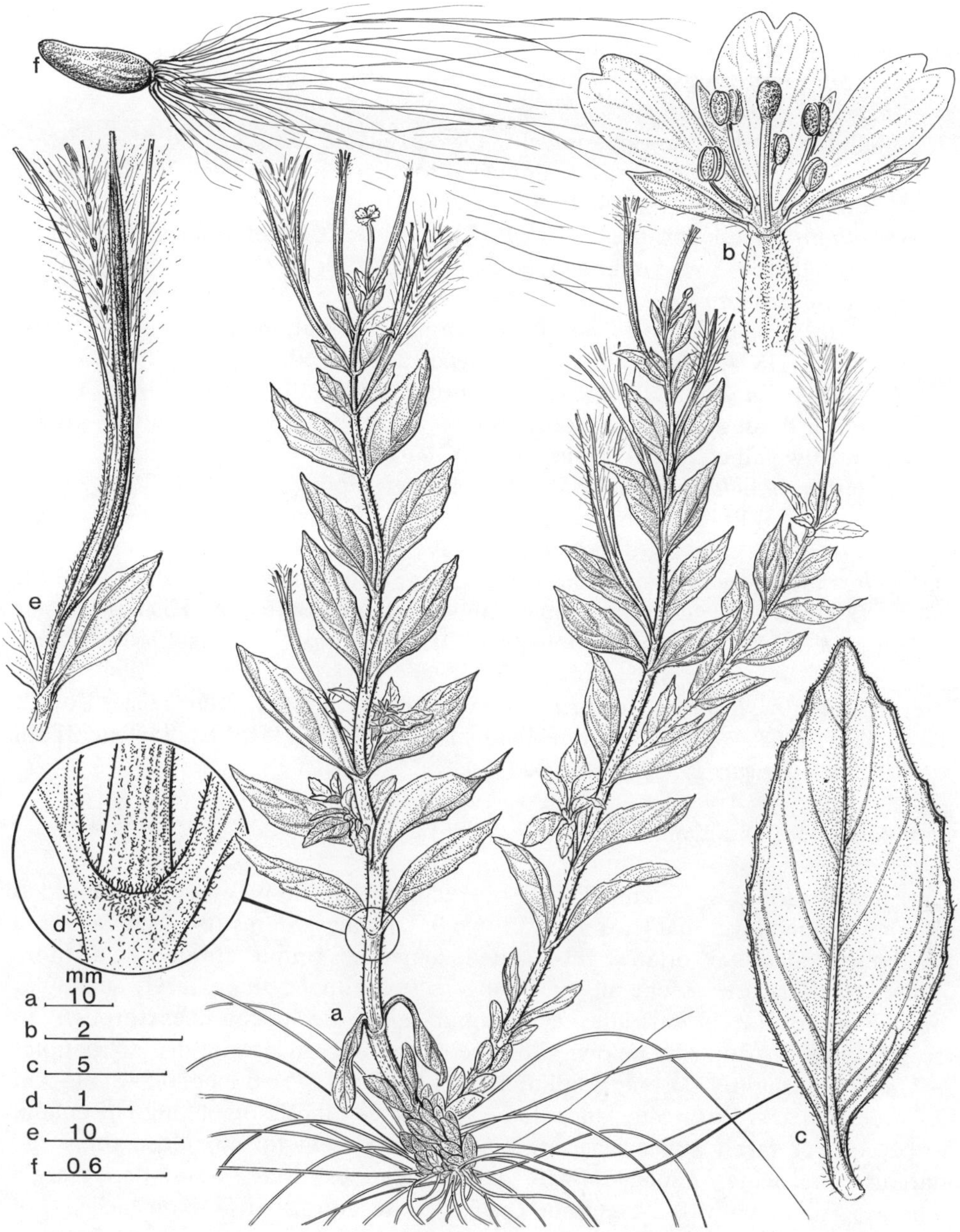

FIG. 36. *Epilobium taiwanianum*. a. Habit. b. Flower (one petal and three stamens removed). c. Mid-cauline leaf. d. Axils from lower stem. e. Capsule and subtending bract. f. Seed with coma attached. (Based on *Hoch & C.-I Peng 3406*, type.)

Hausskn. (India, Himachal Pradesh, Kanawar, *Thomson s.n.*, K; =*E. royleanum*), differing clearly only in seed size, somewhat in leaf shape and attachment, and in the fleshiness and condensation of the basal offshoots. These distinctive populations from Taiwan might be viewed as disjunct outliers of *E. royleanum* (the closest station is Fanjingshan, Guizhou, more than 1000 km distant), or perhaps as a subspecies of that widespread taxon, but the combination of geographical disjunction, ecological specialization, and morphological discontinuity suggests that recognizing them at the species level is more consistent with the treatment of the rest of the genus.

17. Epilobium royleanum Haussknecht, Oesterr. Bot. Zeitschr. 29: 55. 1879.—TYPE: "North-west India" (unplaced locality), s.d., *Royle s.n.* (lectotype, designated by Raven 1962a: K!).

Epilobium roseum var. *indicum* C. B. Clarke in J. D. Hooker, Fl. Brit. India 2: 584. 1879.—TYPE: unknown (see Raven 1962a).

Epilobium roseum var. *dalhousieanum* C. B. Clarke in J. D. Hooker, Fl. Brit. India 2: 584. 1879.—TYPE: INDIA. Himachal Pradesh, Dalhousie, 2100 m, 11 Sep 1874, *Clarke 22190* (lectotype, designated by Raven, 1962a: K!).

Epilobium lividum Haussknecht, Monogr. Epil. 201. 1884.—TYPE: INDIA. Uttar Pradesh, Kumaun; seed from near Deoband, 2400 m, *Duthie,* grown at Kew and collected 30 Jun 1880, *Nicholson s.n.* (holotype: JE!).

Epilobium himalayense Haussknecht, Monogr. Epil. 213. 1884.—TYPE: INDIA. Himachal Pradesh, ascent to Werang Pap, Kanawar, 19 Aug 1847, *Thomson s.n.* (lectotype, designated by Raven, 1962a: K!).

Epilobium royleanum f. *glandulosum* Raven, Bull. Brit. Mus. (Nat. Hist.) Bot. 2: 360. 1962.—TYPE: Jammu and Kashmir, Baltistan, Hispar Glacier snout, 3200 m, growing by spring, damp ground, 21 Aug 1960, *Polunin 6362* (holotype: BM!; isotype: E!).

Epilobium royleanum f. *glabrum* Raven, Bull. Brit. Mus. (Nat. Hist.) Bot. 2: 361. 1962.—TYPE: Kashmir, Pahlgam (c. 34°02′N, 75°20′E), 2450 m, 31 Jul 1945, *Stewart 21523* (holotype: K!).

Common name: Duangeng Liuyecai (Li 1986).

Erect or ascending perennial herb, overwintering by fleshy soboles produced at or below ground level that leave dark brown basal scales. Stems 10–60 cm tall, (1–) 1.5–5 mm in diameter, often well branched, sometimes simple, strigillose and more or less glandular-pubescent all over especially on upper stem, rarely glandular throughout, or rarely subglabrous, without raised decurrent lines through internodes. Leaves opposite below, alternate on inflorescence, with distinct petioles 2–7 mm long, slightly clasping; cauline blades 1.5–5.5 (–7) cm long, 0.5–2.5 (–3.3) cm wide, narrowly ovate to lanceolate, sometimes elliptic or oblong-lanceolate, herbaceous or rarely subcoriaceous, subglabrous except for strigillose veins and margins, apex acute or subacuminate, base cuneate or rarely subrounded, margin densely serrulate with 10–24 teeth on each side, the lateral veins 4–6 on each side of the midrib. Inflorescence erect. Flowers erect; buds 3.5–5 mm long, 2–2.5 mm in diameter, oblong-ovoid, apiculate; ovaries 1.8–2.5 cm long, densely strigillose, often with an admixture of glandular hairs; pedicels 0.3–0.8 cm long; floral tube 0.6–1 mm long, 1.3–2 mm in diameter, usually with a ring of spreading hairs at mouth within; sepals 3.8–6 mm long, 0.8–1.6 mm wide, oblanceolate, apiculate, strigillose, usually with an admixture of glandular hairs; petals 5–7.2 mm long, 2.5–3.8 mm wide, pink to rose-purple, narrowly obcordate, the notch 0.6–1 mm deep; filaments of longer stamens 2.8–4.5 mm long, those of shorter ones 1.8–2.7 mm long; anthers 0.6–0.8 mm long, 0.4–0.5 mm wide, oblong-ovate; style 2–3.2 mm long, erect, usually glabrous; stigma capitate or sometimes broadly clavate, 1.6–2 mm high, 1.2–1.5 mm in diameter, at least longer stamens shedding pollen directly onto stigma at anthesis. Capsules 3.5–7 cm long, strigillose mixed with some scattered glandular hairs, borne on pedicels 0.4–1 cm long. Seeds 0.9–1.2 mm long, 0.3–0.4 mm wide, oblong-obovoid, with short chalazal collar, light brown, the surface papillose; coma 5–6 mm long, dull white, readily detaching. Chromosome number: n = 18. Fig. 37.

Phenology. Collected in flower from July to September, in fruit from August to October.

Distribution (Fig. 22). From central and western China throughout the Himalaya to northeastern Afghanistan (Nuristan); in China in Henan, Shaanxi, Gansu, Qinghai, Xinjiang, Hubei, Sichuan, Guizhou, Yunnan, and Xizang; frequent in southwestern China, rare in central China, western Xizang, and southwestern Xinjiang; wet weedy places in valleys, along roads and streams, sometimes on seasonally flooded grassland or near springs; 1400–3240 (–4300) m.

ADDITIONAL SPECIMENS EXAMINED. **China.** HENAN: Hui Xian to Ling Xian, *Henan Exped. 4204* (PE); Lushi, *J. Q. Fu 2108* (KUN, PE), *K. M. Liou 4703* (PE); Xixia, *Henan Exped. 1519* (PE).—SHAANXI: Zhongnan Shan, *F. T. Wang et al. 538* (PE); Shanyang, *T. P. Wang 16361* (KUN, PE); Taibaishan, *K. S. Hao 4325* (PE), *T. N. Liou & P. C. Tsoong 1554* (KUN, NAS), *Y. Y. Pai 1552* (PE), *Taibaishan Exped. 375* (LZU), *F. T. Wang 600* (PE), *F. T. Wang et al. 107* (PE), *337* (PE), *618* (PE); Huxien, *P. C. Kuo 452* (KUN, PE); Fuping, *P. C. Kuo 1623* (PE); Shanyang, *T. P. Wang 16361* (PE); “Makiapou,” *Licent 328* (PE).—GANSU: Lanzhou, 1965, *Lanchou Exped. s.n.* (LZU), *J. G. Wang 84* (LZU); Maijishan, *M. Q. Zhang et al. 420* (LZU); Minxien, *G. D. Zhang et al. 707-233* (LZU); Kang Xian, *Z. Y. Zhang 16246* (PE), *17042* (PE); Wen Xian, *W. L. Chen & C. J. Chen 8801* (MO, PE), 1959, *Lanzhou Univ. Biol. Dept. Wen Xian Exped. s.n.* (LZU).—QINGHAI: Zeku, *L. H. Chou & L. N. Sun 1941* (HNWP); Xunhuamengda, *P. C. Kuo 9728* (HNWP); Nangqen, *Y. C. Yang 1136* (NWIP, PE); no locality, *Qinghai-Gansu Exped. 1849* (PE).—XINJIANG: Yecheng, *T. N. Liou L.5538* (PE).—HUBEI: Shennongjia, *1980 Sino-Amer. Exped. 318* (GH, KUN, PE).—SICHUAN: Chengkou, *T. L. Dai 102585* (NAS), *104681* (NAS, PE), *105906* (NAS), *106863* (NAS), *107074* (CDBI, PE); Qingxi, Fuling, *Z. P. Huang et al. 2336* (KUN); Barkam, *Z. Y. Chang & H. F. Zhou 22770* (PE), *W. L. Chen & C. J. Chen 7569* (MO, PE), *X. Li 72373* (PE); Xiaojin, *K. Y. Lang et al. 1488* (KUN, PE), *Sichuan Coll. Agr. 6273* (CDBI); Zamatang, *W. L. Chen & C. J. Chen 7507* (MO, PE), *7509* (MO, PE); Erlang Shan, *Sichuan Econ. Exped. 831* (PE); Leibo, *T. S. Ying 4507* (PE); Mt. Emei, *X. B. Peng 6117* (PE); Kangding, *S. Jiang & C. L. Jing 2759* (PE), *K. C. Kuan et al. 649* (PE), *H. S. Kung 6049* (MO, PE), *6268* (MO, PE), *6269* (PE), *K. Y. Lang et al. 805* (PE), *Z. A. Liu 2304* (CDBI); Garze, *S. Jiang 9629* (PE); Yuexi, *Sichuan Med. Exped. 25370* (NAS); Yajian, *H. S. Kung et al. 40* (MO, PE); Qianning, *T. S. Ying et al. 9857* (PE), 1974, *s.c. 5680* (CDBI); Muli, *W. L. Chen & C. J. Chen 5692* (PE), *Ministry of Forest. Exped. 22* (PE), *Qinghai-Xizang Exped. 14169* (KUN, PE), *T. T. Yu 14353* (BM, E, GH, PE), *Q. S. Zhao et al. 8272* (CDBI); Batang, *Soulie 3172* (P), *3636* (P), *Qinghai-Xizang Exped. 5186* (PE), *5204* (KUN, PE), *Xizang Exped. 2912* (NHWP); Daocheng, *Qinghai-Xizang Exped. 4310* (KUN, PE), *4322* (PE), *5904* (PE), *5948* (PE), *Sichuan Exped. 2324* (PE), *2424* (CDBI, KUN, PE); Xiangcheng, *Qingzang Exped. 3124* (PE), *3682* (PE), *3905* (PE), *4521* (KUN, PE), *4551* (KUN, PE), *4821* (PE), *Sichuan Veget. Exped. 3538* (CDBI, KUN, PE); Derong, *Qingzang Exped. 3369* (PE).—GUIZHOU: Fanjing Shan, *Steward et al. 396* (F, S); Hezhang, *P. H. Yu 1371* (KUN, PE).—YUNNAN: Qiaojia, *B. X. Sing 888* (YNU); Huize, *Dongchuan Exped. 63-074* (KUN); Kunming, *T. N. Liou 14379* (PE); Fumin, *B. Y. Qiu 596286* (KUN); Dali, *T. N. Liou 17641* (PE); Lijiang, *C. Z. Bao 63-20285* (KUN), *Forrest 2894* (BM, E); Tuomulang, *C. Y. Wu 4129* (KUN); Zhongdian, *T. T. Yu 12609* (GH, KUN, PE), *Qingzang Exped. 1582* (KUN, PE), *3078* (PE), *3089* (KUN, PE); Weixi, *Hengduanshan Exped. 01817* (PE), *H. T. Tsai 59641* (PE), *59909* (KUN), *C. W. Wang 68072* (GH, KUN, PE); Deqen, *Qingzang Exped. 2248* (PE), *2657* (PE), *C. W. Wang 69744* (GH, KUN, NAS, PE), *69931* (GH, NAS, PE), *T. T. Yu 9955* (KUN, PE); Northwestern Yunnan, *T. T. Yu 9523* (KUN); no locality, *Maire 3289* (NY).—XIZANG: Zayu, *Xizang Exped. 1228* (HNWP); Mainkung, *Kingdon-Ward 10034* (BM); “Nyug Numagang,” 1951, *Aufschnaiter s.n.* (BM); Bomi, *T. S. Ying & D. Y. Hong 650412* (PE); Lulang, Tangmai to

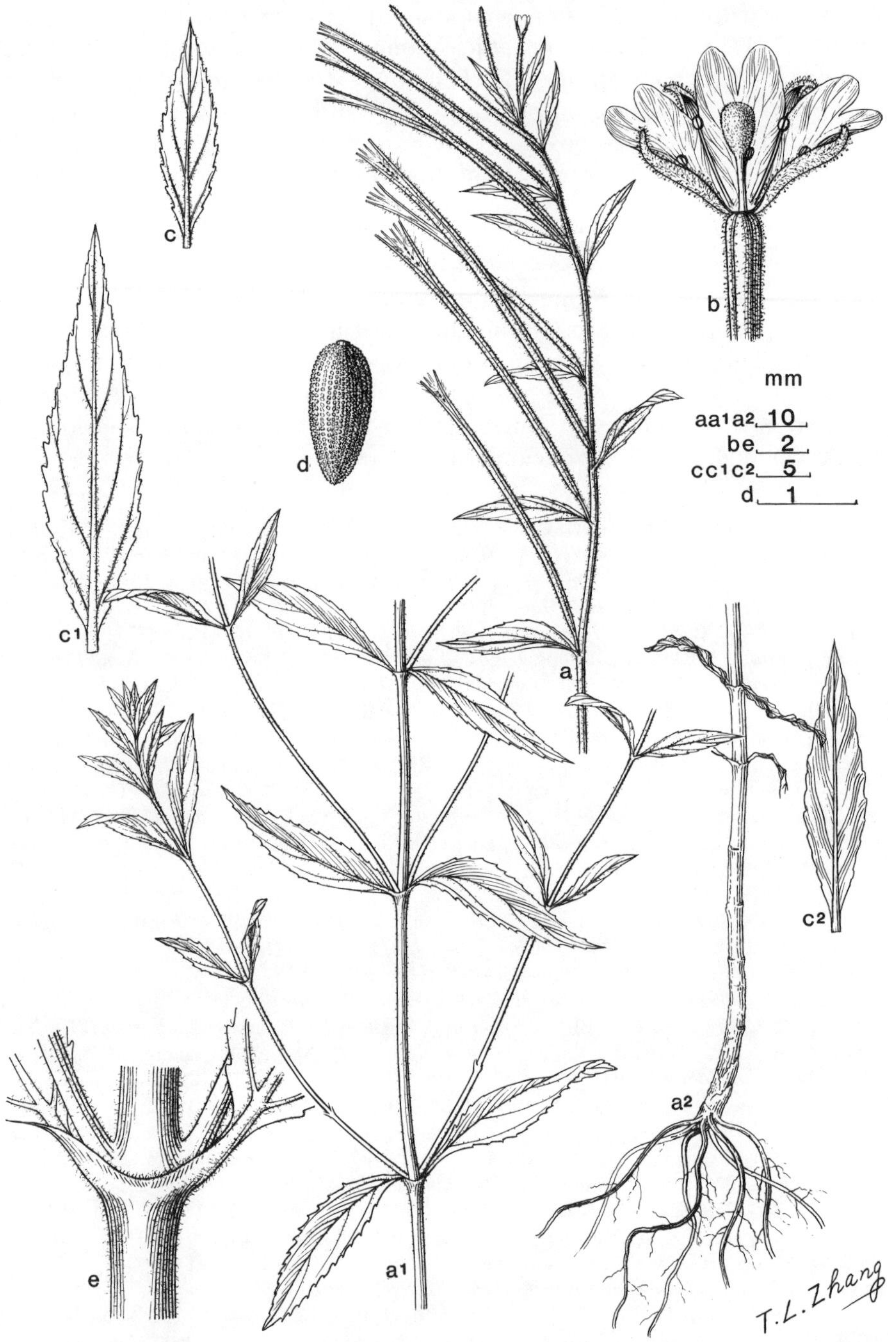

FIG. 37. *Epilobium royleanum*. a. Habit: inflorescence. a[1]. Portion of mid-stem. a[2]. Base of stem. b. Flower (one petal and three anthers removed). c. Leaf from inflorescence. c[1]. Mid-cauline leaf. c[2]. Basal leaf. d. Seed. e. Axils from mid-stem. (Based on *C. W. Wang 69744*.)

Nyingchi, *Y. T. Zhang & K. Y. Lang 951* (PE); Mainling, *Y. T. Chang & K. Y. Lang 951* (PE), *Qinghai-Xizang Exped. 75087* (PE), *Xizang Med. Exped. 3609* (PE); Baxoi, *Qinghai-Xizang Exped. 12196* (PE); Cona, *Qinghai-Xizang Exped. 2907* (PE); Lhunzhub, *Xizang Med. Pl. Exped. 1977* (PE); Lyingchi, *Z. Y. Qin 10258* (HNWP), *Y. T. Zhang & J. Y. Lang 1067* (PE); Lhasa, 1946–1950, *Aufschnaiter s.n.* (BM); Lhasabeishan, *B. Q. Zhong 5589* (PE); "Drukla Gompa," *Ludlow et al. 15516* (BM); Nielamu, *N. Jiang & C. F. Zhao 218* (PE); Gyzrong, *Qingzang Exped. 7053* (PE); Alizhada, *Qingzang Exped. 76-8090* (PE); Rutog, *Qinghai-Xizang Exped. 76-9120* (PE); no locality, *T. P. Wang 16361* (PE).

Epilobium royleanum, a distinctive species of the Himalaya and southwestern China, is variable in stature, leaf shape and size, pubescence, and stigma shape. The combination of stems strigillose-pubescent all over and lacking raised decurrent pubescent lines, conspicuously petiolate leaves with cuneate bases, subcapitate stigma, and relatively short pedicels clearly marks this species. Although *E. royleanum* is similar in some respects to *E. minutiflorum,* it differs from that species in its leaf shape, longer petioles, larger flowers, broader stigma, and seeds with a very short, inconspicuous chalazal collar. In addition, these two species mostly occur in different regions; *E. minutiflorum* extends farther south and west than *E. royleanum,* although the two co-occur in the western Himalaya. Some collections of *E. royleanum* are similar to *E. brevifolium* subsp. *trichoneurum,* but that species has less-branched stems, broader and shorter subsessile leaves with a more rounded base, and larger flowers.

A number of collections from western Sichuan and northwestern Yunnan have predominantly glandular pubescence on the stems. Raven (1962a) called attention to plants from Jammu and Kashmir that are similar to these Chinese collections by describing them as *Epilobium royleanum* f. *glandulosum.* We can find no other morphological differences that correlate with this variation in vestiture, however, and no longer consider that these plants should be segregated formally from typical *E. royleanum.* Such anomalous plants occur both in the western Himalaya and at the eastern end of the range in China, which suggests that the plants may represent local extremes and not a persistent pattern recognizable at either the specific or infraspecific level.

Raven (1962a) also described *Epilobium royleanum* f. *glabrum* in order to call attention to a series of collections, mostly from Jammu and Kashmir, that differ from the typical form by lacking the characteristic dense strigillose pubescence covering the stem. A single collection from eastern Xixang (Baxoi, *Qinghai-Xizang Exped. 12196*) shares this distinctive vestiture pattern, but otherwise does not differ from typical specimens of *E. royleanum.*

In previous treatments, the distribution of *E. royleanum* has been limited to the Himalayan region from Afghanistan to Yunnan (summary in Raven, 1962a). Our study of additional specimens extends this range substantially to the northeast in China through Sichuan to Qinghai, Gansu, Shaanxi, Henan, and Hubei. At the western end of the distribution, *E. royleanum* occurs in the western Kun Lun Shan region of southwestern Xinjiang; it does not appear to extend farther north.

18. Epilobium kermodei Raven, Bull. Brit. Mus. (Nat. Hist.) Bot. 2: 364, pl. 33B. 1962.—TYPE: BURMA. Myitkyina District, near Hpimaw (approx. 25°45′N, 97°23′E), 2100 m, by roadside, 25 Apr 1938, *Kermode 17233* (holotype: K!).

Common name: Ruichi Liuyecai.

Robust erect perennial herb, producing fleshy basal stolons 1–12 cm long, extending horizontally underground and terminating in thick fleshy buds. Stems 40–120 (–200) cm tall, 3–10 (–15) mm in diameter, simple or sometimes sparingly branched, covered entirely with glandular and some strigillose pubescence, especially on inflorescence, with barely raised, indistinct hairy lines decurrent from margins of petioles on lower stem. Leaves opposite, alternate on the inflorescence, subsessile on upper stem, on petioles 1–6 mm long below; cauline blades 3.5–8 (–11) cm long, 1.5–4 (–4.5) cm wide, narrowly ovate to lanceolate, densely strigillose on veins, especially on lower surface, otherwise subglabrous, apex acute, base broadly cuneate to subrounded, margin sharp-serrulate with 28–42 (–60) teeth on each side, the lateral veins 5–6 on each side of midrib. Inflorescence erect, subcorymbose initially, later elongated, often densely glandular-pubescent, leafy, the bracts more than half the length of the ovaries they subtend. Flowers erect; buds 3–5 mm long, 2–3 mm in diameter, narrowly ovoid; ovaries 2–5 cm long, strigillose with an admixture of glandular hairs; pedicels 0.3–1.2 cm long; floral tube 1.2–2 mm long, 1.5–2.5 mm in diameter, a ring of white spreading hairs at the mouth within; sepals 5–8 mm long, 1.5–2 mm wide, lanceolate, keeled, strigillose and glandular-pubescent; petals 7–15 (–18) mm long, 4–11 (–15) mm wide, rose or rose-purple, broadly obcordate, the notch 1–2 mm deep; filaments of longer stamens 4–10 mm long, those of shorter ones 2.5–6 mm long; anthers 0.8–1.2 mm long, 0.4–0.5 mm wide, oblong; style 4–10 mm long, erect, usually with a few spreading hairs near base; stigma capitate to broadly clavate, 1.7–2 mm high, 1.4–1.8 mm in diameter, at least longer stamens shedding pollen directly onto stigma at anthesis. Capsules 7–11 cm long, strigillose and glandular, borne on pedicels 0.7–1.5 cm long. Seeds 0.8–1.2 mm long, 0.35–0.45 mm in diameter, obovoid, with very short chalazal collar, dark brown, the surface coarsely papillose; coma 5–6 mm long, white, detaching easily. Chromosome number: n = 18. Figs. 2e, 6a, 8d–f, 38.

Phenology. Collected in flower from May (rarely from February) to July, in fruit from July (rarely from May) to September.

Distribution (Fig. 10). Central to southwestern China and northern Burma; in China in northern Guangxi, western Hunan, western Hubei, Sichuan, Guizhou, and Yunnan; often in open weedy meadows at edges of forests or in disturbed places along roads, stream banks, and boggy areas; at 400–1400 m in central China to 1800–2800 (–3800) m in Sichuan and Yunnan.

Additional Specimens Examined. **China.** Guangxi: Huaping, *L. F. Liu & S. F. Yang 5514* (NAS).—Hunan: Wugang, *T. H. Wang 56* (C, E, W); "Patakun Shan," *H. J. Li 3652* (PE).—Hubei: Yichang, *Henry 1747* (GH, K, P), *4012* (K); Changyang, *Wilson 1139* (HK, K, W); Western Hubei, *Wilson 1368* (NY, P, W); Shennongjia, *Shennongjia Pl. Exped. 10582* (PE); Batung, *Henry 6141* (BM, GH, K); Laifeng, *F. W. Yang 335* (HIIM).—Sichuan: Wushan, *Wilson 1368* (E, K); Shizhu, *Y. Chen 990* (CDBI), *2025* (CDBI), *W. H. Wang 214* (CDBI), *253* (CDBI), *1283* (CDBI); Pengshui, *F. D. Fu & Y. L. Cao 125* (CDBI, PE), *149* (CDBI, PE); Wulong, *F. D. Fu & Y. L. Cao 282* (CDBI, PE); Nanchuan, *C. J. Chen & Z. Y. Liu 83091* (PE), *C. H. Chu 1336* (PE), *K. L. Chu 1115* (PE), *W. P. Fang 1075* (E, GH, PE), *C. H. Hsiang et al. 93309* (PE), *G. F. Li 63533* (PE), *W. China Acad. Sci. 3212* (PE), *3215* (PE), *X. Y. Huo 5031* (NAS, PE), *Jingfo Shan Exped. 1013* (PE), *K. C. Kuan et al. 1150* (PE), *2220* (PE), *Kunming Inst. Bot. 64601* (KUN), *G. F. Li 61735* (KUN, PE), *62249* (KUN), *62580* (KUN, SZ), *Pl. Geogr. Exped. 557* (CDBI, PE), *J. H. Xiong & Z. L. Zhou 91142* (KUN, PE); Qijiang, *W. P. Fang 1309* (E, GH, NY, PE); Guan Xian, *F. T. Wang 20815* (KUN, PE); Songpan, *W. L. Chen & C. J. Chen 8415* (PE); Nanping, *W. L. Chen & C. J. Chen 8846* (MO, PE), *8853* (MO, PE); Pingshan, *T. T. Yu 3042* (PE), *T. T. Yu 3042a* (PE); Mabian, *T. H. Tu 5608* (PE); Leibo, *Z. T. Guan 8277* (PE), *8991*

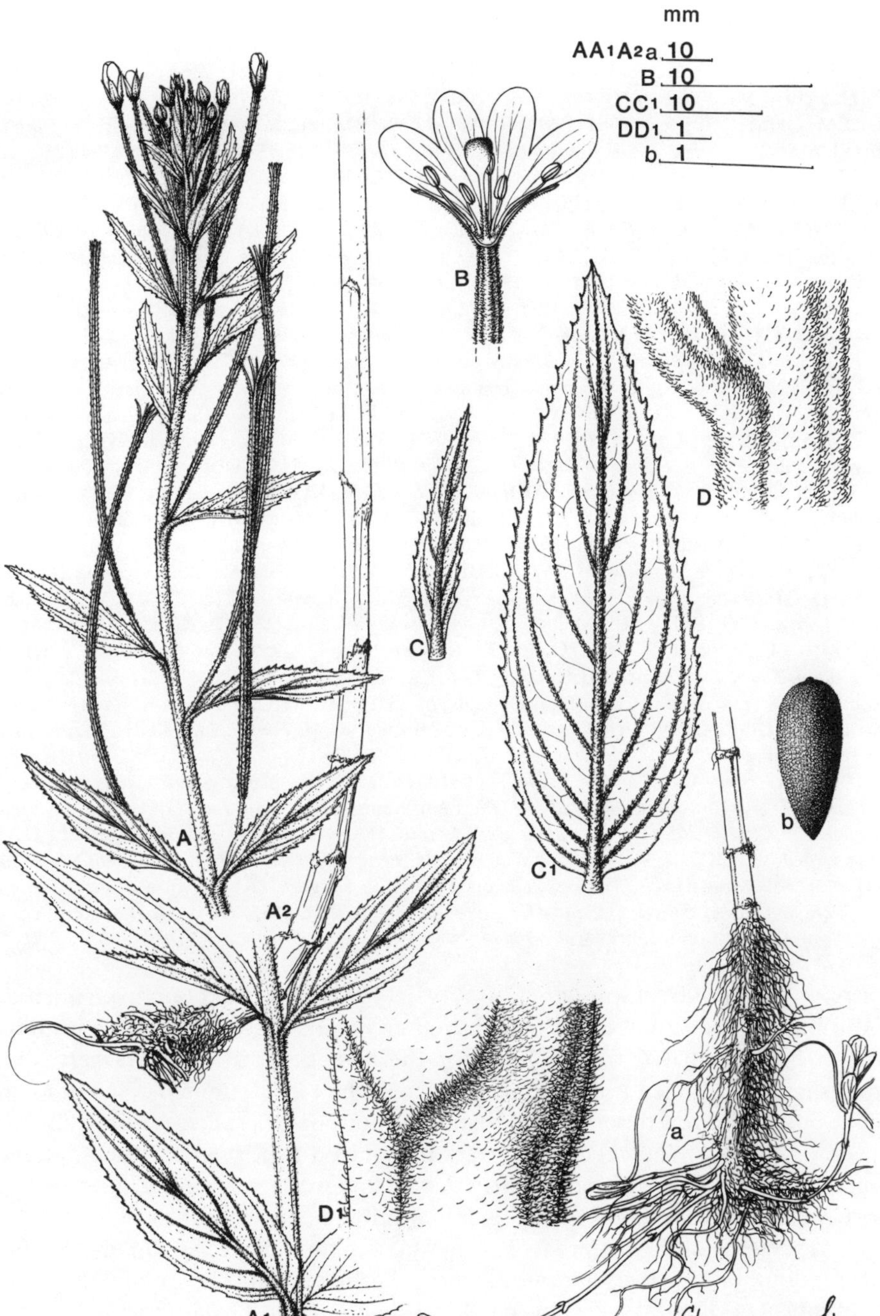

FIG. 38. *Epilobium kermodei*. A. Habit: inflorescence. A[1]. Portion of mid-stem. A[2]. Base of stem. B. Flower (two petals and three stamens removed): note hairs at mouth of floral tube and on lower part of style. C. Leaf from inflorescence (bract). C[1]. Mid-cauline leaf. D. Axil from inflorescence. D[1]. Axils from mid-stem. a. Base of stem showing root system and stolons terminating in leafy buds. b. Seed. (A–D based on *T. P. Soong 38763*; a–b based on *Raven et al. 26117*.)

(PE), *9665* (PE); Liang Shan, *Sichuan Econ. Pl. Exped. 460* (PE), *1290* (PE); Yuexi, *Sichuan Veg. Exped. 3629* (PE); Meigu, *Sichuan Veg. Exped. 13074* (CDBI, PE); Butuo, *Sichuan Veg. Exped. 13739* (CDBI, PE); Simian, *C. C. Hsieh 41462* (PE); Yingjing, *C. W. Yao 2240* (PE); Ebian, *C. W. Yao 4352* (PE), *4353* (PE); Mt. Emei, *H. C. Chow 7672* (GH), *9930* (GH), *12190* (GH), *W. P. Fang 2884* (GH), *12811* (BM, GH, KUN, PE), *12985* (BM, UC, W), *12998* (BM, PE, W), *17580* (PE), *W. P. Fang et al. 20452* (PE), *K. S. Hsiang & K. Y. Lang 41* (PE), *L. C. Hu 50148* (SZ), *Raven et al. 26104* (PE), *26110* (MO, PE), *26112* (MO, PE), *26115* (MO, PE), *26116* (MO, PE), *26117* (MO, PE), *26121* (PE), *26122* (MO, PE), *26123* (MO, PE), *26127* (PE), *26129* (PE), *26131* (MO, PE), *26134* (MO, PE), *Sichuan Med. Exped. 12413* (NAS), *Sino-U.S.S.R. Bot. Exped. 1947* (PE), *2109* (PE), *S. C. Sun & K. Chang 739* (GH), *789* (GH), *S. L. Sun 2210* (KUN), *2415* (KUN), *T. H. Tu 148* (PE), *F. T. Wang 23354* (GH, PE), *S. X. Wang 450* (PE), *Wilson 4914* (BM, HK, K), *C. Y. Wu et al. 53* (KUN, PE), *73* (KUN, PE), *103* (KUN, PE), *J. H. Xiong et al. 30977* (PE), *G. H. Yang 54745* (KUN, NAS, PE), *55114* (KUN, NAS, PE); Hongya, *T. H. Tu 282* (PE), *C. W. Yao 2362* (PE); *3691* (PE); *3943* (PE); Baoxin, *S. S. Chang 5713* (PE); *K. L. Chu 2929* (BM, PE, W), *3182* (BM, PE), *T. P. Soong 38763* (PE), *38867* (KUN, PE), *Smith 2108* (UPS), *10117* (S, UPS); Erlang Shan, Tianquan, *S. S. Chang et al. 5114* (CDBI), *W. G. Fu 34751* (PE), *B. Z. Go et al. 20177* (KUN), *W. G. Hu & Z. He 10181* (PE), *10200* (PE), *S. Jiang & C. L. Jing 1869* (PE), *Sichuan Agr. Coll. 667* (CDBI, KUN), *F. C. Tai & C. M. Teng 5019* (KUN), *H. C. Tsiang 37986* (PE); Luding, *Z. P. Huang et al. 809* (KUN), *Q. P. Wang 22543* (CDBI); Kangding, *Z. P. Huang et al. 1341* (PE), 1981, *H. S. Kung s.n.* (CDBI), *K. Y. Lang 918* (KUN, PE), *Pratt 433* (K), *723* (BM, K), *Smith 11535* (UPS), *C. S. Liu 1277* (PE); Jiulong, *Q. Q. Wang 21004* (CDBI); Zamtang, *Chengdu Inst. Biol. 9302* (CDBI); Western Sichuan, *Henry 59* (K), *8945* (P), *T. Tang 23354* (KUN, PE).—GUIZHOU: Fangjing Shan, *Fangjing Shan Exped. 362* (GZBI), *907* (GZBI), *C. P. Tsien et al. 30924* (GZ); Leishan Shan, *C. P. Tsien et al. 50357* (GZBI, KUN, PE); Tongzi, *Y. Tsiang 5127* (NAS, NY, W); Shiqian, *S. Guizhou Exped. 2641* (KUN, PE); Maitan, *N. Guizhou Exped. 2548* (GZBI); Kaili, *S. Guizhou Exped. 1640* (GZBI, PE), *2108* (PE), *2152* (GZBI, KUN, PE); Nayong, 1959, *Bijie Exped. s.n.* (PE); Pang Xian, *Beijing Youth Bot. Exped. to Guizhou 146* (PE).—YUNNAN: Suijiang, *B. X. Sun 347* (YNU); Yongshan, *H. T. Tsia 51027* (KUN, PE); Daguan, *B. X. Sun 815* (KUN, YNU); Kunming, *Murata & Yamazaki 8902* (BM); Qiaojia, *Ducloux 5639* (P); Wenshan, *K. M. Feng 22332* (KUN); Mengzi, *Henry 10638* (E); Jinping, *Sino-U.S.S.R. Bot. Exped. 1571* (PE); Chuxiong, *H. T. Tsia 52278* (KUN, PE), *52670* (PE); Dali, *Sino-American Bot. Exped. 860a* (KUN, MO, PE); Yangbi, *R. C. Ching 22610* (KUN, PE), *99610* (KUN); Tengchong, *Howell 222* (E); Longchuan, 1921, *Maire s.n.* (G); Lushui, *Hengduang Shan Exped. 59* (PE), *365* (PE), *Nujiang Exped. 1804* (KUN); Bijiang, *Bijiang Exped. 18* (KUN), *Nujiang Exped. 991* (KUN), *992* (KUN); Weixi, *K. M. Feng 4410* (KUN, PE), *Hengduan Shan Exped. 1263* (PE), *C. W. Wang 63699* (GH, PE); Deqen, 1979, *J. S. Yang s.n.* (KUN); Gongshan, *Lujiang Bot. Exped. 79-0006* (KUN), *Qinghai-Xizang Exped. 8372* (KUN, PE), *8555* (KUN, PE); western Yunnan, 1896, *Orleans s.n.* (P); no locality, *Maire 10044* (NY).

Epilobium kermodei was based by Raven (1962) on a collection from northeastern Burma; no additional collections have been assigned to this species until now, and it is here recorded from China for the first time. Handel-Mazzetti (1933) treated these plants as a probable new species when he examined collections from Hunan, but he did not describe it. Most collections of this species from China have been determined as *E. tanguticum,* now recognized as a synonym of *E. wallichianum.* Modern collections, particularly those from western Yunnan, have provided a more accurate understanding of the delimitation and distribution of *E. kermodei.*

This species is quite distinct from other taxa in the same region in its tall, simple habit, predominantly glandular pubescence, sharp-serrulate leaves, and larger flowers and capsules. This taxon seems most closely related to *E. sikkimense,* but differs from it by lacking fleshy basal scales, and having stems with glandular and strigillose pubescence all over and with indistinct hairy lines, leaves with an acute apex, and coarsely papillose seeds. In addition, *E. sikkimense* is distributed farther west and at higher elevations (subalpine zone from 2700–4500 m), whereas *E. kermodei* occurs farther east and south at lower elevations (400–2800 m).

Natural hybrids involving *E. kermodei* and other species apparently occur, and often can be distinguished by intermediate morphology and reduced seed set and pollen stainability. Apparent hybrids between *E. kermodei* and *E. amurense* subsp. *cephalostigma* [e.g., Guangxi, Longsheng, *L. F. Liu & S. F. Yuang 5514* (NAS); Sichuan, Shimian, *C. C. Hsieh 41462* (PE); Sichuan, Kangding, *Pratt 433* (BM, K); Sichuan, *Henry 8945* (P), *8948* (K)] have more branches, fewer glandular hairs, elliptic-lanceolate leaves with cuneate base, strongly reduced seed set, and pollen stainability of 0–9%. A set of specimens from Sichuan [Kangding, *Pratt 723* (BM, K, P)] with unusually large petals (to 1.8 cm long), stems with distinct hairy lines, and pollen stainability of 0%, might be of hybrid origin between *E. kermodei* and *E. sikkimense*.

19. Epilobium pannosum Haussknecht, Oesterr. Bot. Zeitschr. 29: 54. 1879 (Feb.). *Epilobium khasianum* C. B. Clarke in J. D. Hooker, Fl. Brit. India 2: 585. (May) 1879, nom. superfl. *Epilobium brevifolium* subsp. *pannosum* (Haussknecht) Raven, Bull. Brit. Mus. (Nat. Hist.) Bot. 2: 363. 1962.—Type: India. Assam, Khasi Hills, below Pomrang, 1500 m, 10 Sep 1850, *J. D. Hooker & Thomson 2322* (lectotype, designated by Raven 1962a: K!).

Common name: Yinmao Liuyecai (Wu 1984).

Robust erect perennial herb, sprouting from caudex or forming leafy basal rosettes. Stems 20–120 cm tall, 2–8 mm in diameter, simple below, sparsely branched above, covered all over with dense appressed tomentose pubescence, sometimes with an admixture of glandular hairs especially on inflorescence. Leaves opposite, or sometimes alternate on inflorescence, sessile and often clasping; cauline blades 1–4.8 cm long, 0.5–1.7 cm wide, elliptic to lanceolate or ovate, subcoriaceous, covered on both surfaces with dense appressed tomentose pubescence, apex acuminate to acute or subobtuse, base subrounded, the margins remotely denticulate with 3–15 teeth on each side, sometimes revolute, the lateral veins inconspicuous, 5–8 on each side of midrib. Inflorescence nodding prior to anthesis, later erect. Flowers nodding to suberect, up to 17 mm long; buds 6–8 mm long, 3–3.5 mm in diameter, oblong-ovoid, acute or apiculate; ovaries 2–3.5 cm long, very densely tomentose mixed with glandular hairs; pedicels 1–2 cm long; floral tube 0.8–1.2 mm long, 1.8–2.8 mm in diameter, a conspicuous ring of long white hairs at the mouth within; sepals 5.5–8 mm long, 1.5–2.5 mm wide, oblong-lanceolate, covered with dense tomentose and glandular pubescence; petals 8–16 mm long, 6.4–10 mm wide, pink to rose-purple, broadly obcordate, the notch 1–2 mm deep; filaments pinkish, those of longer stamens 4–5.4 mm long, those of shorter ones 3–3.4 mm long; anthers 2–2.7 mm long, 0.5–0.9 mm in diameter, oblong, apiculate; style 5–8.5 mm long, erect, white, subglabrous or with a few glassy hairs; stigma cylindric to broadly clavate, 1.5–3.5 mm high, 1–1.8 mm in diameter, elevated above anthers at anthesis. Capsules 3.5–6.5 cm long, tomentose and with scattered glandular hairs, borne on pedicels 1.2–2.8 cm long. Seeds 0.9–1 mm long, 0.5–0.55 mm wide, broadly obovoid, with very short chalazal collar, dark brown, the surface finely papillose; coma 7–8 mm long, dull white, detaching easily. Chromosome number: n = 18. Figs. 4i, j, 5j–l, 6d, i, 39.

Phenology. Collected in flower from August (rarely July) to October, in fruit from September to November.

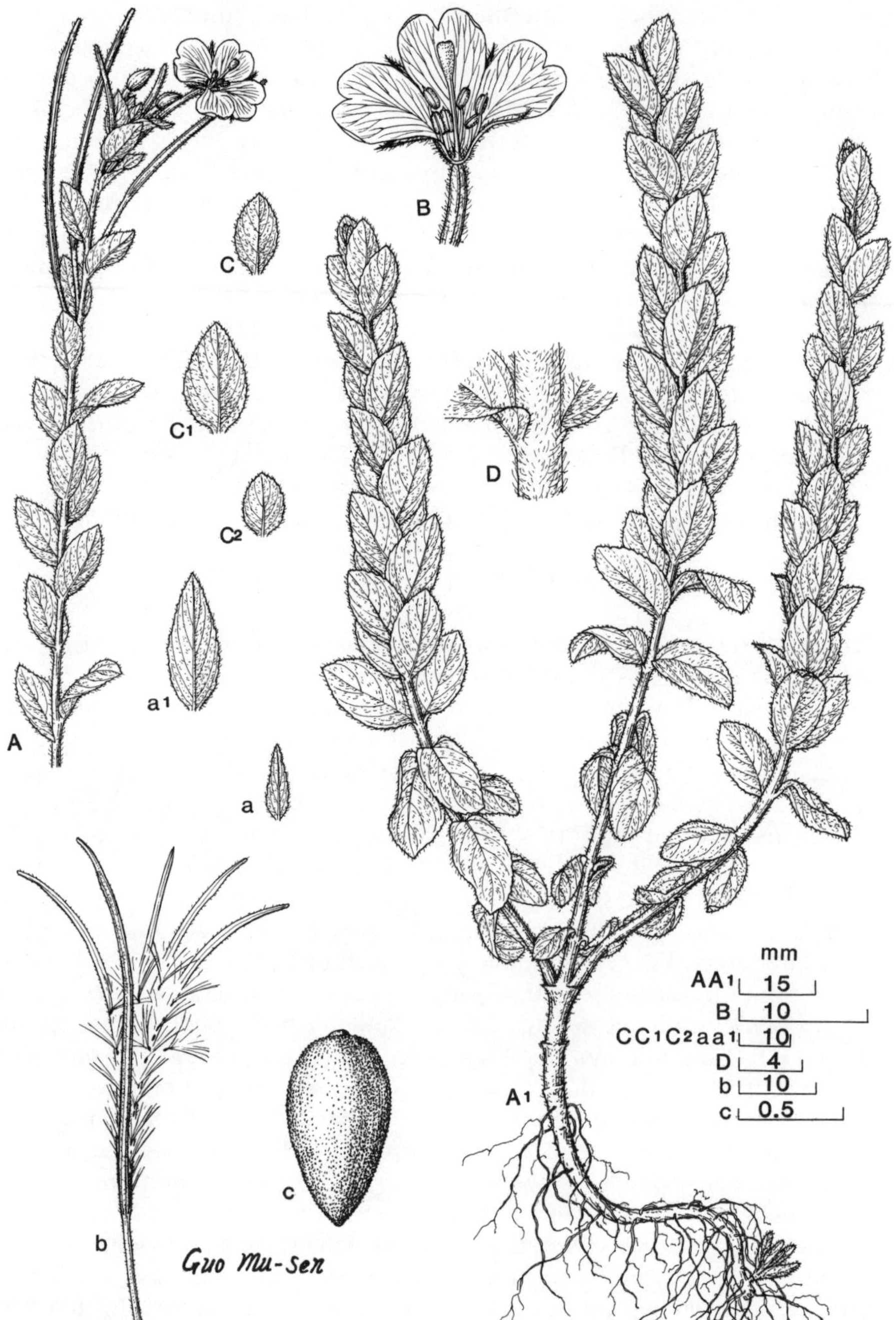

FIG. 39. *Epilobium pannosum*. A. Upper stem and inflorescence. A[1]. Habit (inflorescence excluded); note leafy rosette at base of stem. B. Flower (one petal and two stamens removed); note hairs at mouth of floral tube and near base of style. C. Leaf from inflorescence. C[1]. Mid-cauline leaf. C[2]. Lower leaf. D.·Axils from mid-stem. a. Leaf from inflorescence. a[1]. Mid-cauline leaf. b. Capsule. c. Seed. (A–D based on *K. K. Tsoong 2296*; a–c based on *Y. Tsiang 12405*.)

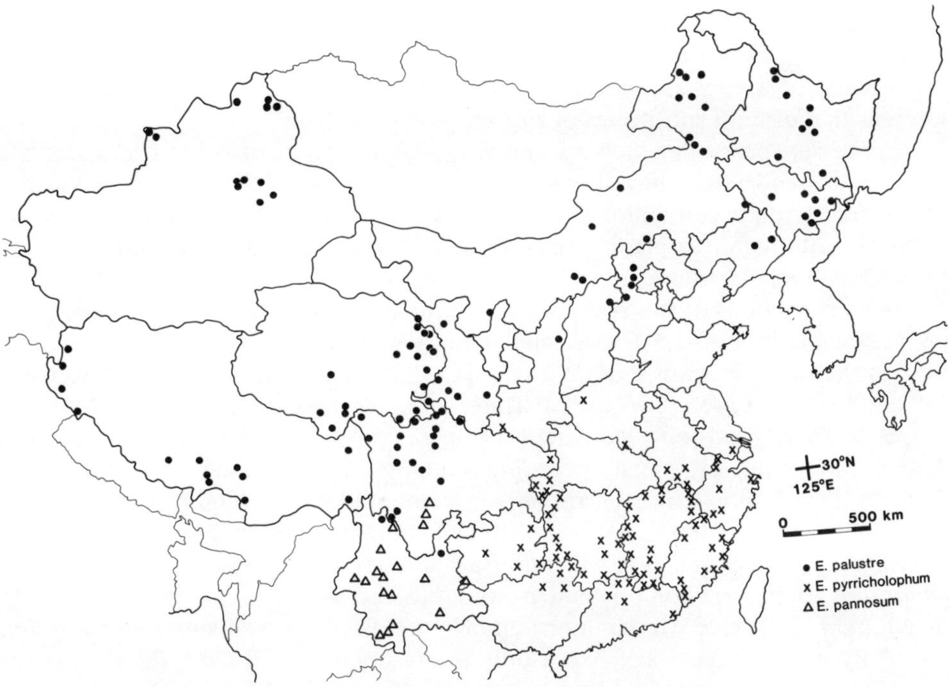

FIG. 40. Distribution in China of *Epilobium palustre*, *E. pannosum*, and *E. pyrricholophum*.

Distribution (Fig. 40). Southwestern China, northern Vietnam, northeastern Burma, and Assam (India); in China in southwestern Sichuan, southwestern Guizhou, and Yunnan; in open moist pastures, shaded wet weedy places by streams, or in valleys in evergreen broad-leaf forests; (760–) 1500–2100 m.

ADDITIONAL SPECIMENS EXAMINED. **China.** SICHUAN: Yuexi, *Sichuan Med. Exped. 26797* (NAS); Ganluo, *Chengdu Inst. Exped. 14330* (CDBI, PE); Xichang, *Legendre 1359* (P).—GUIZHOU: Jiangdi, *Cavalerie 4123* (P).—YUNNAN: Kunming, *K. H. Liang 260* (KUN), *T. N. Liou 14292* (PE), *14656* (PE), *20277* (PE), *20664* (KUN, PE), *B. Y. Qiu 50223* (KUN, PE), *H. K. Tang 80* (KUN); Mengzi, *Henry 12627* (NY), *13457* (NY); Dayao, s.d., *Simeon s.n.* (C), eastern flank of Lijiang, *Forrest 2975* (E, P); Heqing, *K. M. Feng 811* (KUN); Dali to Cangshan, *T. N. Liou 17441* (PE); Dali, *Forrest 425* (E), *T. N. Liou 17441* (PE), *20967* (KUN, PE), *Schneider 2532* (G), *2619* (G, GH), *1984 Sino-American Exped. 745A* (A, KUN, PE), *Tsiang 11343* (NAS), *K. K. Tsoong 2296* (PE), *C. W. Wang 305* (KUN), *358* (KUN), *H. C. Wang 1509* (KUN), *2935* (KUN); Cangshan, *Koyama et al. 1187* (KYO); Weishan, *Forrest 9275* (E); Fengqing, *T. T. Yu 17639* (A, E, KUN, PE); Jingdong, *M. K. Li 413* (KUN, PE), *Y. Tsiang 12405* (NAS, PE); Tengchong, *Delavay 2* (P), *2003* (P), *2004* (P), *Howell 83* (E), *W. Q. Yin 60-1265* (KUN); Simao, *Henry 12627* (K, NY), *13457* (BM, E, HK, K); Puwen, *P. Y. Mao 6759* (KUN); Xishuangbanna, *P. Y. Mao 6759* (PE); western Yunnan, *Lianda 1072* (KUN), *Delevay 2937* (P), *Forrest 2975* (BM).

This species is restricted to low or middle elevations in areas of high rainfall on the southern slope of the eastern Himalaya and the adjacent portion of the Hengduang Shan range. Although its distribution overlaps with that of *E. brevifolium,* especially of subsp. *trichoneurum,* and some morphological characteristics of the two species are similar, *E. pannosum* is remarkably distinct by virtue of its tomentose indumentum. *Epilobium pannosum* also differs from *E. brevifolium* by having rosettes rather than soboles, longer pedicels, and a stigma elevated above the anthers at anthesis. The mean pollen diameter of *E. pannosum* is 90 μm, whereas that of both subspecies of *E. brevifolium* is 73 μm.

Epilobium pannosum flowers and fruits later than most other species of the genus; many collections have been made in October in full flower, and in November in fruit. Even in cultivation in St. Louis, *E. pannosum* is the last species to begin to flower, although it is planted and germinates simultaneously with other species from China and elsewhere.

Most collections of this species have dense appressed velvety tomentum covering the plants and also a few glandular hairs scattered on the inflorescence and ovaries. Some collections from Yunnan [e.g., Kunming, *H. K. Tang 80* (KUN); Jingdong, *M. K. Li 413* (KUN, PE); Tengchong, *Delavay 2* (P), *2004* (P); Weishan, *Forrest 9275* (E); and some cultivated specimens], including also the lectotype from Assam, have relatively sparse spreading tomentose and glandular (sometimes predominantly so) pubescence on the upper stem and ovaries. There is no apparent geographical pattern to this variation. Raven (1962a) considered similar apparently intermediate collections in Assam and Burma to indicate a close relationship of *E. pannosum* to *E. brevifolium,* and he reduced the former to a subspecies of *E. brevifolium.* However, the greater number of collections available for this study, especially from China, clarifies not only the distinctive morphology of *E. pannosum,* but also its unusual and largely unique ecogeographical range. It seems preferable now to recognize it as a distinct species, albeit closely related to *E. brevifolium,* as noted by Raven and others.

The collection from Guizhou [Jiangdi, *Cavalerie 4123* (P)] and two from Yunnan [Fengqing, *T. T. Yu 17639* (A, E, KUN) and Tengchong, *W. Q. Yin 60-1265* (KUN)] have stems with four elevated pubescent lines decurrent from the margins of the petioles and sparse but predominantly glandular hairs all over the upper stems. These may be similar to plants described informally by Haussknecht (1884: 209) as forma *pilosa,* although he cited no specimens. The pollen fertility of these three collections is somewhat reduced (respectively 11%, 81%, and 59%), and several or all of them may be hybrids between *E. pannosum* and *E. wallichianum* or another species. The large flowers with exserted stigmas suggest that *E. pannosum* outcrosses predominantly, providing the mechanism for hybridization when other species are sympatric with it.

The flowers of this species are particularly large and attractive, and the plants are sometimes cultivated in gardens in Yunnan.

20. Epilobium brevifolium D. Don, Prodr. Fl. Nepal. 222. 1825.—TYPE: NEPAL. Near Katmandu, "in Nepalia superiore," 1802, *Hamilton s.n.* (not located; see Raven 1962a).

Common name: Duanye Liuyecai (Li 1980).

Erect or ascending perennial herb, forming fleshy basal soboles that leave loose brownish scales at base of stem. Stems 15–90 cm tall, simple or branched, strigillose all over, sometimes with an admixture of glandular hairs on the inflorescence, lacking raised lines decurrent from the nodes. Leaves mostly opposite below, alternate on the inflorescence, subsessile or on petioles 1–4 mm long; cauline blades 1.5–5 (–8) cm long, 0.5–2.2 (–3) cm wide, broadly ovate to ovate or broadly lance-elliptic, strigillose on veins, the remainder subglabrous, apex subobtuse to acute, base subcordate to cuneate, the margins more or less sharply denticulate with 7–22 teeth on each side, the lateral veins often prominent beneath, 5–6 on each side of the midrib. Inflorescence erect to slightly nodding. Flowers erect; buds 4–7 mm long, 2.5–3.5 mm in diameter, narrowly ovoid, apiculate; ovaries 1.8–4.5 cm long, strigillose or mixed with glandular hairs; pedicels 0.5–1.4 cm long; floral tube 1–1.4 mm long, 2–2.5 mm in diameter, with a few spreading hairs at the mouth within; sepals 4.5–6.5 mm long, 1–1.2 mm wide, lance-oblong, keeled, strigillose and glandular-pubescent; petals 7–11 mm long, 3–5 mm wide, pink to reddish purple, obcordate, the notch 1–2 mm deep; filaments of longer stamens 3–4.5 mm long, those of shorter ones 1.5–3 mm long; anthers 1–1.4 mm long, 0.4–0.6 mm wide, oblong; style 3–5 mm long, erect, glabrous; stigma broadly clavate to clavate, 1.5–3.5 mm high, 0.8–1.3 mm in diameter, subequal with stamens. Capsules 3.5–7 cm long, strigillose, sometimes with an admixture of glandular hairs, borne on pedicels 0.4–1.5 cm long. Seeds 0.9–1.1 mm long, 0.4–0.5 mm wide, oblong-obovoid, with very short, inconspicuous chalazal collar, dark brown, the surface coarsely papillose; coma 5–10 mm long, dull white, detaching readily. Chromosome number: n = 18.

Raven (1962) treated *E. brevifolium* as a broadly circumscribed and common Himalayan species that comprises three subspecies with partially overlapping distributions and partially intergrading morphologies. We here retain this treatment in part, but on the basis of additional material of *E. pannosum* from China we have restored that taxon to specific status, as discussed under that species. Intermediate specimens between *E. brevifolium* and *E. pannosum* are rare, and the morphological and ecological differences are substantial.

Consequently, we now treat *E. brevifolium* as consisting of two subspecies, subsp. *trichoneurum* and subsp. *brevifolium.* These two taxa intergrade broadly, primarily in leaf shape and texture, and to a lesser degree in branching pattern and stigma shape. They replace one another geographically (cf. fig. 42; Raven 1962a, fig. 8). Subspecies *brevifolium* has a more western range and barely reaches Xizang and Yunnan from the west, whereas subsp. *trichoneurum* is common throughout the warm temperate and subtropical areas of eastern and southern China. Subspecies *brevifolium* is rare in China, and we have not seen intermediates with subsp. *trichoneurum* from China, although they are more frequent in Sikkim, Assam, and Burma (Raven 1962a).

For the most part, *E. brevifolium* is distinct from *E. royleanum* over their large common range by virtue of its capitate stigma and distinct petioles. However, some collections that we identify as *E. brevifolium* subsp. *trichoneurum* have broadly clavate (rather than narrower) stigmas and short but definite petioles. They are few enough in relation to the thousands of unambiguous collections that they do not cast doubt on the distinctness of the two species, and may in fact represent hybrids or backcrosses between these predominantly self-pollinating taxa.

20a. Epilobium brevifolium subsp. **brevifolium.**

Epilobium trichoneurum var. *brachyphyllum* Haussknecht, Oesterr. Bot. Zeitschr. 29: 54. 1879.—Type: "Himalaya bor. occid.," unplaced locality in

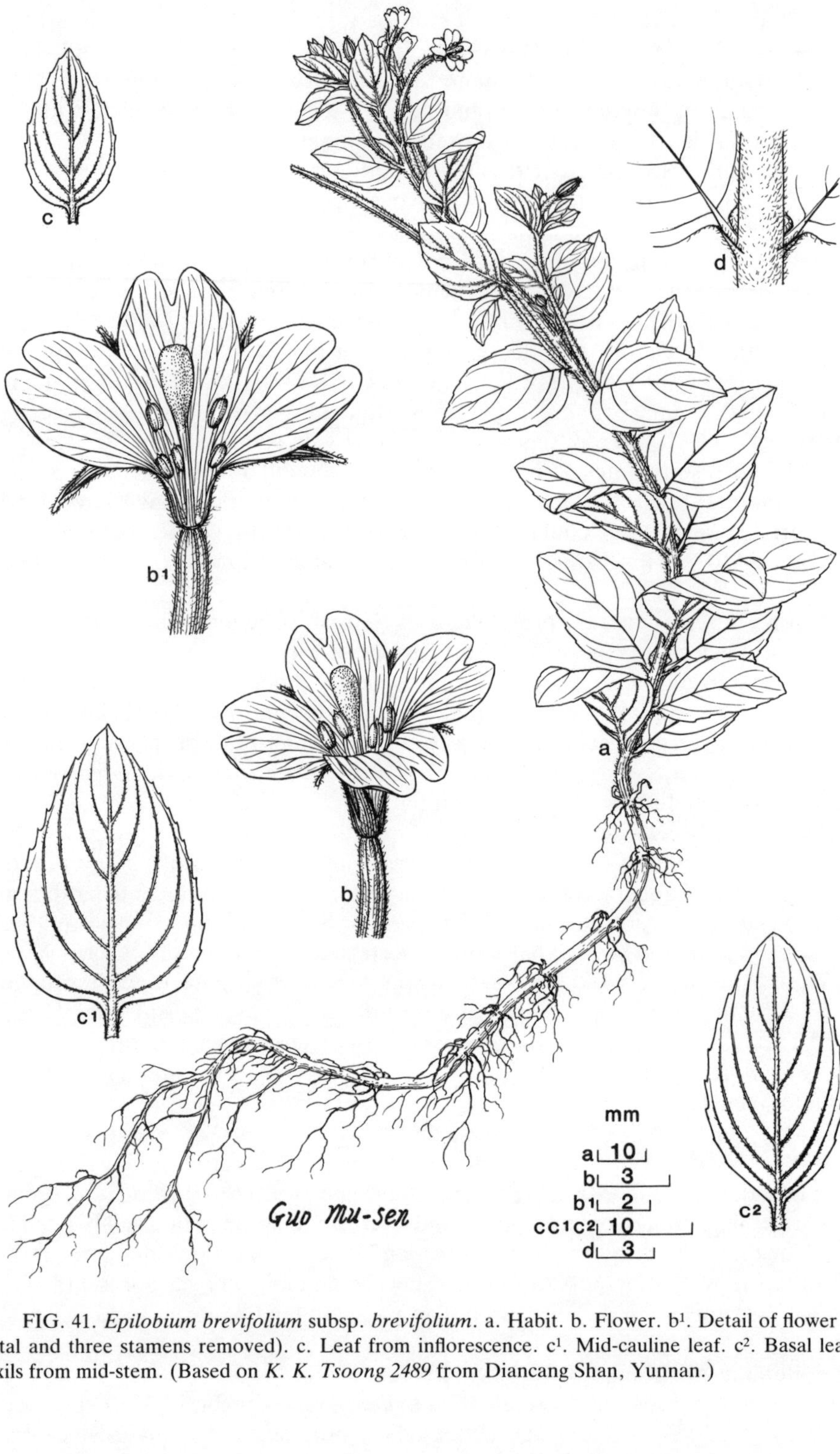

FIG. 41. *Epilobium brevifolium* subsp. *brevifolium*. a. Habit. b. Flower. b¹. Detail of flower (one petal and three stamens removed). c. Leaf from inflorescence. c¹. Mid-cauline leaf. c². Basal leaf. d. Axils from mid-stem. (Based on *K. K. Tsoong 2489* from Diancang Shan, Yunnan.)

northwest Himalaya, 1800–2400 m, 1847–1848, *Thomson s.n.* (lectotype, designated by Raven 1962a: K!).

Perennial herb. Stems 25–60 cm tall, 2.5–4 mm in diameter, simple or sparingly branched, strigillose all over, with a few scattered glandular hairs especially on the upper part, rarely predominantly glandular. Leaves subsessile or with petioles 1–2 mm long; cauline blades 2.5–4.5 cm long, 1.5–2.2 cm wide, broadly ovate or ovate, strigillose and rarely with a few glandular hairs on both sides, especially on the veins, herbaceous, apex acute or subobtuse, base subcordate, the margins sharply or weakly serrulate with 15–22 teeth on each side. Ovaries 2–3 cm long; pedicels 0.5–0.8 cm long; petals 9–11 mm long, 4–5 mm wide, pink to rose-purple, the notch 1.2–2 mm deep; filaments of longer stamens 4–5 mm long, those of shorter ones 2.5–3.5 mm long; style 4–5 mm long; stigma broadly clavate or clavate, 2–3.2 mm long, 1–1.6 mm in diameter. Capsules 5–7 cm long, borne on pedicels 0.4–1.5 cm long. Seeds 0.9–1.1 mm long, 0.4–0.5 mm wide, the surface papillose. Chromosome number: $n = 18$. Fig. 41.

Phenology. Collected in flower from June to July, in fruit from August to September.

Distribution (Fig. 42B). Himalayan region, from Himachal Pradesh, India, eastward to Nepal, and barely to southwestern China; in China rare in southeastern Xizang (Tibet) and western Yunnan; in grassy places by streams in the mountains; 1700–2100 m.

ADDITIONAL SPECIMENS EXAMINED. **China.** YUNNAN: Diancang Shan, *K. K. Tsoong 2489* (PE).—XIZANG: Zayu, *Z. C. Ni et al. 428* (PE); Mt. Nanjiabawa, Medog, *P. S. Li & S. Z. Cheng 1420* (PE), *3206* (PE); Cayu, *Xizang Exped. 1199* (HNWP); Shangzayu, *Z. C. Ni 801* (PE).

This taxon is restricted in China to low elevations in a small area of the Himalaya and adjacent mountains. Two of the collections from Xizang (*Z. C. Ni et al. 428* and *801*) agree with typical specimens of subsp. *brevifolium,* but have an unusual abundance of glandular hairs mixed with the strigillose pubescence of the stem and inflorescence. Otherwise, the collections from China are indistinguishable from those obtained farther west, although they are disjunct geographically from most other reported localities. Raven (1962a) reported the range of subsp. *brevifolium* only as far east as Nepal, but he had not seen these collections from China.

20b. Epilobium brevifolium subsp. **trichoneurum.** (Haussknecht) Raven, Bull. Brit. Mus. (Nat. Hist.) Bot. 2: 362. 1962. *Epilobium trichoneurum* Haussknecht, Oesterr. Bot. Zeitschr. 29: 54. (Feb) 1879. *Epilobium hookeri* C. B. Clarke in J. D. Hooker, Fl. Brit. India 2: 585. (May) 1879, nom. superfl.—TYPE: INDIA. Assam, Khasi Hills, 1500–2100 m, 1850, *J. D. Hooker & Thomson s.n.* (lectotype, designated by Raven 1962a: K!; isolectotype: BM!).

Epilobium esquirolii H. Léveillé, Bull. Herb. Boiss., sér. 2. 7: 590. 1907.—TYPE: CHINA. Guizhou, Pingfa, "Kouy-Tcheou," *Esquirol 607* (holotype: G!).

Epilobium cavalieri H. Léveillé, Bull. Herb. Boiss., sér. 2. 7: 590. 1907.—TYPE: CHINA. Guizhou, Pingfa, "Kouy-Tcheou," 31 Aug 1902, *Cavalerie 352* (lectotype, here designated: G!; isolectotypes: E! K!).

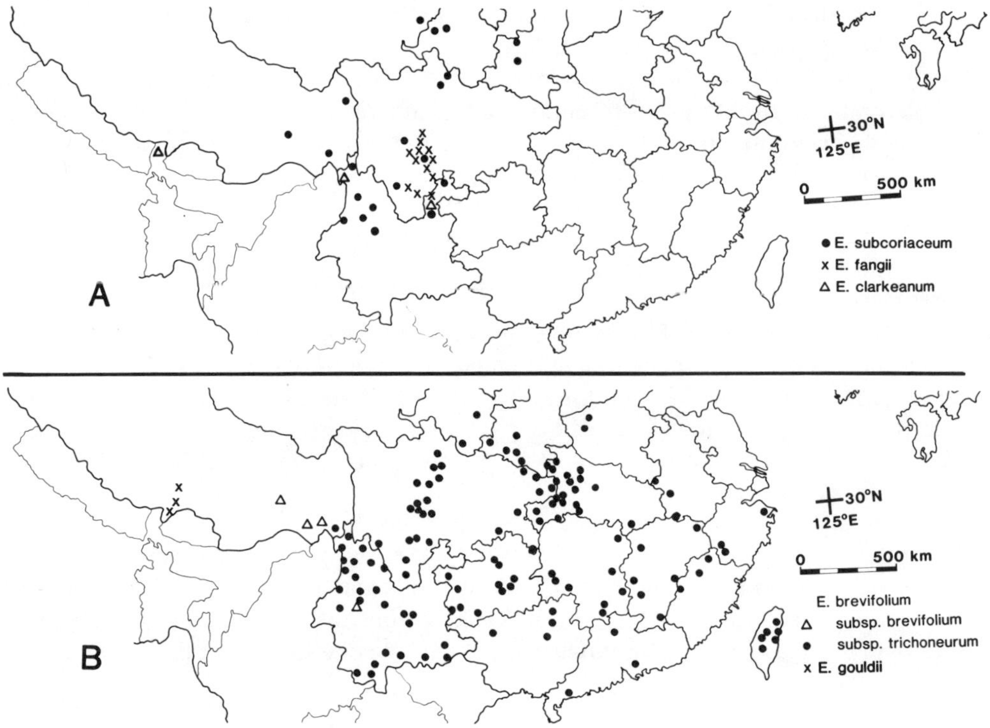

FIG. 42. Distribution in China of: A. *Epilobium clarkeanum*, *E. fangii*, and *E. subcoriaceum*; B. *Epilobium brevifolium* subsp. *brevifolium*, *E. brevifolium* subsp. *trichoneurum*, and *E. gouldii.*

Epilobium cordouei H. Léveillé, Repert. Spec. Nov. Regni Veg. 6: 110. 1908.—TYPE: CHINA. Guizhou, Pingfa, "Kouy-Tcheou, Majo," rare, 5 Sep 1907, *Cavalerie 3151* (not located).

Epilobium philippinense C. B. Robinson, Philipp J. Sci., Sect. C, 3: 209. 1908.—TYPE: THE PHILIPPINES. Luzon, District of Lepanto, dry grassy slopes in pine forest on Mt. Data, 3 Nov 1905, *Merrill 4484* (lectotype, here designated: K!). The holotype at the herbarium of the Bureau of Science in Manila was destroyed in 1945; we choose the only known isotype as the lectotype.

Common name: Xianjing Liuyecai (Li 1986).

Erect or ascending perennial herb, sometimes woody at the base. Stems 15–90 cm tall, 1.5–6 mm in diameter, simple to well branched above, strigillose all over, sometimes with a sparse admixture of glandular hairs especially on upper part. Leaves on petioles 1–4 mm long or rarely sessile; cauline blades 1.5–5 (–8) cm long, 0.5–2 (–3) cm wide, lanceolate or elliptic to narrowly ovate, very rarely broader, strigillose on veins of both surfaces, subcoriaceous, green or red-purplish beneath and along margins, apex subobtuse or acute, base rounded or cuneate, margins distinctly denticulate, with 7–17 teeth on each side. Ovaries 1.8–4.5 cm long; pedicels 0.6–1 cm long; petals 7–10 mm long, 3–4 mm wide, dark pink to rose-purple, the notch 1–1.5 mm deep; filaments of longer stamens 3–4.5 mm long, those of shorter ones 2–3.5 mm long; style 3–5 mm long; stigma clavate, 1.5–3.5 mm long, 0.5–1 mm in diameter. Capsules 3.5–7 cm long, borne on pedicels 0.4–1.4 cm long. Seeds 1–1.1 mm long, 0.4–0.5 mm wide, the surface papillose. Chromosome number: n = 18. Fig. 43.

Phenology. Collected in flower from July to September (rarely to October), in fruit from September to October.

Distribution (Fig. 42B). Eastern Himalaya from Bhutan to Assam and Burma, southern China, northern Vietnam, and northern Luzon; in China in Henan, southern Shaanxi, southeastern Gansu, southern Anhui, Zhejiang, Jiangxi, Fujian, Taiwan, Guangdong, northern Guangxi, western Hunan, western Hubei, Sichuan, Guizhou, Yunnan, and southeastern Xizang (Tibet); in open grassy places, wet areas in valleys, and disturbed places in the mountains; 600–1800 m in eastern and central China, 900–2500 (-3600) m in western China.

ADDITIONAL SPECIMENS EXAMINED. **China.** HENAN: Song Xian, *Henan Forestry Depart. 951* (PE); Luanchuan, *Henan Exped. 21087* (PE).—SHAANXI: Taibai Shan, *K. T. Fu 9867* (WUK); Xixiang, *K. T. Fu 11483* (WUK); Lueyang, *K. T. Fu 5778* (PE); Ziyang, *P. Y. Li 4759* (PE); Hanzhong, *J. W. Wang et al. 332* (PE).—GANSU: Tianshui, *C. C. Wu 20258* (PE); Wen Xian, *W. L. Chen & C. J. Chen 8834* (MO, PE), *J. Q. Fu 2889* (WUK), *Z. Y. Zhang 13975* (PE), *14229* (WUK).—ANHUI: Baimazhai, Dabei Shan, *Y. L. Wu 16* (HSFS); Yuexi, *E. China Bot. Station 6881* (PE), *6886* (NAS); Laozhuling, *Zhou et al. 583* (PE).—ZHEJIANG: Ningbo, *Hancock 42* (K); Longquan, *S. Y. Zhang 336* (PE); S. W. Zhejiang, *R. C. Ching 2490* (K, UC).—JIANGXI: Yushan, *M. X. Nie 6081* (PE); Zhang Gou Shan, *R. C. Ching 4506* (E, UC); Lu Shan, *Carles 28* (E, K), *Steward 4706* (E, UC); Yifeng, *S. K. Lai et al. 343* (LBG); Wugong Shan, *Jiangxi Pl. Exped. 1021* (PE), *Y. G. Xiong 8551* (LBG); Shicheng, *C. M. Hu 4848* (LBG); Suichuan, *J. Xiong 2417* (LBG).—FUJIAN: Chong'an, *Wuyi Shan Exped. 830* (FJSI); Shaowu, *H. C. Zhou 5105* (IBSC); Ninghua, *C. H. Huang 83-001* (PE).—TAIWAN: Taibei Xian, *S. Suzuki 6219* (TAI); Yilan Xian, *C. C. Hsu 5916* (TAI), *M. T. Kao 8601* (TAI); 1928, *S. Suzuki s.n.* (TAI); 1925, *Y. Yamamoto s.n.* (TAI); Taiping Shan, *Kanehira 419* (UC); Xinzhu Xian, *G. Masamune 1289* (TAI); 1932, *N. Fukuyama s.n.* (TAI); 1932, *S. Sasaki s.n.* (TAI); Hualian Xian, *C. C. Chuang & M. T. Kao 4450* (TAI); 1936, *Fukuyama & Suzuki-Tokio s.n.* (TAI), *M. T. Kao 7772* (H, PH, TAI), *C. S. Kuo et al. 7054* (TAI); 1937, *Suzuki-Tokio et al. s.n.* (TAI); Taizhong Xian, *T. G. Huang 7080* (TAI), *7123* (TAI), *7126* (TAI), *7130* (TAI), *T. C. Huang & C. F. Hsieh 7174* (TAI), *C. M. Kuo 3983* (H, TAI); 1936, *K. Mori s.n.* (TAI); 1930, *S. Onuma s.n.* (TAI); Nantou Xian, *C. C. Hsu & R. Hsu 4009* (TAI); 1967, *C. C. Hsu & R. Hsu s.n.* (TAI), *Tamura et al. 22034* (SHIN), *Wu et al. 7728* (TAI); Taihokuo, *Suzuki 6219* (MICH); Jiayi, *C. C. Hsu & R. Hsu 3950* (TAI), *C. C. Hsu 6238* (TAI), *C. S. Kuoh 1255* (TAI); 1935, *Simada-Hidetaro s.n.* (TAI); Manda Shan, 1908, *T. Kawakami & U. Mori s.n.* (TAI); Yushan ("Mt. Niitaka"), *Kanehira & Sasaki 21782* (UC).—GUANGDONG: Ruyuan, *Y. X. Liu 28854* (IBSC); Luofu Shan, *Merrill 10658* (GH, HK, NA, NU, UC); 1920, *Merrill et al. s.n.* (K); Maoming, *L. Deng 2382* (IBSC, PE).—GUANGXI: Guangyang, *Z. Z. Chen 52432* (KUN); Longsheng, *Y. F. Tang et al. 700565* (IBSC, NAS); Guilin, *W. T. Tsang 28090* (IBSC, UC); Sanjiang, *A. Steward & C. C. Cheo 987* (NY); Linyun, *Y. X. Liu 28675* (IBSC).—HUNAN: Pingjiang, *J. Xiong 6403* (LBG), *6470* (LBG); Hengshan, *K. K. Tsoong 4051* (PE); Ling Xian, *B. M. Yang 27079* (HTC), *L. H. Liu 11151* (HTC); Shimen, *Huping Shan Exped. A25* (PE), *A48* (PE), *A109* (PE); Yizhang, *G. Z. He 4862* (HTC), *L. H. Liu s.n.* (KUN, YNU), *283* (HTC), *757* (HTC, YUN); Xinning, *Hunan Med. Exped. 63476* (HTC), *L. H. Liu & G. X. He 15144* (HTC, KUN, PE), *P. X. Tan 63622* (HTC), *Ziyun Shan Bot. Exped. 529* (PE); Dongkou, *L. H. Liu & G. Z. He 16621* (PE), *P. X. Tan 62916* (HTC); Qianyang, *Anjiang School of Agr. 1339* (PE), *Z. T. Li 1601* (IBSC); "Mang Shan," *G. Z. He 4861* (HTC), *M. X. Huang 11986* (HTC, IBSC), *112191* (IBSC), *L. H. Liu 575* (HTC, PE); Yun-wu Shan, *S. S. Sin 471* (IBSC); "Yunshang," *H. T. Chang 4716* (IBSC); no locality, *Hunan Exped. 12235* (PE).—HUBEI: Tongcheng, *J. Xiong 6050* (LBG); Nanzhang, *D. Q. You 6935* (HIIM); Shennongjia, *H. J. Li 1490* (PE), *Y. Chen 692* (UC), *Shennongjia Pl. Exped. 22500* (PE), *22840*(PE); Zhuxi, *B. C. Yang 312* (HIIM); Xingshan, *W. J. Wang 289* (HIIM); Badong, *Q. L. Chen et al. 1820* (PE), *M. Z. Chien 1487*(PE), *L. Q. Kong 1526* (HIIM), *M. Z. Qian 1628* (PE); Yidu, *S. L. Xu 556* (HIIM); Changyang, *S. X. He 403* (HIIM); Jianshi,

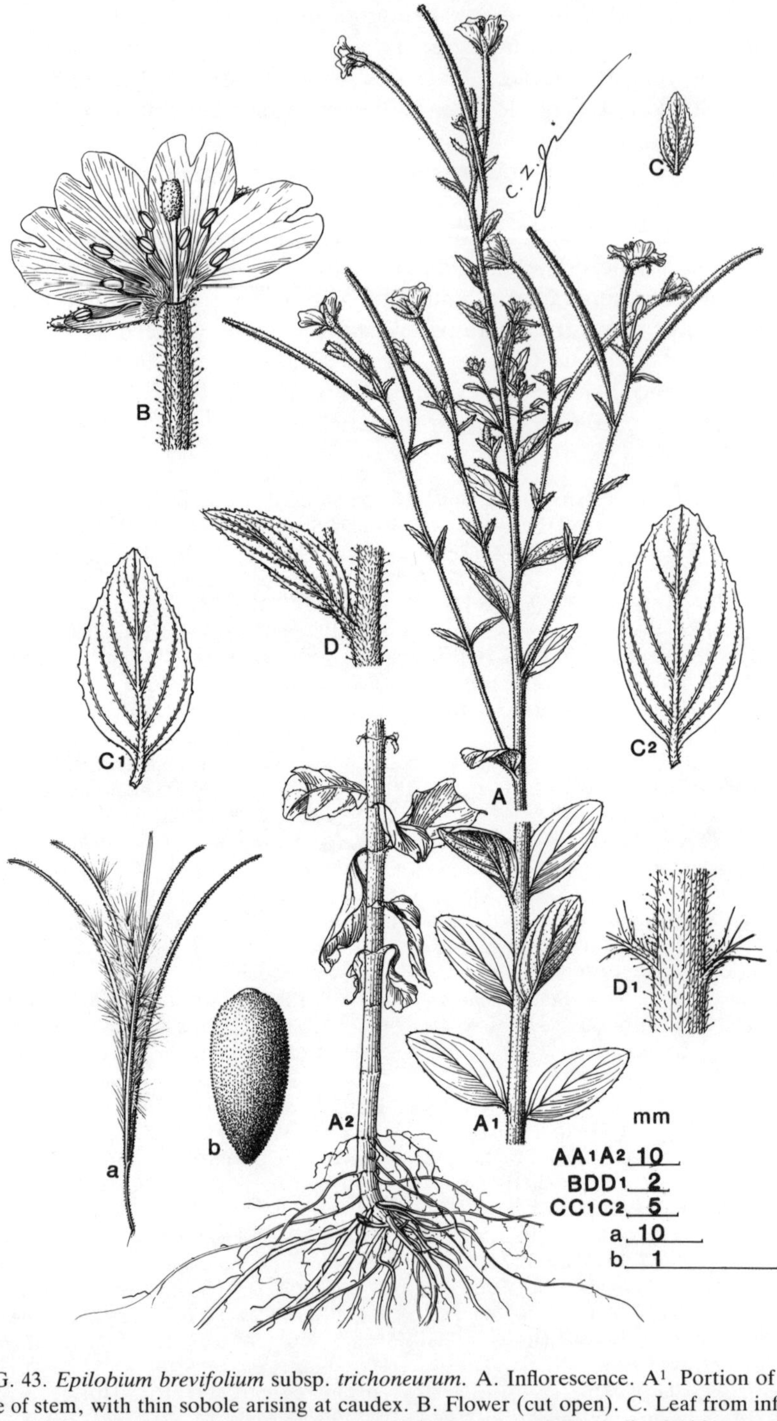

FIG. 43. *Epilobium brevifolium* subsp. *trichoneurum*. A. Inflorescence. A¹. Portion of mid-stem. A². Base of stem, with thin sobole arising at caudex. B. Flower (cut open). C. Leaf from inflorescence (bract). C¹. Mid-cauline leaf. C². Basal leaf. D. Axil of inflorescence. D¹. Axils of mid-stem. a. Capsule. b. Seed. (A–D¹ based on *J. H. Xiong & Z. L. Zhou 93500;* a–b based on *L. Y. Dai & C. H. Qian 1135.*)

1951, *L. Y. Dai & C. H. Qian s.n.* (PE); Xianfeng, *H. J. Li 9078* (NAS); Lichuan, *L. Y. Dai & C. H. Qian 1135* (PE), *G. X. Fu & Z. S. Zhang 1589* (PE), *J. P. Ma s.n.* (HIIM); Western Hubei, *Wilson 1324* (NY); Hefeng, *H. J. Li 5967* (PE), *Y. J. Ma 4274* (HIIM); Jinyao Shan, Xishan, *S. S. Li 535* (HIIM).—SICHUAN: Wushan, *H. F. Zhou & H. Y. Su 108976* (PE); Fengjie, *H. F. Zhou & H. Y. Su 109962* (PE, SZ), *110948* (NAS); Chengkou, *T. L. Dai 101179* (PE), *102673* (NAS, PE, SZ), *107406* (PE, SZ); 1895–1897, *Farges s.n.* (P); Wanyuan, *P. Y. Li 6132* (PE); Nanchuan, *C. J. Chen & Z. Y. Liu 83074* (PE); *Kunming Inst. Bot. 64602* (KUN), *G. F. Li 64602* (PE), *J. H. Xiong & Z. L. Zhou 93500* (PE, SZ); Shizhu, *Y. Chen 593* (CDBI); Xiushan, *E. Sichuan Exped. 3952* (PE); Eastern Sichuan, *Henry 173* (NY), *285* (NY); Songpan, *K. T. Fu 1911* (PE); Zheming Shan, *X. Li & J. X. Zhou 74297* (PE); Heishui, *X. Li & J. X. Zhou 73353* (PE, SZ); Mao Xian, *Z. He & Z. L. Zhou 13554* (PE); Maowen, *Maowen Pl. Exped. 11099* (CDBI, PE); Li Xian, *D. P. He 46043* (SZ); Jinchuan, *Sichuan Med. Exped. 23026* (NAS); Tianquan, *K. C. Kuan et al. 2306* (PE); Mt. Emei, *C. Y. Chiao & C. S. Fan 330* (GH), *Faber 173* (K), *285* (K), *L. Hamet-Ahti 3231* (H), *K. C. Kuan et al. 1561* (PE), *Raven & C. J. Chen 26144* (MO, PE), *26147* (PE), *C. Y. Wu et al. 49* (KUN, PE), *C. W. Yao 3289* (PE); Luding, *H. Chen 22133* (CDBI), *Sichuan Econ. Pl. Exped. 55* (CDBI); Lushan, *K. L. Chu 3994* (PE); Shiangling Shan, *T. P. Wang 9442* (PE); Qianglin, *T. S. Ying 9857* (PE); Meigu, *Sichuan Exped. 13071* (PE); Mianning, *S. K. Wu 1775* (PE); Dechang, *S. F. Zhu 20120* (PE), *20310* (PE); Xichang, *T. T. Yu 1274* (GH, PE), *Q. S. Zhao et al. 5001* (CDBI); Huili, *Chengdu Inst. Biol. 12304* (CDBI, PE), Muli, *Qinghai-Xizang Exped. 14186* (PE), *S. K. Wu 2577* (KUN); Western Sichuan, *K. L. Chu 6358* (PE); no locality, *M. Chen 2104* (KUN), *T. T. Li 516* (PE).—GUIZHOU: Pingfa, *Cavalerie 352* (E, K), *3036* (P); Suiyang, *Esquirol 3007* (P); Zunyi, *Sichuan-Guizhou Exped. 1275* (PE); Longli, *Cavalerie 3152* (P); Huangping, *S. Guizhou Pl. Exped. 2641* (GZBI, NAS, PE); Kaili, *S. Guizhou Pl. Exped. 3456* (GZBI, NAS, PE); Nayong, *Bijie Exped. 693* (PE); Hezhang, *Hezhang Exped. 453* (GZBI); Pu-an, *Anshun Exped. 1807* (PE); Panxian, *Anshun Exped. 1051* (PE); Wangmo, *Esquirol 2501* (P); Xingren, *P. Dang 574* (PE); "Shin-guy-fou (Eul lang Keou)," *Esquirol 84* (HK); no locality, s.d., *Cavalerie s.n.* (P); *Guizhou Exped. 335* (GZBI); 1858, *Perny s.n.* (P).—YUNNAN: Songming, *Ducloux 7263* (P), *7269* (P), *R. Z. Long et al. 40* (YNU), *Y. B. Zhang 7* (KUN, PE); Fuming, *B. Y. Qiu & B. X. Sun 596147* (KUN); Kunming, *K. M. Feng 10681* (KUN), *A. R. Li 7923* (PE), *K. H. Liang 231* (KUN), *T. N. Liou 14252* (KUN, PE), *16981* (PE), *B. Y. Qiu 50658* (KUN), *77597* (KUN), *B. X. Sun 64479* (YNU), *64526* (YNU), *64543* (YNU), *64544* (YNU), *64549* (YNU), *64598* (YNU), *64608* (YNU), *64624* (YNU), *64633* (YNU), *64634* (YNU), *64846* (YNU), *Yunnan Univ. 155* (YNU); Malipo, *K. M. Feng 12669* (GH, KUN, PE); Wenshan, *K. M. Feng 11180* (GH, KUN, PE), *H. Li 248* (KUN), *H. T. Tsai 588168* (KUN); Mengzi, *Henry 9642* (K, NY); Luquan, *D. D. Dao 225* (KUN), *1126* (KUN), *B. X. Sun 64187* (YNU); Yufu, Lijiang, *C. Z. Lu 63-20096* (KUN); "Long-ky," s.d., *Maire s.n.* (P); Deqen, *K. M. Feng 23681* (KUN), *C. W. Wang 69336* (GH, PE); Gongshan, *K. M. Feng 24605* (KUN), *X. F. Deng 791221* (KUN), *Qinghai-Xizang Exped. 8030* (KUN, PE), *9033* (KUN, PE), *9807* (KUN, PE), *C. W. Wang 66859* (GH, KUN, NAS, PE), 1980, *J. S. Yang s.n.* (KUN); Zhongdian, *Handel-Mazzetti 7646* (E, WU); Ninglang, *B. Y. Qiu 61181* (KUN), *T. T. Yu 13677* (PE); Lijiang, *C. Z. Bao 63-20096* (KUN), *Forrest 2490* (E), *Kunming Inst. Bot. 21272* (KUN, PE), *22454* (KUN, PE), *Rock 4905* (E, NY), *A. L. Zhang & S. W. Yu 100918* (KUN); Weixi, *T. H. Tsai 59641* (GH, KUN, NAS, PE); Fugong, *H. T. Tsai 54633* (GH, KUN, NAS, PE), *54727* (KUN, PE); Dayao, *P. Zhai 60-083* (KUN); Lanping, *H. T. Tsai 56008* (GH, KUN, NAS, PE); Eryuan to Heqing, *Delavay 932* (P), *1224* (P), *2146* (P); 1885, *Delavay s.n.* (P); Binchuan, *B. X. Sun 35* (YNU), *83* (YNU), *149* (YNU), *351* (YNU); Dali, 1882, *Delavay s.n.* (P), *Delavay 2005* (P), *T. N. Liou 21102* (PE), *Y. Tsiang 11372* (NAS), *H. C. Wang 1124* (KUN, PE), *4405* (PE), *W. T. Wang 43* (KUN); Yangbi, *K. M. Feng et al. 80-64* (KUN); Pianma, *Bijiang Exped. 1545* (KUN); Bijiang, *Z. G. He 559* (KUN); Tengchong, *Howell 221 (=Forrest 2490;* E); Jingdong, *M. K. Li 484* (KUN), *B. Y. Qiu 52546* (KUN, PE), *S. G. Xu 5310* (KUN), *C. W. Wang 78483* (GH, KUN, NAS, PE); Yaunjiang, *Y. H. Li 5847* (KUN), *W. Q. Yin 2072* (KUN); Zhenkang, *T. T. Yu 17121* (KUN, PE); Western Yunnan, *Forrest 2490* (E, K); no locality, *Cavalerie 7622* (E), *Lianda 1063* (KUN), *1065* (KUN), *1067* (KUN), *Maire 2215* (E).—XIZANG: Mainkung Nu Jiang Divide, *T. T. Yu 22701* (E, GH, KUN, PE).

Epilobium brevifolium subsp. *trichoneurum* is very widespread in southern China, and occurs as an element of the warm temperate flora of eastern Asia. Indeed, this taxon occurs in warmer areas than almost any species of *Epilobium* in eastern Asia, and is one of only two species (the other is *E. platystigmatosum*) to reach the Philippines (Luzon; Raven 1967a). It is extremely variable in stature, shape and size of the leaf and its posture on the stem, and degree of pubescence, but these features are not, for the most part, correlated with one another or geographically. There are, however, some unusual collections from Taiwan [Hualian, *M. T. Kao 7772* (H, PH, TAI); Taizhong, *C. M. Kuo 3983* (H, TAI); Taibei, *Suzuki 6219* (TAI); and Nantou, *C. C. Hsu & R. Hsu 4009* (TAI)] with much-branched stems and smaller and thinner leaves with a cuneate base.

21. Epilobium pyrricholophum Franchet & P. A. L. Savatier, Enum. Pl. Jap. 1: 168. 1875, nomen nudum; 2: 370. 1879.—TYPE: JAPAN. Honshu, Kanagawa pref., near Yokosuka ("Yokoska"), and on Mt. Hakone, in moist shady places in montane areas, *Savatier 410* (holotype: not located).

Epilobium tetragonum var. *japonica* Miquel, Ann. Mus. Bot. Lugduno-Batavum 3: 94. 1867. *Epilobium japonicum* (Miquel) Haussknecht, Oesterr. Bot. Zeitschr. 29: 56. 1879. *Epilobium pyrricholophum* var. *japonicum* (Miquel) H. Hara, J. Jap. Bot. 18: 236. 1942.—TYPE: JAPAN (not found; at least one syntype [*Siebold*] should be at M but cannot be located).

Epilobium japonicum β. *glanduloso-pubescens* Haussknecht, Oesterr. Bot. Zeitschr. 29: 56. 1879.—TYPE: JAPAN. Hokkaido, Oshiwa Pref., Hakodate, *Albrecht s.n.* (holotype: LE, not located).

Epilobium oligodontum Haussknecht, Oesterr. Bot. Zeitschr. 29: 58. 1879.—TYPE: JAPAN. Hokkaido, Oshiwa Pref., Hakodate, *Albrecht s.n.* (holotype: LE, not located).

Epilobium rouyeanum H. Léveillé, Bull. Acad. Int. Geogr. Bot. 9: 210. 1900. *Epilobium pyrricholophum* var. *curvatopilosum* H. Hara, J. Jap. Bot. 18: 237. 1942.—TYPE: JAPAN. Honshu, Fukushima Pref., at the foot of Mt. Bandai-san, 7 Sep 1898, *Faurie 1317* (lectotype, designated here: G!; isolectotypes: KYO!, 2 sheets). Hara designated his variety as a *nom. nov.,* but did not state clearly what his name replaced; he was free to use any name, however, because no name existed at the varietal level. He included a lengthy synonymy in his treatment of this taxon, and we have placed his taxon under the earliest validly published basionym that he cited.

Epilobium hakkodense H. Léveillé, Bull. Acad. Int. Geogr. Bot. 10: 34. 1901.—TYPE: JAPAN. Honshu, Aomori Pref., Mt. Hakkoda-san, Aug 1899, *K. Koriba s.n.* under *Faurie 3731* (holotype: not located; isotype: TI!).

Epilobium quadrangulum H. Léveillé, Bull. Soc. Agric. Sarthe 40: 72. 1905.—TYPE: JAPAN. Hokkaido, Aomori Pref., Mt. Osorezan, 2 Sep 1902, *Faurie 5101* (lectotype, here designated: KYO!).

Epilobium makinoense H. Léveillé, Bull. Soc. Agric. Sarthe 40: 73. 1905.—TYPE: JAPAN. Hokkaido ("Yezo"), Ochiai, in humid forests, Sep 1904, *Faurie 6314* (holotype: not located; isotypes: KYO! 2 sheets).

Epilobium chrysocoma H. Léveillé, Bull. Herb. Boiss., sér. 2, 7: 589. 1907.—TYPE: JAPAN. Hokkaido, Aomori, along streams in the mountains, 7 Sep 1882, *Faurie 1098* (lectotype, here designated: G!).

Epilobium arcuatum H. Léveillé, Bull. Herb. Boiss., sér. 2, 7: 589. 1907.—TYPE: JAPAN. At the foot of "Norikosa," 21 Aug 1901, *Faurie 6663* (lectotype, here designated: G!).

Epilobium prostratum H. Léveillé, Bull. Soc. Bot. France 54: 520, 522. 1907, non *Epilobium prostratum* Warburg, 1893.—TYPE: JAPAN. Miyadzu, in wet forests, Oct 1901, *Faurie 4896* (lectotype, here designated: KYO!).

Epilobium kiusianum Nakai, Bot. Mag. (Tokyo) 22: 84. 1908.—TYPE: JAPAN. Chikuzen Prov., Kiusiu, *K. Nagano s.n.* (holotype: TI?).

Epilobium pyrricholophum var. *anuoleucholophum* H. Léveillé, Repert. Spec. Nov. Regni Veg. 6: 330. 1909.—TYPE: JAPAN. Honshu, Tokyo Pref. (Prov. Musashi, "Funabashi"), Mt. Takao, Oct 1908, *C. Funabash s.n.* (holotype: TI!).

Epilobium nakaianum H. Léveillé, Icon. Gen. Epil. 166, t. 140. 1910.—TYPE: JAPAN (holotype: TI?).

Epilobium axillare Franchet ex Koidzumi, Fl. Symb. Orient.-Asiat. 85. 1930.—TYPE: JAPAN. Honshu, Kanagawa Pref., Yokosuka ("Yokoska"), along streams and wet places, Sep 1867, (holotype: P "Herb. Drakeano," not located; isotype: K!).

Epilobium myokoense Koidzumi, Fl. Symb. Orient.-Asiat. 86. 1930.—TYPE: JAPAN. Myokosan, 27 Jul 1897, *Faurie 10* (lectotype, here designated: P!).

Common name: Changzi Liuyecai (Institute of Botany, Academia Sinica 1972; Jiangsu Institute of Botany 1982).

Erect perennial herb, forming basal filiform stolons with small, widely spaced, subentire, suborbicular, obtuse leaves. Stems 25–80 cm tall, 2.5–7 mm in diameter, terete, usually well branched in larger plants or simple in smaller ones, strigillose all over with a conspicuous admixture of glandular hairs, pubescence especially dense on the upper part. Leaves opposite, alternate in the inflorescence, usually densely arranged and exceeding the internodes, subsessile; cauline blades 2–5 cm long, 0.5–2 cm wide, ovate to broadly oblong, the upper ones narrowly ovate to lanceolate, strigillose especially on veins on both sides, sometimes with an admixture of glandular hairs in upper pairs, apex acute or subobtuse, base obtuse, rounded, or subcordate, the margin usually sharply serrulate with 7–15 teeth on each side, the lateral veins 4–6 on each side of the midrib, usually prominent beneath. Inflorescence erect, densely glandular and strigillose. Flowers erect; buds 4–8 mm long, 2.5–5 mm in diameter, narrowly ovoid, apiculate; ovaries 1.5–3 cm long, mostly glandular-pubescent; pedicels 0.4–0.7 cm long; floral tube 1–1.2 mm long, 1.8–3 mm in diameter, a ring of spreading white hairs at the mouth within; sepals 4–7 mm long, 1–1.2 mm wide, lance-oblong, apiculate, strigillose and glandular-pubescent; petals 6–8 mm long, 3–4.5 mm wide, pink to purple, obovate to obcordate, the notch 1–1.4 mm deep; filaments of longer stamens 2.5–3.5 mm long, those of shorter ones 1.8–2.5 mm long; anthers 0.7–1.3 mm long, 0.3–0.6 mm wide, ovoid, apiculate; style erect, 2.8–4 mm long, glabrous; stigma clavate or subcapitate, 2–3 mm long, 1–2.3 mm in diameter, sometimes elevated above anthers at anthesis or subequal with them. Capsules 3.5–7 cm long, predominantly glandular-pubescent, borne on pedicels 0.7–1.5 cm long. Seeds 1.5–1.8 mm long, 0.35–0.5 mm wide, narrowly obovoid, with a conspicuous chalazal collar 0.08–0.1 mm long, 0.25–0.3 mm wide, brown, the surface finely papillose; coma 7–12 mm long, reddish brown, rather persistent. Chromosome number: $n = 18$. Figs. 4c–e, 7b, 44.

FIG. 44. *Epilobium pyrricholophum.* a. Upper stem and inflorescence. a[1]. Portion of mid-stem. a[2]. Base of stem, with filiform stolon. b. Flower (one petal and three stamens removed). c. Leaf from inflorescence (bract). c[1]. Mid-cauline leaf. c[2]. Basal leaf. d. Capsule. e. Seed. f. Axils from inflorescence. (Based on *Guizhou Exped. 3625.*)

Phenology. Collected in flower from July to September, in fruit from August to November.

Distribution (Fig. 40). Japan, Ussuri area of the Soviet Union and eastern to southern China; in China in eastern Shandong, Henan, Shaanxi, southern Anhui, southern Jiangsu, Zhejiang, Jiangxi, Fujian, Guangdong, northern Guangxi, Hunan, western Hubei, eastern Sichuan, and Guizhou; in wet places along streams and valley bottoms, in moist weedy places and hillsides in mountainous regions; (150–) 300–1770 m.

ADDITIONAL SPECIMENS EXAMINED. **China.** SHANDONG: Muping, *T. N. Liou & K. M. Liou 1214* (PE), *1343* (PE).—HENAN: Luanchuan, *Henan Exped. 21087* (KUN); Jigong Shan, *K. K. Tsoong 4051* (PE).—SHAANXI: Nanzhen, *K. T. Fu 10943* (KUN).—ANHUI: Jing Xian, *Anhui Exped. 703* (NAS), *East China Station of Botany 6052* (NAS); Qianshan, *Anhui Exped. 7198* (NAS).—JIANGSU: Suzhou, 1933, *Migo s.n.* (NAS).—ZHEJIANG: Ningbo, 1888, *Faber s.n.* (HK); Shouchang, *B. L. Qiu 687* (HZBG); Longquan, *Longquan Exped. 6648* (PE), *R. H. Shan 5709* (NAS, PE), *Zhejiang Museum 3625* (KUN), *D. X. Zuo et al. 23496* (NAS); West Tianmu Shan, *X. Y. Ho 1088* (NAS); Mt. Tiantai, *Y. X. He 27953* (PE); Southwest Zhejiang, *R. C. Ching 2490* (BM, PE), *T. S. Gao 1511* (GH).—JIANGXI: Yushan, 1958, *M. X. Nei s.n.* (LBG); Shangrao, *M. X. Nei 4943* (LBG, PE); Yanshan, *M. X. Nie 4307* (KUN, LBG, PE); Jiujian, *Hemsley 897* (BM), *898* (BM); Lu Shan, *R. C. Ching 10173* (LBG), *Y. Lin 13695* (IBSC); 1941, *Migo s.n.* (NAS), *Schindler 352* (BM, E, G, K, NAS, W), *H. Y. Shan et al. 75161* (LBG), *Y. Tsiang 10760* (NAS), *M. J. Wang 991* (LBG), *Y. G. Xiong 6863* (LBG, NAS), *7382* (LBG), *9655* (LBG), *G. F. Zhu 43* (LBG); King Yuan, *R. C. Ching 2490* (E); Jing'an, *Y. G. Xiong 1412* (LBG); Tonggu, *S. S. Lai 3704* (PE); Xiushui, *S. S. Lai 2911* (PE), *J. Xiong 5946* (LBG), *Y. G. Xiong 4179* (LBG); Ningdu to Jian, *T. H. Wang 482* (WU); Pingxiang, *Y. G. Xiong 8406* (LBG); Anfu, *J. S. Yuo et al. 3287* (KUN, NAS, PE); Lianhua, *S. S. Lai 1640* (LBG, PE); Xingguo, *Jiangxi Pl. Exped. 552* (PE); Jinggang Shan, *Jiangxi Pl. Exped. 890* (PE), *S. S. Lai et al. 4540* (KUN, LBG), *J. S. Yue et al. 5123* (NAS, PE); Jiulian Shan, *Jiulian Shan Tree Farm 781304* (IBSC); Xunwu, *Y. Ling 15199* (IBSC, LBG, PE), *J. S. Yue et al. 2035* (KUN, NAS); Anfu, *J. S. Yue 3287* (KUN, NAS); Shangyou, *Jiangxi Pl. Exped. 222* (PE); Suichuan, *J. S. Yue 4225* (KUN, NAS, PE); Longnan, *Jiangxi Pl. Exped. 838* (PE).—FUJIAN: Pucheng, *G. Ho 2859* (FJTU); Chong'an, 1961, *Y. X. Huang s.n.* (FJTU), *Y. X. Huang 2279* (FJTU), *Wuyi Shan Exped. 428* (FJSI), *00540* (FJSI), *790032* (FJSI), *790835* (FJSI); Shaowu, *H. C. Zhou 6604* (IBSC); Luoyuan, *Ling 5439* (PE); Fuzhou, 1937, *Migo s.n.* (NAS); Dehua, *Fujian Exped. 377* (FJTU); 1931, *K. K. Tsoong s.n.* (PE); Liancheng, *D. S. Wang 1001* (FJTU, PE); Longyan, *Q. Q. Zhang 328* (FJTU); no locality, *R. Ling 3175* (PE).—GUANGDONG: Shantou, 1900, *Dalziel s.n.* (E); 1901, *Dalziel s.n.* (BM); Fengshun, *C. J. Chen & H. Y. Gu 88050* (PE), *88055* (PE); Renhua, *L. Deng 7686* (IBSC, KUN, NAS, PE); Lechang, *T. S. Chu 326* (IBSC); Wengyuang, *S. K. Lau 2636* (IBSC), *X. X. Liu 24173* (IBG, IBSC); Ruyuan, *J. X. Zhong 10947* (IBSC), *X. X. Liu 29173* (IBSC), *Y. G. Liu 619* (PE).—GUANGXI: Ziyuan, *X. Q. Wang 78034* (NAS); Xin'an, *J. Z. Chen 51413* (KUN, PE); Lingui, *X. F. Deng 181* (IBG); Damio Shan, *S. H. Chun 16631* (KUN, NAS, PE).—HUNAN: Ling Xian, *C. J. Chen 89522* (PE); Nanyue, *Y. Liu 512* (PE), *Herb. Hunan Teachers' Coll. 9248* (HTC); Yizhang, *S. H. Chun 2478* (IBSC), *B. H. Liang 83532* (IBSC); Dao Xian, *Inst. Wild Plants 7422* (HTC); Hengyang, *Yang 5726* (PE); Baoqing, *L. H. Liu 9746* (HTC, IBSC, KUN, PE); Yuaning, *L. H. Liu 8372* (IBSC); Cili Xian, *W. Hunan Exped. 480* (PE); Anjiang, *Anjiang Agr. School 36* (PE); Dongkou, *L. H. Liu & G. Z. He 16621* (HTC, IBSC); Xinning, *C. S. Fan & Y. Y. Li 627* (BM, G, GH, P, W); Zhijiang, *Wuling Exped. 1890* (PE).—HUBEI: Shennongjia, *Shennongjia Pl. Exped. 11481* (PE); Lichuan, *G. X. Fu & Z. S. Zhang 1661* (KUN, NAS, PE), *Lichuan Med. Exped. 1064* (HIMC); Hefeng, *H. Z. Li 5967* (KUN); Xianfeng, *Y. M. Wang 6469* (PE).—SICHUAN: Fengjie, *H. F. Zhou & H. Y. Su 110674* (NAS, PE), *110948* (PE, SZ), *111052* (NAS, PE).—GUIZHOU: Fanjing Shan, *Fanjing Shan Exped. 362* (GZBI), *Guizhou Exped. 3625* (PE), *Sino-Amer. Guizhou Bot. Exped. 1190* (A, PE), *1674* (A, PE), *Steward et al. 396* (E, K, PE); Shiqian, *Wuling Shan Exped. 2361* (PE); Liping, *K. Z. Yang 255* (GZBI); Dushan, *Y. Tsiang 6582* (NAS, NY, W); Nayong, *Bijie Exped. 438* (GZBI).

This species, one of the most characteristic of the East Asian *Epilobium* flora, is superficially similar to *E. brevifolium* subsp. *trichoneurum* and has been confused with it in earlier treatments. *Epilobium pyrricholophum* differs by its stolons, larger seeds with longer reddish brown comas, and the strong admixture of glandular hairs on the stem and inflorescence. This species is not as morphologically variable in China as it appears to be in Japan, where many names have been proposed; however, even in Japan, opinion now seems to favor a broader circumscription of the species (Hara 1965).

22. Epilobium amurense Haussknecht, Oesterr. Bot. Zeitschr. 29: 55. 1879. *Epilobium origanifolium* var. *pubescens* Maximowicz, Prim. Fl. Amur. 105. 1859.—TYPE: Amur River [Heilongjiang River, border between the Far East region of the U.S.S.R. and Heilongjiang Province, China], "Um die Bai de Castries, an sonnigen Mergelabhängen", 14 Jul 1854, *Maximowicz s.n.* (holotype: LE!; isotypes: G! K!).

Erect perennial herb. Stems (10–) 20–150 cm tall, simple or branched, covered with dense strigillose pubescence, often mixed with glandular hairs on upper parts, with 2 (or 4) raised, pubescent lines decurrent from margins of the petioles, more or less conspicuous through the internode, or very rarely stem subglabrous, subterete to somewhat two-edged. Leaves mostly opposite, alternate in the inflorescence, sessile or with petioles 1–6 mm long; cauline blades 2–9 cm long, 0.5–2.5 cm wide, ovate to narrowly so, or rarely lance-oblong, subglabrous except for sparsely strigillose veins and margins, apex acute or acuminate, base rounded or cuneate, the margins serrulate or denticulate, the lateral veins 4–6 on each side of the midrib, often prominent beneath. Inflorescence erect to somewhat nodding in bud; flowers suberect in bud; buds 3–4.8 mm long, 1.5–2.4 mm in diameter, elliptic-ovoid, apiculate; ovaries 1.2–4 cm long, strigillose, often mixed with glandular hairs, rarely subglabrous; pedicels 0.2–0.8 cm long; floral tube 0.6–1 mm long, 1.3–1.8 mm in diameter, with a ring of villous pubescence at the mouth within; sepals 3.5–6 mm long, 0.8–1.9 mm wide, lance-oblong, often keeled; petals 4.5–8 mm long, 2.4–4.5 mm wide, white, pink, or rose-purple, obovate, the notch 0.8–1.5 mm deep; filaments of longer stamens 2.8–4 mm long, those of shorter ones 1.2–2.8 mm long; anthers 0.4–0.7 mm long, 0.3–0.4 mm wide, ovoid; style 2–4.7 mm long, erect, glabrous or with a few glassy hairs at the base; stigma capitate or rarely broadly clavate, 1–1.5 mm long, 1–1.4 mm in diameter, anthers of longer stamens shedding pollen directly onto stigma at anthesis. Capsules 1.5–7 cm long, scattered strigillose or glabrescent, rarely glabrous, borne on pedicels 0.3–1.3 cm long. Seeds 0.8–1 mm long, 0.3–0.4 mm in diameter, obovoid, with a short inconspicuous chalazal collar, brown or dark brown, the surface coarsely papillose; coma 6–9 mm long, dull white, easily detaching. Chromosome number: $n = 18$.

Epilobium amurense is a conspicuously complex species that occurs nearly throughout eastern Asia; within it, we are recognizing here two broadly circumscribed subspecies, subsp. *amurense* and subsp. *cephalostigma.* The former occurs in Kamchatka, Japan, Korea, northeastern to southwestern China and Taiwan (excluding eastern and southern China), westwards along the Himalayan range to Kashmir and Hunza (Raven 1962a), whereas subsp. *cephalostigma* occurs in the Soviet Far East, Japan, Korea, northeastern and eastern China except Taiwan. Subspecies *amurense* grows at somewhat higher elevations (1310–3700 m) than does subsp. *cephalostigma* (800–2100 m), though they overlap considerably. Morphologically, the two taxa are similar in floral, fruit, and seed characters, and in having stems with two lines of hairs decurrent from the nodes. They differ primarily in stem and floral tube pubescence and leaf shape: subsp. *amurense* has conspicuous hairy lines on a subglabrous stem rather than the faint lines on a more uniformly strigillose stem found in subsp. *cephalostigma;* mixed glandular and strigillose pubescence on the inflorescence rather than strictly strigillose pubescence; tufts of hairs between the bases of the sepals on the floral tube rather than uniformly distributed hairs; and ovate rather than oblong to broadly lanceolate leaves.

In the region of overlap between the two taxa in Japan, Korea, the Soviet Far East, and northeastern and central China, most populations can be distinguished easily. Nevertheless, collections with mixed or intermediate leaf and pubescence characters can be found in many areas of range overlap, for example especially in Anhui (Huang Shan, *S. S. Chien 1284,* PE; "Jung-mo-pongo," *P. C. Tsoong 4093,* PE), Guangxi (Damiao Shan, Rongshui, *Q. H. Lu 2485,* IBK, PE; *Liuzhou Exped. 2594,* IBK), Hunan (Xuefeng Shan, *Z. T. Li 2870,* PE), Guizhou (Kaili, *S. Guizhou Pl. Exped. 1640, 2152,* NAS; Dafang, *Y. Tsiang 8954,* NY), and Yunnan (Mengzi, *Henry 10638,* HK). All individuals of *E. amurense* that we have examined in the field and in the greenhouse are modally autogamous, and this persistent self-pollination undoubtedly contributes to the maintenance of the two entities as distinct over a rather wide sympatric range.

We have not seen enough material of *E. amurense* from Japan to comment precisely about the degree of differentiation between the two subspecies there. Hara's treatment in the *Flora of Japan* (Hara 1965) indicates a very close relationship and distinguishes the taxa by the same characteristics that we are using here. In view of Hara's treatment and our examination of a representative sample of Japanese collections, as well as collections from Korea and the Soviet Far East, we feel justified in reducing *E. cephalostigma* to a subspecies of *E. amurense.*

Epilobium amurense is closely related to *E. wallichianum,* from which it can be distinguished by having a capitate rather than clavate stigma, the stem usually with two rather than four decurrent lines from the nodes, and ovate to lance-oblong leaves with acute or acuminate apex. The two species overlap substantially in the Himalayan region and much of southwestern China, but virtually all mature specimens are clearly one or the other species, with very few intermediates.

22a. Epilobium amurense subsp. **amurense.**

Epilobium laetum Wallich ex Haussknecht, Monogr. Epil. 218. 1884. [*Epilobium laetum* Wallich, Numer. List 216, n. 6329. 1832, nomen nudum, quoad A.] *Epilobium amurense* subsp. *laetum* (Wallich ex Haussknecht) Raven, Bull. Brit. Mus. (Nat. Hist.) Bot. 2: 367. 1962.—TYPE: NEPAL. Towards Gosainthan, 1821–1822, *Wallich 6329A* (lectotype, here designated: K!; isolectotypes: BM! in part, G!). The holotype noted by Haussknecht was destroyed in Berlin in 1945; we select the lectotype from the isotypes.

Epilobium nepalense Haussknecht, Oesterr. Bot. Zeitschr. 29: 53. 1879.—TYPE: NEPAL (without definite locality), 1821–1822, *Wallich s.n.* (lectotype, designated by Raven 1962a: C!). As noted by Raven (1962a), this unnumbered sheet may be a duplicate of *Wallich 6329A,* the type collection of *Epilobium laetum* Wallich ex Haussknecht.

Epilobium tenue Komarov, Trudy Imp. S.-Peterburgsk. Bot. Sada 25: 95. 1905.—TYPE: KOREA (no specific locality), *Komarov s.n.* (lectotype, designated by Steinberg 1949: LE).

Epilobium yabei H. Léveillé, Bull. Soc. Agric. Sarthe 40: 72. 1905.—TYPE: JAPAN. Hokkaido (Yezo), Junsainuma, Aug 1903, *Faurie 5396* (lectotype, here designated: KYO!).

Epilobium gansuense H. Léveillé, Bull. Herb. Boiss., sér. 2, 7: 590. 1907.—TYPE: JAPAN. "Sommet du Gansu," 12 Aug 1898, *Faurie 1307* (lectotype, here designated: G!; isolectotypes: K! KYO!).

Epilobium miyabe H. Léveillé, Repert. Spec. Nov. Regni Veg. 5: 8. 1908.—TYPE: JAPAN. Hokkaido, Sapporo, 9 Jul 1891, *Miyabe s.n.* (holotype: MO!).

Epilobium ovale Takeda, J. Linn. Soc., Bot. 42: 466. 1914.—TYPE: U.S.S.R. Kurile Islands. Shikotan, Anama, in wet places near the sea, Jul 1909, *Takeda s.n.* (not located).

Common name: Maomai Liuyecai (Li 1986).

Erect perennial herb, in autumn forming short leafy soboles, elongated rosettes, or rarely fleshy stolons. Stems (10–) 20–50 (–80) cm tall, 1.5–4 mm in diameter, simple or sparingly branched, moderately strigillose on upper part with admixture of glandular hairs, the lower part sparsely strigillose with two raised densely strigillose lines decurrent from margins of petioles, or rarely the stem subglabrous. Leaves subsessile or the lower pairs with petioles 1–4 mm long; cauline blades 2–7 cm long, 0.5–2.5 cm wide, ovate, rarely oblong or lanceolate to obovate below, apex acute, sometimes subacuminate or subobtuse, base rounded or broadly cuneate, the margin usually sharply serrulate with 6–25 teeth on each side. Inflorescence nodding to suberect, strigillose and sometimes with an admixture of glandular hairs. Flowers suberect; buds sparsely strigillose with scattered glandular hairs; ovaries 1.5–2.8 cm long, strigillose with an admixture of glandular hairs; floral tube 0.6–0.9 mm long, 1.5–1.8 mm in diameter; sepals 3.5–5 mm long, 0.8–1.9 mm wide, sparsely strigillose, with definite tuft of hairs at junctures of sepal bases; petals 5–8 (–10) mm long, 2.4–4.5 mm wide, white to rose or purple; style 2–4.7 mm long, sometimes with a few spreading hairs near the base; stigma subcapitate, 1–1.5 mm long, 1–1.3 mm in diameter. Capsules 1.5–7 cm long, sparsely pubescent to glabrescent, borne on pedicels 0.3–1.2 cm long. Seeds 0.8–1 mm long, 0.3–0.4 mm wide, dark brown; coma 6–9 mm long. Chromosome number: $n = 18$. Figs. 3a, 4a, b, 45.

Phenology. Collected in flower from July (rarely from May) to August, in fruit from August (rarely from June) to October (rarely to December).

Distribution (Fig. 25). Kamchatka, Soviet Far East region, Kurile Islands, Japan, northern Korea, northeastern to southwestern China, westwards through the Himalaya to Kashmir and Hunza; in China in Jilin, Nei Monggol, Hebei, Shanxi, Shangdong, Henan, Shaanxi, eastern Gansu, Qinghai, Taiwan, northern Guangxi, Hunan, Hubei, Sichuan, Guizhou, Yunnan, and southern Xizang; in wet places by streams and springs, on moist grassy slopes, and in disturbed places in the mountains; 1300–2000 m in northern China, 1800–4200 m in western China.

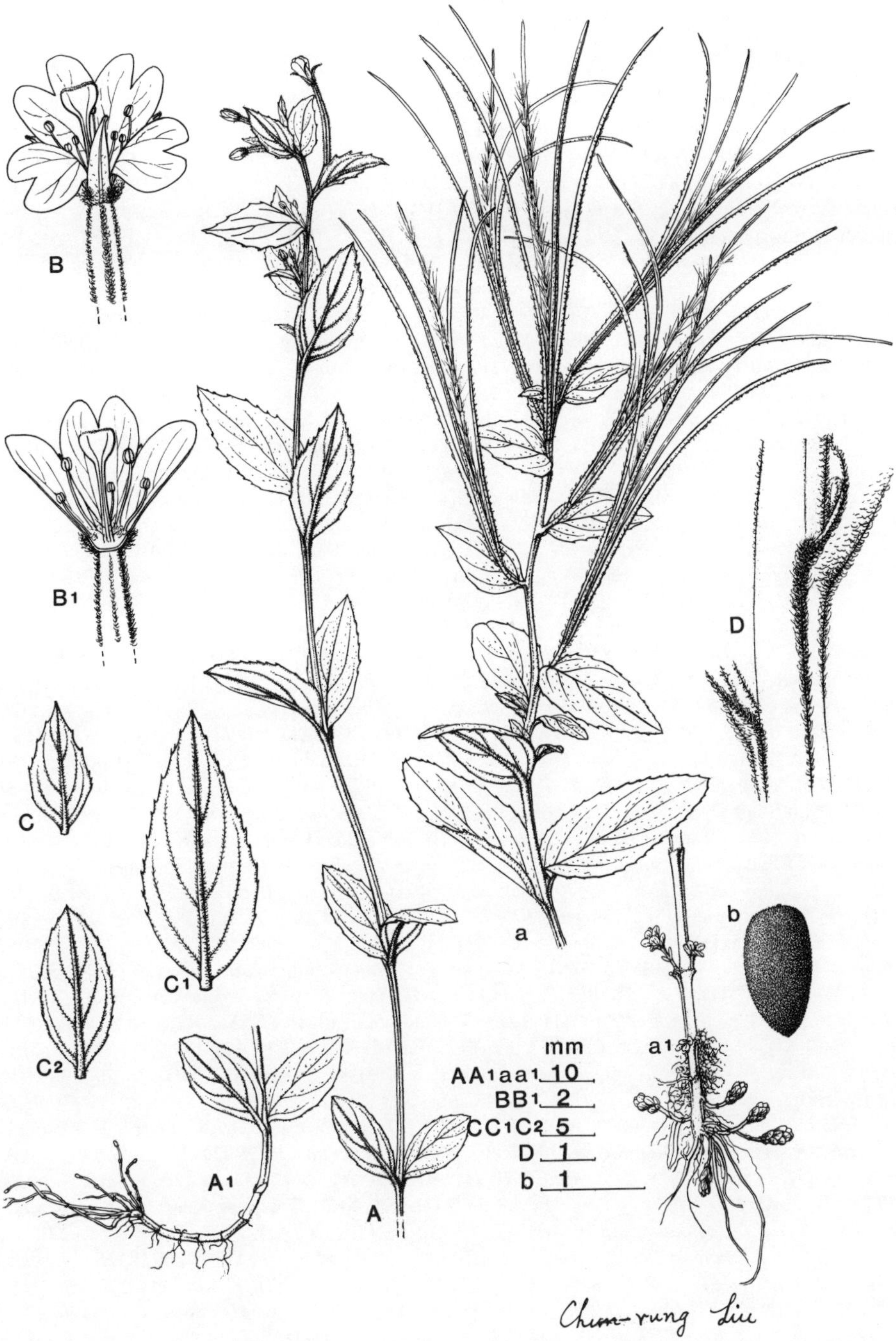

FIG. 45. *Epilobium amurense* subsp. *amurense*. A. Upper portion of stem and suberect inflorescence. A¹. Base of stem with filiform stolons. B. Flower, showing tuft of hairs at insertion of sepals on floral tube. B¹. Flower (one petal and three stamens removed); note hairs at mouth of floral tube and near base of style. C. Leaf from inflorescence (bract). C¹. Mid-cauline leaf. C². Basal leaf. D. Axils from inflorescence. a. Habit: upper stem and inflorescence. a¹. Base of stem with short stolons terminating in fleshy turions. b. Seed. (A–D based on *G. H. Yang 58825;* a–b based on *S. S. Chang & Y. X. Ren 7165*.)

ADDITIONAL SPECIMENS EXAMINED. **China.** JILIN: Changbai Shan, *Changbai Shan Exped. 954* (PE), *T. N. Liou 1516* (IFP, PE); 1963, *W. Wang et al. s.n.* (IFP), *W. Z. Zhang 434* (NETU); Fusong, *Y. L. Zhang et al. 414* (IFP).—NEI MONGGOL: Wula Shan, *Q. R. Wu et al. 96* (NMU).—HEBEI: Mt. Wuling, *Z. F. Fang 959* (IFP), *W. X. Zhang et al. 246* (PE); Dongling, *C. G. Yang 1371* (PE); Chicheng, *J. W. Wang 57* (PE); Beijing, *J. C. Liu 1113* (UC), *1131* (UC), *D. F. Jin 369* (PE); Xiaowutai Shan, *H. W. Kung 1216* (NY), *Smith 134* (S, UPS), *136* (UPS), *137* (UPS), *1083* (UPS), *1172* (UPS), *C. W. Wang 62525* (GH, PE), *Xiaowutai Shan Exped. 1872* (PE), *5041* (PE); Laiyuan, *Hebei Univ. Agr. 42252* (PE), *43613* (PE).—SHANXI: Mt. Hongtas, *K. C. Kuan & Y. L. Chen 800* (PE); Wutai Shan, *Hancock 69* (K); 1959, *K. C. Kuan & Y. L. Chen* (PE); Ningwu, *W. Y. Hsia 1575* (PE), *Shanxi Exped. 1965* (PE); Wuzhai, *Huang He Exped. 2903* (PE); "Yuan-chu dist.: Shui-wang-ping," *Smith 6541* (UPS); Luya Shan, *Smith & Chin 8108* (UPS), *Smith 8296* (UPS); Jiaocheng, *Smith 7232* (UPS), *7305* (UPS); Xi Xian, *T. P. Wang 3178* (PE); "Ho Schan," *Licent 12121* (W); Guandi Shan, *X. Y. Liu 21497* (PE); N.E. Shanxi, *T. Tang 1087* (PE).—SHANDONG: Tai Shan, *Clemens 1448* (E).—HENAN: Lingbao, *J. Q. Fu 108* (NAS); Lushi, *K. M. Liou 4584* (PE), *5124* (PE); Luanchuan, *Henan Exped. 20164* (PE).—SHAANXI: Hu Xian, *B. Z. Guo 632* (PE), *827* (PE), *K. T. Fu 14392* (WUK), *14556* (WUK); Qinling, *Limgreft 2721* (WU), *2825* (WU); Taibai Shan, *T. N. Liou & P. C. Tsoong 766* (PE), *943* (LZU), *1522* (PE), *Z. X. Peng & X. Z. Zhang 647* (LZU), *1020* (LZU), *1237* (LZU), *Z. X. Peng 459* (LZU), *518* (LZU), *558* (LZU), *716* (LZU), *Qin Ling Exped. 401* (PE, WUK); Foping, *B. Z. Guo 1728* (CDBI, PE), *T. H. Liou & P. C. Tsoong 3229* (KUN, PE); Yang Xian, *K. T. Fu 5169* (PE), *J. X. Yang 1479* (PE, WUK); Lueyang, *K. T. Fu 5819* (KUN, PE, WUK).—GANSU: Tianshui, *J. M. Liu 10481* (KUN, PE), *Z. W. Zhang 192* (NAS, WUK), *Tianshui Exped. 151* (LZU), *K. L. Chen 184* (LZU), *T. H. Lu 274* (LZU); Xinglong, *S. B. Xie 47* (LZU), *Lanzhou Univ. Exped. 177* (LZU), *223* (LZU), *C. J. Wang & Y. S. Lian 10174* (PE), *S. R. Wang 010* (LZU); Hui Xian, *Z. X. Peng 5508* (LZU), *5847* (LZU), *7380* (LZU); Kang Xian, *Z. Y. Zhang 16652* (PE); Wen Xian, *W. L. & C. J. Chen 8818* (PE), *8829* (MO, PE), *8743* (MO, PE), *Wen Xian Exped. 2204* (LZU), *Y. G. Zhu et al. 0431* (LZU); Xiaohe, *T. P. Wang 5728* (PE), *T. P. Wang 7155* (PE); Xigu, *T. P. Wang 14493* (PE), *14577* (PE); Yuzhong, *Huang He Exped. 3154* (PE), *3322* (HNWP), *5981* (KUN, PE), *H. P. Tsui et al. 80-369* (PE); near "Chuoni," *T. P. Wang 5285* (PE), *5351* (PE); Min Shan, *T. P. Wang 7664* (PE); Lintan, 1950, *Lanzhou Univ. s.n.* (LZU), *Z. G. Song 453* (LZU); "South of Old Taochow," *R. C. Ching 909* (E, NY, US); Lianhua Shan, 1950, *Lanzhou Univ. Lianhua Shan Exped. s.n.* (LZU); Tianshui, *C. W. Zhang 192* (PE); Pingfen, *R. C. Ching 575* (E, US); Liancheng, *Qinghai-Gansu Exped. 1440* (PE), *3527* (PE).—QINGHAI: Ledu, *Qinghai & Gansu Exped. 1440* (HNWP); Menyuan, *P. C. Tsoong 10007* (PE); Qinghai Lake, *Y. R. Ling et al. 80-158* (PE); Datong, *F. Liang 505* (HNWP), *K. M. Liou 6139* (PE); Zekog, *L. H. Zhou & L. N. Sun 1974* (HNWP), *He Exped. 3154* (PE), *3322* (PE), *5981* (PE).—TAIWAN: Taibei Xian, *Suzuki-Tokio 8711* (NAS); 1932, *Suzuki-Tokio s.n.* (TAI); 1933, *Suzuki-Tokio s.n.* (TAI); Yilan Xian, *C. S. Kuoh 3045* (MO, TAI), *C. C. Hsu 5916* (TAI), *S. Suzuki 243* (TAI); 1925, *S. Suzuki s.n.* (TAI), *S. Suzuki 5081* (TAI); 1930, *S. Suzuki s.n.* (TAI), *M. T. Kao 5241* (TAI); 1969, *T. Yamazaki et al. s.n.* (TAI); 1938, *S. Hibino et al. s.n.* (TAI); *S. Suzuki 5643* (PE), 1940, *Masamume et al. s.n.* (TAI), *Tamura & Shimizu 20727* (SHIN), *20728* (SHIN), *Tamura et al. 21377* (SHIN); Xinzhu Xian, *M. T. Kao 8506* (TAI), *C. F. Hsieh 1332* (TAI), *Shimizu 20253* (SHIN), *20358* (S, SHIN), *20379* (SHIN), *Matsuda 18284* (TAIF); Hualian Xian, *C. C. Hsu & R. Hsu s.n.* (TAI), *Steenis 20726* (LE); 1956, *Liu et al. s.n.* (TAI), *M. T. Kao 4125* (TAI); 1956, *Chen & Kao s.n.* (TAI), *C. C. Hsu & R. Hsu 9291* (TAI), *T. C. Huang 4206* (TAI), *Kawakami 81728* (TAI); *C. M. Kuo 8429* (TAI), *C.-I Peng 5872* (HAST, MO); Taizhong Xian, 1937, *Yamamota & Mori s.n.* (TAI), s.d., *Skikamo s.n.* (TAI), *G. L. Liu et al. 311* (TAI), 1923, *S. Sasaki s.n.* (TAIF), *Shimizu & C. C. Chuang 20190* (SHIN), *20204* (SHIN), 1936, *Mori s.n.* (TAI), 1931, *Hosokawa s.n.* (TAI), *T. C. Huang 1343* (SHIN, SMU, TAI), *C. M. Kuo 8332* (TAI), *Kimura et al. 57* (SHIN), *77* (SHIN), *78* (SHIN), *T. S. Liu et al. 290* (GH, TAI); Nantou Xian, *C. C. Hsu 6111* (TAI), *9202* (TAI), *T. C. Huang & M. T. Kao 5731* (MO), *T. C. Huang et al. 5730* (TAI), *5774* (TAI), *Tamura & Koyama 23190* (S), *23270* (S), *23291* (S); 1907, *Mori s.n.* (TAI); 1930, *Kudo & Mori s.n.* (TAI); 1926, *S. Suzuki s.n.* (TAI), *Tamura et al. 22001* (SHIN), *22043* (S), *C. C. Hsu 6111* (GH); Taiheishan, *Gressitt 382* (BM, G, GH, K, MICH, S, U), *Ohwi 2341* (K); 1929, *Suzuki s.n.* (S); Yushan, *Tamura et al. 22152* (S, SHIN), *Tamura et al. 22136* (SHIN); 1969, *Yamazaki & Yamuzaki s.n.* (TAI); Jiayi Xian, *T. C. Huang 8513* (TAI), 1930, *Wakatsuki s.n.* (TAI), *C. C. Hsu & R. Hsu 3950* (TAI), *C. C. Hsu 5368* (TAI), *M. T. Kao 4899* (MO), *C. S. Kuo 5738* (TAI), *80405* (TAI), *Kuo 80293* (TAI), *J. M. Chao & M. T. Kao 6181* (TAI), 1965, *J. M. Chao & M. T. Kao s.n.* (TAI), *C. C. Chuang & Chuma 8862* (TAI), *C. C. Chuang & M. T. Kao 4219* (PH, TAI), *Faurie 561* (G, GH, P), *Gressitt 108* (GH), *140* (GH), 1956, *T. C. Huang s.n.* (TAI), 1957, *T. I. Chuang & Kao s.n.* (TAI), 1932, *Sashioka s.n.* (TAI), 1939, *Nakamura-Taizo s.n.*

(TAI), 1925, *S. Sasaki s.n.* (TAI), *C. C. Chuang 2952* (MO, TAI), *M. T. Kao 7581* (TAI), *C. S. Kuo 5738* (TAI), *Tanaka 501* (BM, C), *Steenis 20863* (L); Taidong Xian, *T. C. Huang 7078* (TAI); Yushan (Mt. Niitaka), 1933, *Y. Hashioka s.n.* (TAI); Tienchi to Yiakou, *T. C. Huang et al. 8906* (TAI); "Bunkito," *Faurie 559* (G, P); "Hassen-zan," *Kanehit 21160* (UC); Gaoxiong Xian, *T. C. Huang 6521* (TAI); 1972, *T. C. Huang s.n.* (TAI); *C.-I Peng 8627* (HAST, MO); 1937, *Yamamoto & Mori s.n.* (TAI).—Guangxi: Ziyuan, *Z. S. Chung 83512* (GH), *X. J. Wang 78080a* (NAS).—Hunan: Yizhang, *B. H. Liang 83783* (IBSC); Badagun Shan, *H. J. Li 4364* (KUN, PE).—Hubei: Shennongjia, *R. H. Huang 142* (PE), *H. J. Li 1506* (PE), *4364* (PE), *Shennongjia Pl. Exped. 10816* (PE), *11062* (PE), *22467* (PE), *31650* (PE), *31784* (PE), *32213* (PE), *32266* (PE), *32274* (PE); Xingshan, *H. J. Li 556* (KUN, PE), *1152* (PE); W. Hubei, *Henry 6141* (E), *6409* (K), *6941* (GH), *7320* (P); Lichuan, *G. X. Fu & Z. S. Zhang 1589* (KUN, NAS, PE); Badong, *Wilson 1324* (E); Xuan'en, *Y. M. Wang 5492* (HIB).—Sichuan: Wuxi, *G. H. Yang 58825* (PE); Wushan, *H. F. Zhou & H. Y. Su 108976* (SZ), *H. F. Zhou & H. Y. Su 109962* (SZ), *110151* (PE, SZ); Kai Xian, *T. L. Dai 101750* (CDBI, PE); Chengkuo, *T. L. Dai 103867* (PE), *105811* (PE, SZ), *Farges 726* (P), *105811* (CDBI, PE), *106863* (PE), *107074* (PE); Shizhu, *Y. Chen 2025* (CDBI), *W. H. Wang 1582* (CDBI), *2354* (CDBI); Qianjiang, *Z. C. Zhao 88-1769* (CDBI, PE); Fuling, *Z. P. Huang et al. 2326* (KUN); Nanchuan, *C. J. Chen & Z. Y. Liu 83087* (PE), *C. H. Chu 1112* (PE), *S. Y. Hao 4171* (NAS), *G. F. Li 61230* (KUN, PE, SZ), *J. H. Xiong & Z. L. Zhou 92459* (PE); Qijiang, *W. P. Fang 1309* (K); Pingwu, *W. P. Fang 4190* (GH, K); Nanping, *W. L. Chen & C. J. Chen 8547* (MO, PE), *8548* (MO, PE), *8549* (MO, PE), *8584* (PE), *8585* (PE), *8695* (MO, PE), *J. T. Li & X. C. Zao 2893* (PE); Songpan, *W. L. Chen & C. J. Chen 8315* (MO, PE), *8480* (MO, PE), *Smith 2649* (UPS), *2844* (UPS); Heishui, *X. Li 73359* (NAS), *73415* (NAS, PE), *73485* (NAS), *73487* (PE), *73775* (PE); Aba, *W. L. Chen & C. J. Chen 7910* (MO, PE); Shuajinsi, *H. Lee 74573* (PE), *Sichuan Univ. 23603* (PE, SZ); Barkam, *Barkam Pl. Exped. 23144* (PE, SZ), *Sichuan Inst. Forest. 196* (CDBI), *Z. L. Wu 32772* (PE), *Sichuan Grassland Exped. 22458* (NAS), *22482* (NAS); Mao Xian, *F. T. Wang 21813* (GH, PE), *21952* (GH, PE), *Z. He & Z. L. Zhou 14071* (PE), *Maowen Pl. Exped. 11021* (CDBI, PE), *F. T. Wang 22051* (PE), *22052* (KUN), *Chengdu Inst. of Biol. 11047* (CDBI, PE); Wenchuan, *K. Y. Lang et al. 1218* (KUN, PE); Li Xian, *T. L. Chow & C. Ho 13244* (PE), *W. P. Fang & W. K. Hu 53-977* (SZ), *D. P. He 44646* (SZ), *T. P. Soong 46265* (SZ); Xiaojin, *S. S. Chang & Y. X. Ren 6273* (CDBI, PE), *7165* (PE), *7186* (PE), *Xiaojin 6726* (CDBI); Jinchuan, *X. Li 75173* (NAS, PE, SZ), *75175* (NAS, PE, SZ), *75542* (NAS, PE), *76964* (NAS, PE), *77979* (NAS, PE, SZ), *78204* (NAS, PE, SZ), *78275* (NAS, PE), *G. Z. Zhu & X. Li 75743* (PE); Miyaluo, *D. P. He 45581* (SZ), *Z. L. Wu 33369* (PE); Tianquan, *X. L. Jiang 37969* (PE, SZ); Garze, *Sichuan Med. Pl. Exped. 15205* (NAS); Luding, *Luding Pl. Exped. 23113* (CDBI), *G. Y. Xu 23109* (CDBI); Baoxing, *K. L. Chu 3182* (PE, W), *3491* (BM, PE), *S. S. Chang & Y. X. Ren 4832* (PE), *5464* (PE), *7519* (NAS), *7556* (PE), *Baoxing Pl. Exped. 7192* (CDBI, PE), *7246* (CDBI); Kangding, s.d., *Bonvalot & Orleans s.n.* (P); 1923, *Cunningham s.n.* (E), *W. P. Fang 6058* (GH, K, P, PE), *6635* (NAS), *7304* (NAS), *8466* (NAS), *8911* (NAS), *11104* (NAS), *C. S. Liu 1061* (PE), *1062* (PE), *K. C. Kuan et al. 86* (PE), *H. S. Kung 6007* (MO, PE), *6052* (MO, PE), *6093* (MO), *6203* (MO), *Smith 10747* (UPS), *11118* (S, UPS); Yajiang, *Zh. P. Huang et al. 538* (PE); Mt. Emei, *H. C. Chow 12265* (GH), *W. P. Fang 1309* (P), *2647* (E, GH, K, P), *2715* (E, GH, K), *2717* (E, GH, PE), *2823* (E, GH, K, NY, P), *2884* (E, PE), *2885* (E), *12794* (BM, GH, PE, SZ, W), *12819* (BM, GH, PE, SZ, W), *17129* (PE), *17694* (PE), *20383* (PE), *S. Y. Hoo 5917* (NAS), *6056* (NAS), *6152* (NAS), *K. C. Kuan et al. 1622* (CDBI, PE), *1646* (CDBI, PE), *1842* (CDBI, PE), *2085* (PE), *2704* (PE), *T. H. Tu 522* (PE), *F. T. Wang 23343* (GH, P, PE), *Reed 286* (K), *288* (K), *Wilson 4916* (BM, K, P), *W. G. Fu 8257* (PE), *T. N. Liou & C. Wang 1092* (PE), *L. Y. Tai 24* (GH), *G. F. Li 52562* (SZ), *Y. H. Tauo 51751* (SZ), *G. H. Yang 56132* (NAS), *56509* (PE), *Sichuan Medical Pl. Exped. 12731* (NAS), *Raven et al. 26102* (MO, PE), *26105* (MO, PE), *26107* (PE), *26109* (MO, PE), *26113* (MO, PE), *26118* (MO, PE), *26120* (MO, PE), *26121* (MO, PE), *26124* (MO, PE), *26125* (MO, PE), *26128* (MO, PE), *26135* (MO, PE), *26136* (PE), *26140* (MO, PE), *26142* (MO, PE), *C. Y. Wu & C. J. Chen 74* (KUN, PE), *96* (KUN, PE), *C. Y. Wu et al. 100* (KUN, PE); Ebian, *Z. T. Guan 6623* (PE), *C. W. Yao 4398* (PE); Ya'an, *Nanshuibediao Exped. 10112* (PE); Erlang Shan, *W. K. Hu 34657* (PE), *37799* (PE, SZ), *S. Jiang & C. L. Jing 1801* (PE), *1817* (PE), *Yunnan Inst. Bot. 781* (KUN); Shimian, *C. C. Hsieh 41458* (PE, SZ), *42516* (PE, SZ), *42633* (PE), *42677* (PE); Dege, *Y. W. Tsui 4360* (KUN, PE); Ganluo, *Sichuan Econ. Pl. Exped. 4033* (PE), *4161* (PE), *4176* (PE); Mianning, *W. L. Chen & C. J. Chen 6189* (MO, PE), *S. F. Zhu 20430* (PE), *20431* (PE); Yajiang, *K. Huang et al. 633* (KUN), *Sichuan Exped. 2631* (KUN, PE); Hongxi, *Z. T. Guan 6981* (PE), *7477* (PE), *9086* (PE); Meigu, *Sichuan Econ. Pl. Exped. 1727* (PE), *Meigu Pl. Exped. 13016* (CDBI, PE), *13071* (CDBI, PE); Leibo, *Sichuan Econ. Pl. Exped. 738* (PE), *T. T. Yu 3552* (PE); Xichang, *W. L. Chen & C. J. Chen 6128* (MO, PE); Puxiong, *Sichuan Econ. Pl.*

Exped. 4561 (PE); Butuo, *Sichuan Econ. Pl. Exped. 5913* (CDBI, PE), *Butuo Pl. Exped. 13739* (CDBI, PE); Jinyang, *Jinyang Pl. Exped. 13468* (CDBI); Huili, *Sichuan Veget. Exped. 12096* (CDBI, PE); Yanyuan, *W. L. Chen & C. J. Chen 5607* (MO, PE), *Qinghai-Xizang Exped. 12208* (KUN, PE), *12560* (KUN, PE), *12657* (PE); Muli, *W. L. Chen & C. J. Chen 5698* (MO, PE), *Sichuan Forestry Exped. 125* (PE), *Muli Pl. Exped. 7611* (CDBI), *Qinghai-Xizang Exped. 13028* (KUN, PE), *13050* (PE), *14333* (KUN, PE), *Q. S. Zhao 6210* (CDBI, PE), *8662* (CDBI, PE); *Q. S. Zhao et al. 8099* (CDBI), *8276* (CDBI); Jiulong, *Q. Q. Wang 20276* (CDBI), *21528* (CDBI); Daucheng, *Z. G. Liu 285* (CDBI), *Sichuan Veget. Exped. 2525* (CDBI, PE); Xiangcheng, *Qinghai-Xizang Exped. 3095* (KUN, PE), *3724* (KUN, PE), *4852* (KUN, PE), *Sichuan Veget. Exped. 2041* (PE), *3018* (CDBI).—GUIZHOU: Fangjing Shan. *Sino-American Bot. Exped. 419* (A, PE), *432* (A, PE), *481* (A, PE), *Sino-English Fanjing Shan Exped. F066* (PE), *Steward et al. 396* (GH, NY), *C. P. Chien et al. 30758* (GZBI, PE), *30924* (PE), *32107* (GZBI, PE); Dafang, *Bijie Exped. 844* (GZBI, PE), *904* (PE); Pan Xian, *Anshun Exped. 920* (PE).—YUNNAN: Yongshan, 1964, *N.-E. Yunnan Pl. Exped. s.n.* (KUN); "Tchen-fong-chan," 1893, *Delavay s.n.* (P); "San Kong," *Delavay 931* (P); Qiajia, *Ducloux 4158* (P), *N.-E. Yunnan Exped. 1427* (KUN), *B. X. Sing 888* (PE, YNU); Dongchuan, *S. B. Lang 30* (PE), s.d., *Maire s.n.* (S, W); *N. E. Yunnan Exped. 514* (KUN); Songming, *Sino-American Bot. Exped. 1398* (MO, PE), *B. Y. Qiu 54938* (KUN), *54949* (KUN); Kunming, *H. Li 565* (KUN), *Murata & Yamazaki 9203* (BM); "Kiu-tchang-chan," *Maire 3290* (S); S of Yuanjiang, *Henry 9481* (MO, NY); Binchuan ("Piz tchouan"), *Ducloux 7089* (P), *B. X. Sing 11* (YNU), *69* (YNU), *97* (YNU); Xinping, *Xinping Exped. 32* (YNU); Eryuan, *R. C. Ching 23258* (KUN); Dali, *Forrest 4464* (E), *Sino-American Bot. Exped. 715* (PE), *8606* (PE), *H. C. Wang 2462* (KUN), *Yunnan Univ. 51* (YNU), *Zhongdian Pl. Exped. 63-3721* (KUN); Yangbi, *Zhongdian Pl. Exped. 63-4285* (KUN); Tengchong, *Forrest 8298* (E), *Howell 223* (E); Lanping, *H. T. Tsai 54094* (GH, KUN, NAS, PE); Pianma, *S. K. Wu 8069* (KUN), *8260* (KUN); Heqing, *Schneider 2284* (G), *R. C. Ching 24463* (KUN); Lijiang, *Forrest 2589* (BH, E), *2820* (E); *R. C. Ching 21093* (GH, KUN), *22070* (GH), *K. M. Feng 22692* (KUN, PE), *Hengduan Shan Exped. 2579* (PE), *Rock 6434* (UC), *10618* (E), *Schneider 3297* (G, K), *C. W. Wang 71277* (PE), *71653* (GH, KUN, PE), *Yulong Shan Exped. 350* (YNU), *D. L. Zhang & J. W. Yu 100859* (KUN), *Zhongdian Pl. Exped. 648* (KUN, PE, YNU), *R. Y. Xion et al. 610070* (KUN), Lijiang Exped. 100980 (KUN); Weixi, *Hengduang Shan Exped. 1571* (PE), *1612* (PE), *1817* (PE), *H. T. Tsai 53119 GH, KUN), 57659* (GH, KUN, NAS, PE), *58329* (KUN), *59706* (NAS, PE), *59868* (PE), *59909* (GH, KUN, NAS), *63119* (NAS, PE), *C. W. Wang 64197* (GH, KUN, NAS, PE), *64202* (GH, KUN, NAS, PE), *67635* (NAS), *67909* (NAS), *68117* (GH, KUN, NAS, PE), *68132* (GH, KUN, PE), *K. M. Feng 8806* (KUN); Zhongdian, *K. M. Feng 1958* (GH); Gongshan, *X. F. Deng 791435* (KUN), *791456* (KUN), *C. W. Wang 66762* (GH), *67220* (KUN), *K. M. Feng 8402* (KUN); Deqen, *K. M. Feng 5908* (KUN), *23990* (KUN), *Hengduan Shan Exped. 2305* (PE), *2307* (PE), *2344* (KUN, PE), *2562* (PE), *2978* (KUN, PE), *Kunming Bot. Inst. 23990* (PE), *C. W. Wang 69931* (KUN), *T. T. Yu 10163* (KUN); Fugong, *Q. Ling 791675* (KUN), *Qinghai-Xizang Exped. 2978* (PE), *H. T. Tsai 59104* (GH, NAS, PE); Bijiang, *H. T. Tsai 54107* (GH, KUN, NAS, PE), *58328* (PE), *58329* (GH, NAS), *Bijiang Exped. 1689* (KUN); Lushui, *Bijiang Exped. 1503* (KUN); "Hichiung Fu," *Delessent 2284* (G); Jingdong, *C. W. Wang 65285* (NAS): "Yundshi Mtn," *Rock 23465* (NY); S of Red River, *Henry 9481* (E), *Ten 18* (UC); no locality, *Ducloux 522* (K), *Maire 396* (UC), *Y. Tsiang 21952* (NAS).—XIZANG: Yigong, *D. Y. Hong & J. S. Ying 0729* (PE); Gyirong, *Xizang Med. Pl. Exped. 326* (PE); Jomda, *Xizang Exped. 2259* (HNWP); Zayu, *C. W. Wang 651285* (GH, PE), *Xizang Exped. 961* (HNWP), *1216* (HNWP), *1235* (HNWP); East Xizang, *Forrest 2589* (BM); Baxoi, *Soulie 1300* (P); 1909, *Monbeig s.n.* (P); Bomi, *Y. T. Chang & K. Y. Lang 478* (PE), *D. Y. Hong & J. S. Ying 1225* (PE), *Xizang Exped. 810* (HNWP); Medag, *W. L. Chen 10686* (PE); Nyingchi, *Ecol. Group Xizang Exped. 15566* (PE); Mainling, *P. S. Li et al. 6246* (PE), *Ludlow et al. 5371* (BM, E), *Xizang Med. Pl. Exped. 3902* (PE), *Qinghai & Xizang Exped. 74-1942* (PE), *750893* (PE); Gongbo-gyamda (Baiba), *B. Q. Zhong 6760* (PE), *6763* (PE); Longzi, *K. H. Fu 1182* (PE); Cona, *Qinghai-Xizang Exped. 751583* (KUN, PE); Yadong, 1897, *Hobson s.n.* (K), *Qinghai-Xizang Exped. 74-2395* (KUN, PE).

This subspecies is extremely variable in stature, leaf shape, length of petioles, and flower color, but it is distinguished from subsp. *cephalostigma* primarily by its distinctive vestiture. Raven (1962a) recognized *E. amurense* subsp. *laetum* in the Himâlayan region to distinguish the mostly white-flowered plants in that region from the mostly rose-flowered subsp. *amurense* from further north. After seeing more plants from throughout the range, however, he determined that there was no "advantage in recognizing two geographical subspecies" in view of the extreme variability of this character (Raven 1977). We continue to hold that opinion.

The major disjunction in the distribution of *E. amurense* subsp. *amurense* between Taiwan and the western and northeastern provinces on the mainland is most striking, especially when compared with the distribution of subsp. *cephalostigma* in the intervening provinces of southeastern China but not in Taiwan. Similar patterns of disjunct distribution between the central provinces and Taiwan are found also in *E. platystigmatosum* and *E. brevifolium* subsp. *trichoneurum,* in *Circaea glabrescens* (Boufford 1982), and in some other groups. These taxa of *Epilobium* persist in Taiwan in cool montane regions mostly above 2000 m elevation; the closest comparable montane areas on the mainland are in the Luoxiao Shan of eastern Hunan, some 700 km to the west.

A single collection of *E. amurense* subsp. *amurense* from Xiangcheng, Sichuan (*Qinghai-Xizang Exped. 3095,* KUN, PE) is essentially glabrous, which is striking in view of the many typically pubescent collections we examined from Sichuan. We consider it analogous to "*E. nudicarpum*" (=subsp. *cephalostigma*) or *E. royleanum f. glabrum,* in the sense that none of these seem to represent more than minor geographical or even infra-populational variation.

22b. Epilobium amurense subsp. **cephalostigma** (Haussknecht) C. J. Chen, Hoch & Raven, comb. et stat. nov. *Epilobium cephalostigma* Haussknecht, Oesterr. Bot. Zeitschr. 29: 57. 1879.—TYPE: JAPAN. Hokkaido, Hakodate, island of "Jesso," 1–13 Jul 1861, *Maximowicz s.n.* (lectotype, here designated: LE!).

Epilobium calycinum Haussknecht, Monogr. Epil. 196. 1884.—TYPE: JAPAN. Probably Honshu, "Nippon media, Maximowicz. Iter secundum," 1866, *Tschonoski s.n.* (lectotype, here designated: LE!).

Epilobium nudicarpum Komarov, Trudy Imp. S.-Peterburgsk. Bot. Sada 18: 432. 1901. *Epilobium cephalostigma* var. *nudicarpum* (Komarov) H. Hara, J. Jap. Bot. 18: 234. 1942.—TYPE: KOREA. Keng-son Province, Samsu, 5 Aug 1897, *Komarov s.n.* (holotype: LE, not located).

Epilobium angulatum Komarov, Trudy Imp. S.-Peterburgsk. Bot. Sada 18: 432. 1901.—TYPE: NORTH KOREA. Along the banks of streams in the valley of the "Che-czchen-gan" river as it flows into the Yalu River, from Kapsan (41°05′N, 128°17′E) to Samsu (41°17′N, 128°01′E), 2 Aug 1897, *Komarov s.n.* (lectotype, here designated: LE!, photo: K!). Both syntypes mentioned in Komarov's publication are mounted on the same sheet, and that chosen includes three rather than two plants, and more mature material.

Epilobium cylindrostigma Komarov, Trudy Imp. S.-Peterburgsk. Bot. Sada 25: 95. 1905.—TYPE: Border between Korea and China, Yalu River, "Corcadang," 12 Jul 1897, *Komarov s.n.* (holotype: LE!).

Epilobium coreanum H. Léveillé, Bull. Herb. Boiss., sér. 2, 7: 590. 1907.—TYPE: KOREA. "Nai-piang," in wet places, 24 Aug 1901, *Faurie 286* (lectotype, here designated: G!).

Epilobium consimile var. *japonicum* Nakai, Bot. Mag. (Tokyo) 25: 148. 1911.—TYPE: JAPAN. Hokkaido, Aomori, Mutsu prov., Mt. Hakkodayama (40°39′N, 140°51′E), 24 July 1899, *Koriba s.n.* (holotype: TI!).

Epilobium sugaharai Koidzumi, Acta Phytotax. Geobot. 5: 121. 1936.—TYPE: U.S.S.R. Sakhalin, Miho, Kabutoyama, 24 Aug 1931, *Sugahara 8827* (holotype: KYO!).

Common name: Guanghua Liuyecai (Liou et al. 1977).

Erect perennial herb, producing leafy soboles from the base. Stems 25–150 cm tall, 2–8 mm in diameter, usually well branched, strigillose on upper part, completely lacking glandular hairs, with two faint pubescent lines decurrent from margins of petioles below the inflorescence, or rarely stems subglabrous, with raised but subglabrous lines. Leaves subsessile or with petioles 1–6 mm long, especially in lower pairs; cauline blades 3–9 cm long, 0.8–2.5 cm wide, oblong-lanceolate to narrowly ovate or rarely narrowly rhombic, apex acute or acuminate, base cuneate or broadly so, the margins sharp denticulate with 13–35 teeth on each side. Inflorescence suberect, strigillose, very rarely with an admixture of glandular hairs, or very rarely subglabrous. Flowers erect; ovaries 2–4 cm long, scattered strigillose, rarely glabrous; pedicels 0.4–0.8 cm long; floral tube 0.7–1 mm long, 1.3–1.6 mm in diameter; sepals 3.8–6 mm long, 0.8–1.8 mm wide, sparsely strigillose, or rarely subglabrous; petals 4.5–7 mm long, 2.5–4 mm wide, pink or rose-purple, rarely white; style 2.5–4 mm long, subglabrous; stigma subcapitate, 1–1.3 mm long, 1–1.4 mm in diameter. Capsules 3–7 cm long, scattered strigillose or glabrescent, rarely glabrous, borne on pedicels 0.4–1.3 cm long. Seeds 0.8–1 mm long, 0.3–0.4 mm in diameter, brown; coma 6–7 mm long. Chromosome number: n = 18. Fig. 46.

Phenology. Collected in flower from June to August (rarely to September), in fruit from August to September (rarely to October).

Distribution (Fig. 47). Far East region of the Soviet Union, Japan, Korea, and northeastern and eastern China; in China in Heilongjiang, Jilin, Liaoning, Hebei, Shandong, Henan, Shaanxi, eastern Gansu, Anhui, Zhejiang, Jiangxi, Fujian, northern Guangdong, northern Guangxi, Hunan, Hubei, eastern Sichuan, Guizhou and eastern Yunnan; in wet areas along streams and roads at lower elevations or in the mountains in the south; 600–2100 m.

ADDITIONAL SPECIMENS EXAMINED. **China.** HEILONGJIANG: Hailin, *P. Y. Fu et al. 3426* (IFP).—JILIN: Wangqing, *P. Y. Fu et al. 462* (KUN), *C. S. Wang et al. 1488* (IFP); Changbai Shan, *C. C. Chien 462* (KUN), *T. N. Liou et al. 1638* (PE).—LIAONING: Wulongbei, 1909, *Y. Yabe s.n.* (NAS); Fushung, *S. S. Gang 169* (IFP); Liuxi, *Y. C. Zhu 1333* (PE); Kuandian, *S. C. Chu & Y. C. Zhu 422* (IFP, PE); no locality, 1910, *Y. Yabe s.n.* (PE).—HEBEI: Xuanhua, *Serre 9578* (P); Zhuolo, *Hebei Med. Pl. Exped. 121* (PE), *147* (PE); Yu Xian, *Zhangjiakuo Exped. 415* (PE); Laiyuan, *Hebei Univ. Agr. 42252* (PE), *43613* (PE); Wu'an, *Wu'an Exped. 36* (PE); no locality, *Henry 5082* (G).—SHANDONG: Taishan, *Shandong Univ. 53* (KUN, PE).—HENAN: Jiyuan, *Xinxiang Teach. Coll. 6226* (PE); Yiyang, *Henan Exped. 22609* (PE); Song Xian, *Henan Exped. 2365* (PE), *34798* (PE), *Henan Forestry Depart. 1163* (PE); Luanchuan, *Henan Exped. 20135* (PE); Shangcheng, *Beijing Inst. Bot. 10430* (PE).—SHAANXI: Suide, *K. T. Fu 6789* (PE); Hua Shan, *K. S. Hao 4008* (PE); Lounan, *T. P. Wang 15855* (KUN, PE); Shangxian, *T. P. Wang 16029* (KUN), *16124* (KUN, PE), *16228* (KUN, PE); Ziyang, *P. Y. Li 6233* (PE); Foping, *K. T. Fu 4888* (KUN, PE); Taibaishan, *Taibaishan Exped. 334* (LZU), *J. X. Yang 1172* (HTC, PE); "Laoyu," *B. Z. Go 898* (PE, WUK); Yang Xian, *K. C. Fu 5169* (KUN); "Yuan-chu," *Smith 6476* (UPS).—GANSU: Tianshui, *Landa 112* (LZU), *G. L. Zhang & X. Pu 79-125* (LZU), *X. Y. Wang 197* (LZU), *J. Z. Wu 254* (LZU), *W. Y Hsia 5879* (PE); Hui Xian, *T. P. Wang 15120* (PE); Kang Xian, *Z. Y. Zhang 16376* (PE); Wen Xian, *W. L. Chen & C. J. Chen 8840* (MO, PE), *Z. Y. Zhang 7357* (HNWP); Lanzhou, *Y. Q. He 5090* (PE), *5209* (PE), *5356* (PE), *6012* (PE); "Taoping," *Lixian Exped. 697* (LZU).—ANHUI: Huan Shan, *W. C. Cheng 4190* (NAS), *S. S. Chien 1284* (PE, W), *R. C. Ching 4183* (UC), *East China Station Bot. 6404* (KUN, NAS, PE), *K. C. Kuan 75335* (PE), *T. N. Liou & P. C. Tsoong 2623* (K, KUN, PE),

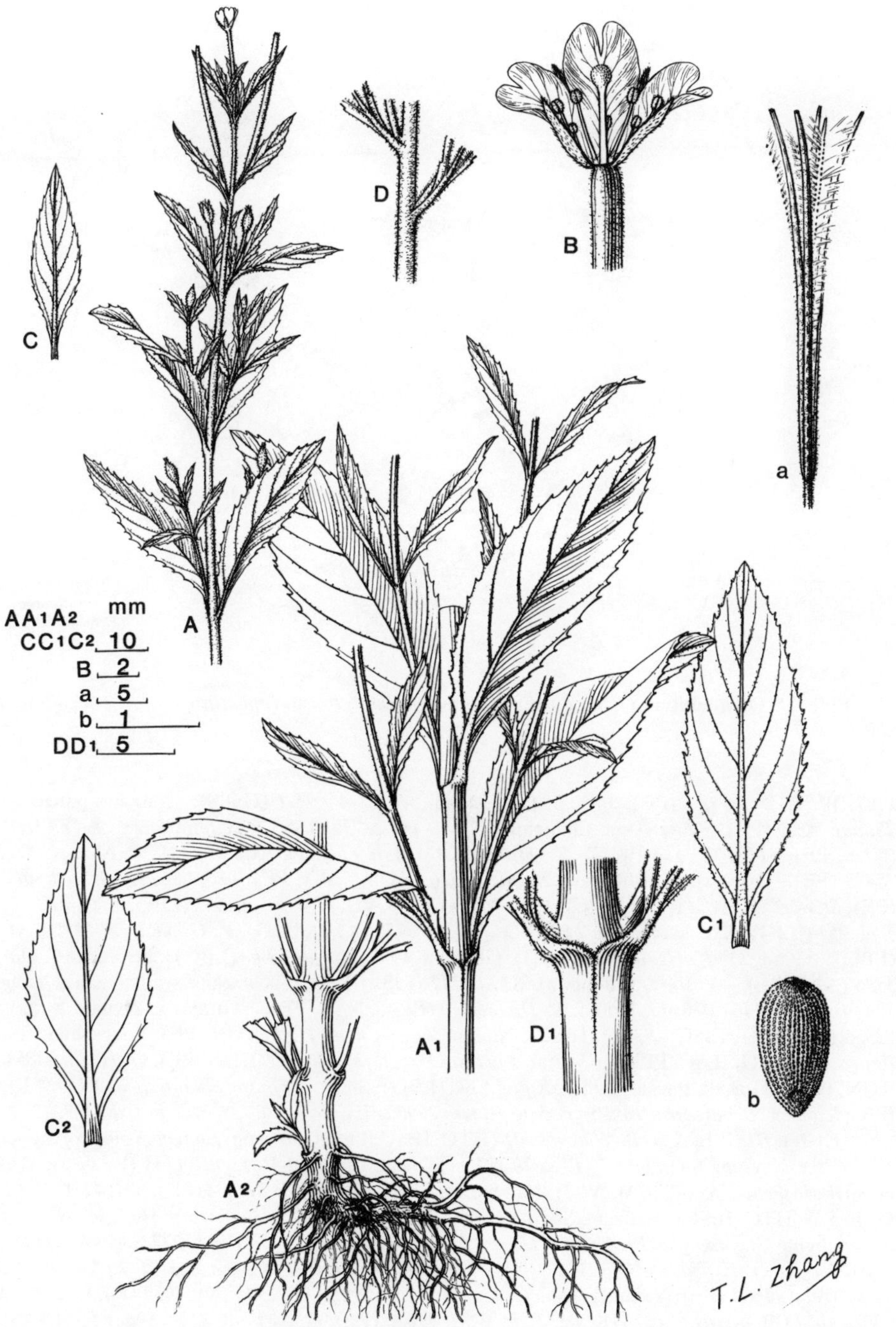

FIG. 46. *Epilobium amurense* subsp. *cephalostigma.* A. Upper stem and inflorescence. A^1. Portion of mid-stem, with lines from margins of petioles inconspicuous and not extending along entire internode. A^2. Base of stem and caudex. B. Flower (one petal and one stamen removed). C. Leaf from inflorescence (bract). C^1. Mid-cauline leaf. C^2. Basal leaf. D. Axils from inflorescence. D^1. Mid-cauline axils. a. Capsule. b. Seed. (A–D^1 based on *C. S. Wang et al. 1488;* a–b based on *J. C. Qiu et al. 3336.*)

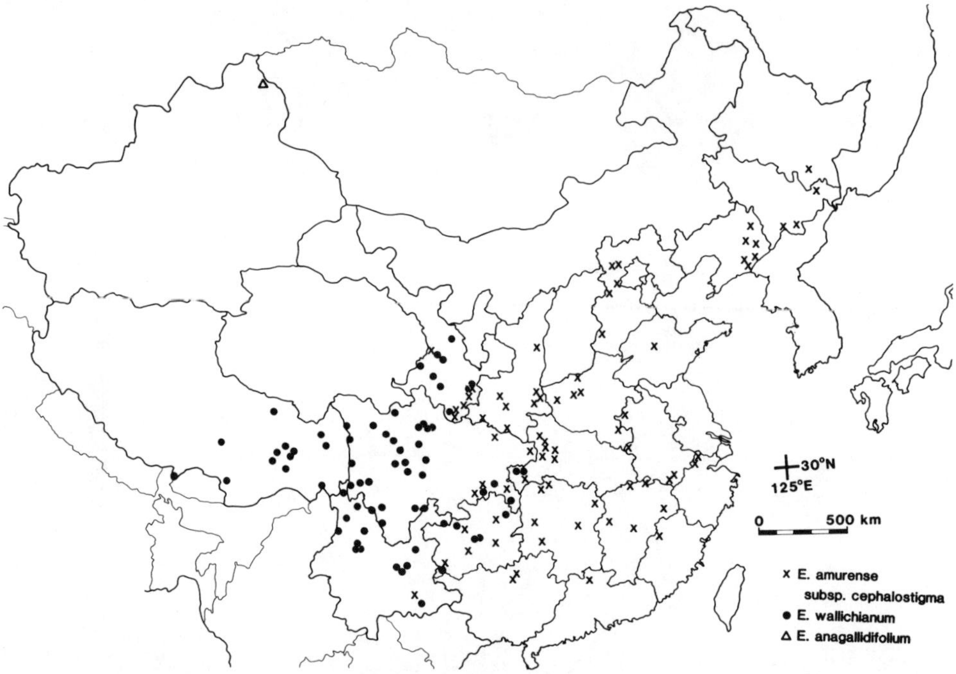

FIG. 47. Distribution in China of *Epilobium amurense* subsp. *cephalostigma, E. anagallidifolium,* and *E. wallichianum.*

1987, *W. T. Wang s.n.* (PE); Jinzhai, Dabei Shan, *Y. L. Wu 78-1* (HSFS); "Jung-mo-pongo," *P. C. Tsoong 4093* (PE).—Zhejiang: Changhua, *X. Y. He 23939* (NAS); Tianmu Shan, *X. Y. He 25426* (NAS, PE), *31174* (PE), 1918, *K. K. Tsoong s.n.* (PE), *Zhejiang Exped. 29342* (NAS), *K. C. Kuan 75433* (PE).—Jiangxi: Yanshan, *M. X. Nie 43526* (KUN, LBG, PE); Lu Shan, *Lu Shan Exped. 188* (PE), *Steward 2450* (C, K, MO, UC); 1941, *Migo s.n.* (NAS), *M. J. Wang 835* (LBG, NAS), *K. C. Kuan et al. 251* (LBG), *K. C. Kuan 74407* (PE), *M. X. Nie 7758* (KUN, LBG), *Y. G. Xiong 5648* (LBG), *5352* (LBG); Wuning, *Y. K. Hsiung 756* (LBG); Gu Xian, 1959, *J. Xiong s.n.* (LBG); Yichun, *J. C. Qiu et al. 3336* (NAS, PE).—Fujian: Taining, *M. S. Li 1211* (FJSI, IBSC).—Guangdong: Ruyuan, *Guangdong 73rd Exped. D1161* (CSFU, IBSC), *L. Deng 5710* (IBSC, NAS, PE).—Guangxi: Ziyuan, *Z. S. Chung 83386* (GH, IBG, IBSC), *83512* (IBSC); Xin'an, *Z. Z. Chen 51173* (KUN, PE); Longsheng, *Guangfu Exped. 899* (IBG, IBSC, PE); Rongshui, *Liuzhou Exped. 2594* (IBG, KUN, PE), *Q. H. Lu 2485* (IBG, KUN, PE).—Hunan: Pinjiang, *J. H. Xiong 6396* (LBG); Shimen, *Foping Shan Exped. A124* (PE), *1301* (PE); Sangzhi, *Beijing Inst. Bot. Exped. to Hunan 2590* (PE), *2924* (PE), *S. Hunan Exped. 3247* (KUN), *3404* (KUN), *B. G. Li & S. B. Wan 750301* (HTC, IBSC, PE); Qianyang, *Jiangan School of Agriculture 1335* (PE); Xuefeng Shan, *Z. T. Li 2870* (PE); Wugang, *Handel-Mazzetti 11233* (E, WU); Wukang, *Handel-Mazzetti 2568* (C, S, W, WU); Nanyue, *K. C. Kuan & B. M. Yang 4* (PE), *80464* (HTC, PE), *B. G. Li 313* (HTC, IBSC); Badagun Shan, *Hunan-Guizhou Exped. 2698* (KUN), *3403* (KUN).—Hubei: Shennongjia, *Shennongjia Pl. Exped. 11796* (PE), *21629* (PE), *21957* (PE), *22723* (PE), *31145* (PE), *1980 Sino-Amer. Exped. 85* (GH, KUN, PE), *119* (GH, KUN, PE); Xingshan, *H. J. Li 556* (PE); Zigui, *T. P. Wang 11804* (PE); Yichang, *Henry 1803* (K, P), *1868* (K, P), *2445* (K), *5082* (K); Badong, *M. Z. Chien 1052* (PE), *Wilson 1324* (K, P), *T. P. Wang 11655* (PE), *G. X. Fu & Z. S. Zhang 907* (NAS, PE); Changyang, *T. P. Wang 11593* (PE); Xuan'en, *H. J. Li 3552* (KUN, PE), *5057* (PE), *H. J. Li 4500* (PE), *Y. M. Wang 4798* (HIB, PE).—Sichuan: Wushan, *G. H. Yang 58956* (KUN, PE); Chengkou, *T. L. Dai 101291* (PE), *102161* (PE), *102585* (PE), *102673* (PE), *105811* (KUN), *105906* (PE); Wanyuan, *P. Y. Li 4363* (PE), *4421* (PE), *6044* (PE), *6188* (PE), *6190* (PE); Qiangjiang, *Z. C. Zhao 88-1736* (CDBI, PE), *88-1934* (CDBI, PE), *88-1981* (CDBI, PE); Shizhu, *Y. Chen 2017* (CDBI), *T. D. Fu & Y. L. Cao 1008* (CDBI, PE); Fengdu, *F. D. Fu & Y. L. Cao 648* (CDBI, PE); Nanchuan, *Bock & van Rosthorn 437* (O),

C. J. Chen & Z. Y. Liu 83069 (PE), *83070* (PE), *83076* (PE), *83081* (PE), *83096* (PE), *Jinfo Shan Exped. 2031* (PE), *G. F. Li 62580* (PE), *K. C. Kuan et al. 2195* (PE), *Pl. Geogr. Exped. 676* (CDBI), *703* (CDBI); 1957, *J. H. Xiong & Z. L. Zhou s.n.* (KUN).—GUIZHOU: Fanjing Shan, *C. P. Tsien et al. 32055* (GZBI, PE); Kaili, *S. Guizhou Pl. Exped. 1640* (NAS), *S. Guizhou Pl. Exped. 2152* (NAS), *2346* (PE); Dafang, *Y. Tsiang 8954* (NY); Pan Xian, *Anshun Exped. 920* (GZBI).—YUNNAN: Mengzi, *Henry 10638* (K).

Some collections of this taxon from Japan, northern Korea, the Ussuri region of the U.S.S.R., and the provinces of Jilin, Liaoning, and Hebei in northeastern China are completely glabrous. These plants are striking in this feature (they have been variously designated as *E. nudicarpum* Komarov or *E. angulatum* Komarov), but they do not differ from typical representatives of subsp. *cephalostigma* in any other discernible character. Although there is some geographical component to this pattern, plants with the more typical pubescence occur throughout the range of the glabrous ones. We have not conducted field studies on any populations in this region, however, so we do not know the exact relationship of these two forms to one another; any intermediates would be noted as the typical form. In view of the much broader pattern of subspecific differentiation we are recognizing in *E. amurense,* we consider that this pattern of variation in vestiture does not warrant formal taxonomic recognition.

23. Epilobium kingdonii Raven, Bull. Brit. Mus. (Nat. Hist.) Bot. 2: 377. 1962.—TYPE: CHINA. Xizang (Tibet), Lhunze, "Chayal Dzong" (approximately 28°18′N, 92°48′E), in pastures, amongst thickets of *Hippophae,* 3350 m, 22 Sep 1935, *Kingdon-Ward 12372* (holotype: BM!).

Common name: Aisheng Liuyecai (Wu 1984).

Low, loosely clumped or matted perennial herb, with fleshy soboles arising 2–6 cm below surface, the caudex with more or less inconspicuous scales. Stems 8–25 cm tall, 1.5–3.5 mm in diameter, simple or rarely branched, strigillose all over on upper part, subglabrous below with 2 raised strigillose lines decurrent from margins of the petioles. Leaves mostly opposite, alternate only on inflorescence, subsessile or with petioles 1–2 mm long; cauline blades 0.8–2 (–2.7) cm long, 0.4–1 (–1.6) cm wide, as long as or longer than the internodes, ovate, glabrous except for sparsely strigillose veins and margins, apex acute, base cuneate to obtuse, the margins obscurely serrulate with 8–15 teeth on each side, the lateral veins 3–5 on each side of midrib. Inflorescence slightly nodding in bud, few-flowered, strigillose, with a few scattered glandular hairs; buds 3–3.5 mm long, 2–2.5 mm in diameter, elliptic, subacute; ovaries 1.4–1.8 cm long, strigillose with an admixture of glandular hairs; pedicels 0.4–1 cm long; floral tube 1–1.2 mm long, 1.8–2 mm in diameter, a ring of long hairs at the mouth within; sepals 4–5 mm long, 1.4–1.5 mm wide, lance-oblong, subacute, mixed glandular- and strigillose-pubescent; petals 7–8 mm long, 4–4.5 mm wide, rose-purple, obcordate, the notch 1–1.5 mm deep; filaments of longer stamens 4–5 mm long, those of shorter ones 2.5–3.5 mm long; anthers 0.5–0.7 mm long, oblong; style 4.5–5 mm long, with a few lax, glassy hairs near base; stigma capitate, 1–1.4 mm long, 0.8–1.3 mm in diameter, surrounded by anthers throughout anthesis. Capsules 3.5–5.5 cm long, subglabrous, borne on pedicels

0.4–1.2 cm long. Seeds 1.4–1.6 mm long, 0.6–0.7 mm in diameter, narrowly obovoid, with short inconspicuous chalazal collar, dark brown, the surface very finely papillose; coma 5–7 mm long, dull white, easily detaching. Chromosome number unknown. Fig. 48.

Phenology. Collected in flower from August to September, in fruit from August to October.

Distribution (Fig. 29B). Endemic to western Sichuan, northwestern Yunnan, and southeastern Xizang in China; in pastures and shaded slopes by rivers; 3300–3700 m.

ADDITIONAL SPECIMENS EXAMINED. **China.** SICHUAN: Kangding, *Smith 10928* (UPS); Dawu, *Smith 11986* (BM, S, UPS).—YUNNAN: Zhongdian, *Zhongdian Exped. 1329* (KUN).—XIZANG: Zogang, 1984, *H. S. Kung et al. s.n.* (MO, PE); Sangagqoiling (Sanga Choling), *Ludlow & Sherriff 1962* (BM); Yadong, *Qinghai-Xizang Exped. 74-2593* (KUN, PE).

This species is found only in a small area in southwestern China. It resembles *E. amurense,* but differs from it in its low stature, much larger seeds with very fine papillae, and floral tube covered with scattered glandular and strigillose pubescence. *Epilobium kingdonii* is also similar to *E. wallichianum,* but that species has a stem with 4 rather than 2 hairy lines; oblong or elliptic acute leaves with densely serrulate margins rather than ovate-obtuse leaves with remotely serrulate margins; and smaller seeds. A collection from Xizang (*Ludlow & Sherriff 1962*), with densely serrulate leaf margins and predominantly glandular inflorescence, might be a hybrid between *E. kingdonii* and *E. wallichianum.*

24. Epilobium wallichianum Haussknecht, Oesterr. Bot. Zeitschr. 29: 54. 1879.—TYPE: SIKKIM. 3 Aug 1849, *Hooker s.n.* (lectotype, designated by Raven 1962a: K!).

Epilobium tanguticum Haussknecht, Oesterr. Bot. Zeitschr. 29: 56. 1879.—TYPE: CHINA. Near border between Gansu and Qinghai, Datong River ("Tetung-gol"), *Przewaldski 109* (holotype: LE!).

Epilobium souliei H. Léveillé, Bull. Herb. Boiss., sér. 2, 7: 588. 1907. *Epilobium wallichianum* subsp. *souliei* (H. Léveillé) Raven, Bull. Brit. Mus. (Nat. Hist.) Bot. 2: 366. 1962.—TYPE: CHINA. Sichuan, Kangding, Tung-o-lo (30°03′N, 101°23′E), "Tongolo, principante de Kiala," 1893, *Soulie 350* (holotype: G!; isotype: G!).

Epilobium duclouxii H. Léveillé, Repert. Spec. Nov. Regni Veg. 6: 110. 1908.—TYPE: CHINA. Yunnan, "San-Chan près Tchen-Hiong," 19 Aug 1905, *Mey 500* (not located; cf. Raven 1962a).

Epilobium mairei H. Léveillé, Repert. Spec. Nov. Regni Veg. 12: 283. 1913.—TYPE: CHINA. Yunnan, lakes on the "plateaux of Ta-Hai," 3200 m, Jul 1912, *Maire s.n.* (not located; cf. Raven 1962a).

Epilobium sykesii Raven, Bull. Brit. Mus. (Nat. Hist.) Bot. 2: 366. 1962.—TYPE: NEPAL. South of Gurjakhani (approximately 28°36′N, 83°13′E), 3050 m, near small streams, 19 Aug 1954, *Stainton, Sykes & Williams 3894* (holotype: BM!; isotype: G!).

Common name: Dianzang Liuyecai (Li 1986).

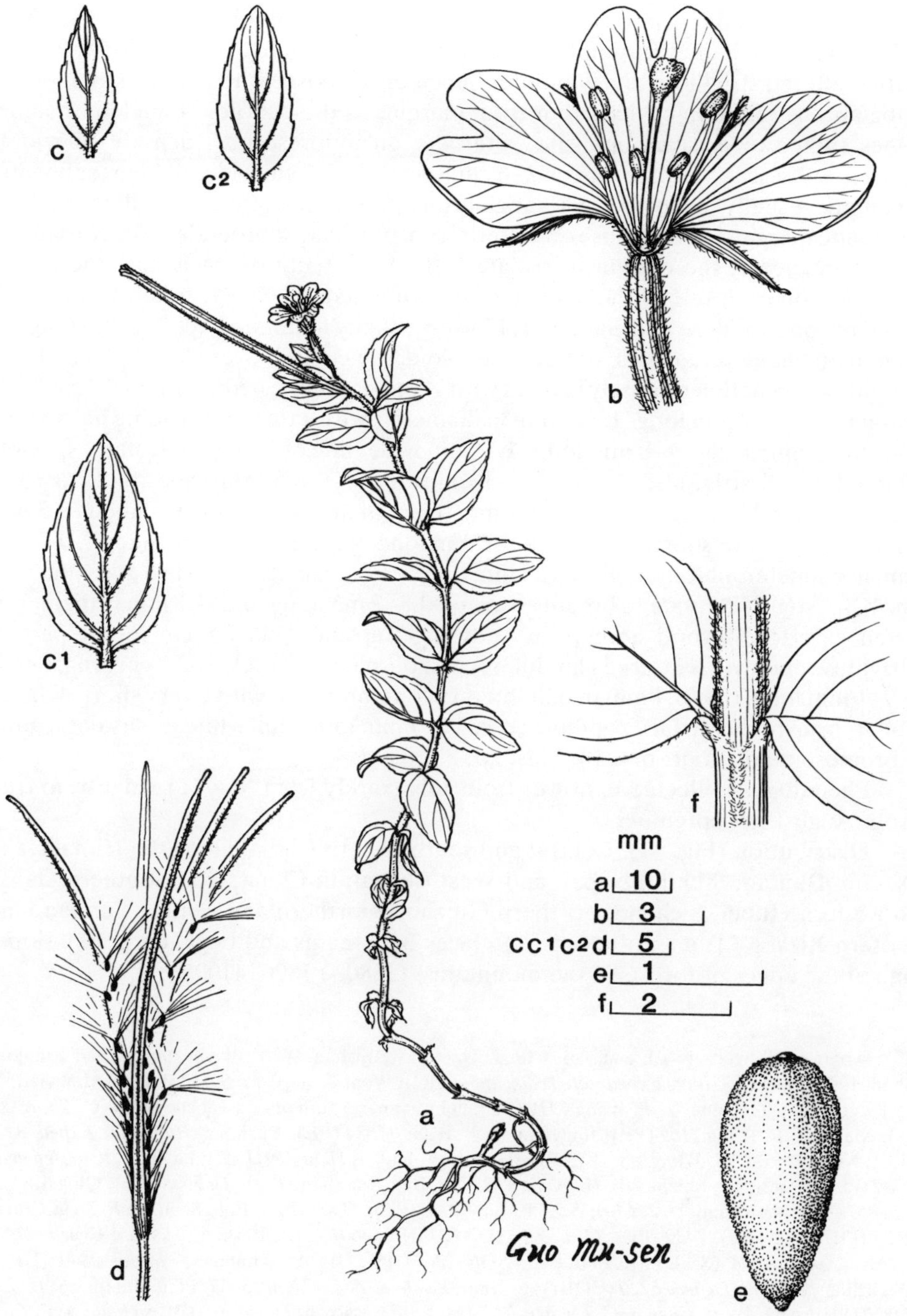

FIG. 48. *Epilobium kingdonii.* a. Habit, with sobole originating from caudex. b. Flower (one petal and three stamens removed): note hairs at mouth of floral tube and near base of style. c. Leaf from inflorescence. c[1]. Mid-cauline leaf. c[2]. Basal leaf. d. Capsule. e. Seed. f. Axils from mid-stem. (Based on *Kingdon-Ward 12372,* type.)

Erect or ascending perennial herb, forming leafy basal soboles. Stems 15–80 cm tall, 2–6 mm in diameter, simple to well branched, subglabrous below inflorescence with scattered strigillose and glandular pubescence above, with (2 or) 4 strigillose or subglabrous raised lines decurrent from margins of the petioles, 4-angled or sometimes subterete. Leaves opposite, alternate on inflorescence, densely or widely spaced, subsessile; cauline blades 2–6 cm long, 0.6–2.5 cm wide, oblong, subovate, or elliptic, usually subcoriaceous, subglabrous except for scattered strigillose hairs on veins and margins, apex obtuse or sometimes acute, base subrounded, subcordate, or broadly cuneate, the margins serrulate with 10–25 teeth on each side, the lateral veins 4–6 on each side of the midrib, often prominent abaxially. Inflorescence nodding, mixed strigillose and glandular. Flowers usually nodding; buds 3–6 mm long, 2–3 mm in diameter, ovoid or globose-ovoid, apiculate; ovaries 1.8–4 cm long, strigillose (sometimes densely) mixed with glandular hairs; pedicels 0.4–1.2 cm long; floral tube 0.8–2 mm long, 1.5–3 mm in diameter, with a ring of spreading hairs at the mouth within; sepals 4–8 mm long, 1–2 mm wide, lance-oblong, apiculate, sparsely glandular and strigillose; petals 5.5–13 mm long, 3–6.5 mm wide, pink to rose-purple, obcordate, the notch 0.7–1.2 mm deep; filaments of longer stamens 3.5–6.2 mm long, those of shorter ones 1.5–3.2 mm long; anthers 0.8–1.4 mm long, 0.3–0.6 mm in diameter, oblong; style 4–8.5 mm long, erect, usually with a few glassy hairs at the base; stigma capitate to broadly clavate, 0.8–2 mm long, 0.8–1.8 mm in diameter, often exserted beyond anthers at anthesis. Capsules 3.8–7.5 cm long, sparsely strigillose and with scattered glandular hairs, borne on pedicels 1–2.5 cm long. Seeds 0.9–1 mm long, 0.3–0.4 mm in diameter, oblong-obovoid, with a very short chalazal collar, brown, the surface papillose; coma 6–7 mm long, dull white, easily detaching. Chromosome number: n = 18. Figs. 2d, 3b, 8a–c, 49.

Phenology. Collected in flower from July (rarely from May) to August, in fruit from August to September.

Distribution (Fig. 47). Central and southwestern China, along the Himalaya to Assam, Bhutan, Sikkim, Nepal, and West Bengal; in China, from southern Gansu to western Hubei, Sichuan, northern Guizhou, northern and central Yunnan, and eastern Xizang (Tibet); often in wet places by streams and bogs, on grassy slopes and at the edges of forests in the mountains; (1380–) 1800–4100 m.

Additional Specimens Examined. **China.** Gansu: Maijishan, *X. Y. Wang 106* (LZU); Xinglong Shan, 1963, *Lanzhou Univ. Exped. 5th Team s.n.* (LZU); Wen Xian, *X. Pu 80152* (LZU); Min Xian, *T. P. Wang 4893* (PE); Jone, *T. P. Wang 5419* (PE); "Gargannar, south of Old Taochow," *R. C. Ching 909* (S); Xiahe, *T. P. Wang 7165* (PE); Jingyuan, *T. P. Wang 17084* (PE); Yuzhong, *Huang He Exped. 3222* (PE); Xigu, 1965, *J. G. Wang s.n.* (LZU).—Hubei: W. Hubei, *Henry 6941* (K); Lichuan, *K. H. Fu et al. 1589* (PE).—Sichuan: Shizhu, *W. H. Wang 1582* (CDBI); Xiu Shan, *T. H. Tu 5608* (PE); Qingxi, *C. S. Liu 467* (PE); Nanchuan, *C. J. Chen & Z. Y. Liu 83062* (PE), *83068* (PE); Jinfo Shan, *Jinfo Shan Exped. 2025* (PE), *2166* (PE), *2439* (PE), *G. F. Li 63533* (KUN), *64601* (PE); Barkam, *Sichuan Exped. 22202* (NAS), *Smith 4694* (S, UPS); Zhoukeji, *Smith 4694* (S, UPS); "Anngpan," *Smith 2849* (UPS); Wenchuan, *Sichuan Exped. 8220* (CDBI); Li Xian, *Z. He & Z. L. Zhou 13662* (PE); Shuajingsi, *H. Lee 74860* (PE); Aba, *W. L. Chen & C. J. Chen 7877* (MO, PE); Baoxin, *David s.n.* (P), *Sichuan Agri. Coll. 7246* (CDBI), *K. L. Chu 3947* (PE, W); Erlang Shan, *W. P. Fang & W. K. Hu 34756* (PE); Emei Shan, *W. P. Fang 12975* (W), *X. Y. He 6055* (NAS), *K. C. Kuan et al. 4008* (PE), *T. H. Tu 230* (PE), *Z. X. Wang 451* (CDBI); Ebian, *C. W. Yao 4352* (NAS); Meigu, *Sichuan Econ. Pl. Exped. 1718* (PE); Zhoajue, *Z. G. Liu et al. 50* (CDBI); Kangding, s.d., *Bonvalot & Henri s.n.* (P), *W. K. Hu & Z. He 10474* (PE), *K. C. Kuan et al. 86* (PE), *H. S. Kung 6007* (MO), *K. Y. Lang et al. 992* (KUN, PE), *C. S. Liu 1000* (PE), *1185* (PE), *Mussot 157* (P), *Soulie 161* (G), *449* (G), *449 p.p.* (K, P); 1893, *Soulie s.n.*

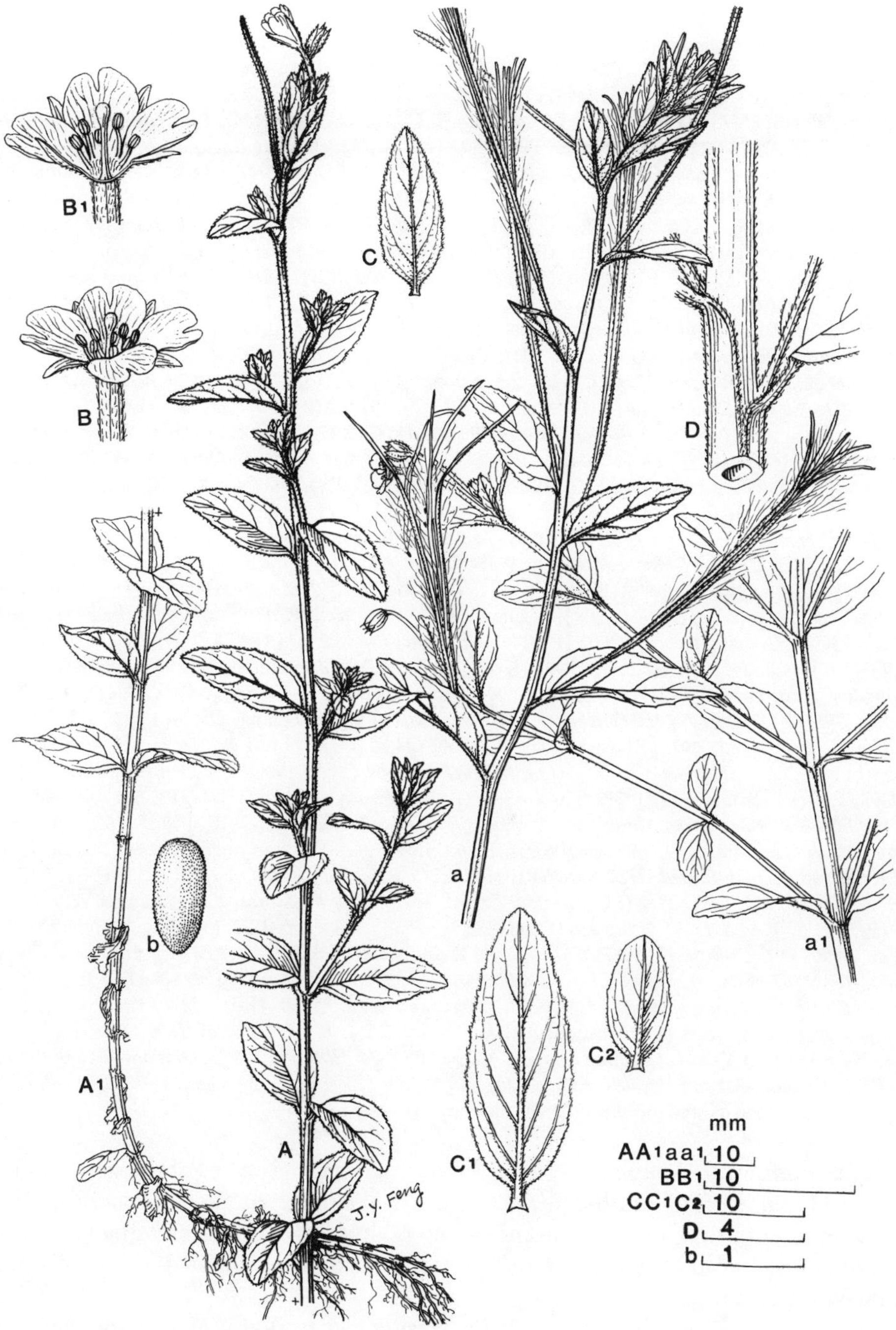

FIG. 49. *Epilobium wallichianum.* A. Upper stem and inflorescence. A¹. Base of stem. B. Flower. B¹. Flower with one petal and three stamens removed. C. Leaf from inflorescence (bract). C¹. Mid-cauline leaf. C². Basal leaf. D. Axils from inflorescence. a. Inflorescence. a¹. Portion of mid-stem. b. Seed. (A–D based on *Xizang Med. Pl. Exped. 1086;* a–b based on *K. K. Tsoong 2489* from Jizu Shan, Yunnan.)

(P), *Smith 10511* (S, UPS), *10928* (S, UPS); Xinduqiao, *Soulie 96* (P); Luhuo, *J. H. Feng 6434* (CDBI, PE); Yajiang, *Z. P. Huang et al. 633* (PE); Qianning, *Qianning Exped. 5549* (CDBI, PE); Dawu, *Dawu Exped. 5859* (PE); Garze, *W. L. Chen & C. J. Chen 6672* (MO, PE), *Sichuan Exped. 15474* (NAS), *Y. W. Tsui 4274* (PE), *Xizang Exped. 2415* (HNWP), *T. S. Ying 4587* (PE); Dege, *F. X. Jiao 2533* (PE), *Y. W. Tsui 4360* (PE), *5137* (PE); Baiyu, *W. L. Chen & C. J. Chen 6856* (MO, PE); Yanguan, *Qinghai-Xizang Exped. 12657* (KUN, PE), *T. S. Ying 4507* (KUN, PE); Muli, *Qinghai-Xizang Exped. 13741* (PE), *Q. S. Zhao et al. 7884* (CDBI); Daocheng, *Daocheng Pl. Exped. 2041* (CDBI, KUN); Jiulong, *Q. Q. Wang 20276* (CDBI), *W. L. Chen & C. J. Chen 6504* (MO, PE); Batong, *Soulie 3978* (P); Hanyuan, *T. P. Wang 8808* (PE); Xiancheng, *K. Y. Lang et al. 2664* (KUN, PE), *Z. G. Liu 1029* (CDBI), *1438* (CDBI), *Qinghai-Xizang Exped. 4852* (PE), *Sichuan Exped. 3018* (CDBI, PE); Derong, *Qinghai-Xizang Exped. 3262* (KUN, PE); "Taining," *K. L. Chu 7488* (PE).—GUIZHOU: Fanjing Shan, *C. P. Chien et al. 30921* (GZBI, PE), *Sino-Amer. Guizhou Exped. 1831* (A, PE); Guiyang, *Cavalerie 992* (P); Qingzhen, *Qinghai-Xizang Exped. 656* (KUN, PE), *S. W. Teng 90339* (GH); Dafang, *Y. Tsiang 8954* (W); Nayong, *Bijie Exped. 683* (GZBI, PE); Pu'an, *Anshun Exped. 1807* (GZBI, PE).—YUNNAN: Yongsheng, *Handel-Mazzetti 3391* (S, WU); Zhenxiong, *H. T. Tsia 52278* (KUN); N.-E. Yunnan, *Maire 1548* (E), *1548a* (E); Dongchuan, 1911, *Maire s.n.* (MO); 1912, *Maire s.n.* (G), *668* (BM, E); "Ta-Hai plateáu," *Maire 119* (E); "Ma-Li-ovan," *Maire 30* (G), *83* (BM, E, P); N. E. Yunnan, *Maire 10043* (C, DD, E, HBG, NY, UC); "Ya-kou," *Maire 10044* (E, UC); Songming, *B. Y. Qiu 54957* (KUN); Fumin, *B. Y. Qiu 596286* (PE); Kunming, 1919–1922, *Cavalerie s.n.* (UPS), *Ducloux 2685* (P), *2686* (P), *Z. H. Yang 101229* (KUN), *B. Y. Qiu 50856* (KUN); "Don dza," *Ducloux 2853* (P); "San Chan," *Ducloux 4378* (P); Wenshan, *Z. H. Tsi et al. 539* (PE); Jizu Shan, Binchuan, *K. K. Tsoong 2489* (PE); Yangbi, *Ducloux 3163* (P), *Handel-Mazzetti 643* (S, W); "Dsolin-ho," *Handel-Mazzetti 782* (W); Dali, *Forrest 4458* (E, K), *4463* (BM, E), *6909* (BM, E, K), *11941* (BM), *H. T. Tsai 53948* (KUN), *H. C. Wang 1418* (KUN, PE), *C. W. Wang 1418* (KUN); "Lanngy tsin," *Ducloux 2687* (P); Luxi, *Manshi Exped. 1056* (KUN, YNU); Eryuan, *Delavay 2007* (P); Heqing, *R. C. Ching 23813* (KUN, PE); Lushui, *C. K. Wu 8069* (KUN); Bijiang, *Nujiang Exped. 991* (KUN), *993* (KUN); Zhongdian, *T. T. Yu 13677* (GH, E), *Zhongdian Exped. 3692* (KUN); Weixi, *C. W. Wang 68471* (GH); Gongshan, *X. F. Deng 791330* (KUN), *791338* (KUN), *791411* (KUN), *K. M. Feng 8402* (PE); Yongning, *Ducloux 7310* (P); Deqen, *Qinghai-Xizang Exped. 2307* (KUN, PE), *C. W. Wang 69070* (GH); "Long gi," *Delavay 5145* (G); "Mt. Che tchutw." 1887, *Delavay s.n.* (P), *Handel-Mazzetti 4904* (S, W, WU); "Echen fong chan," 1894, *Delavay s.n.* (P), *Delavay 4* (P); 1921, *Maire s.n.* (P); no locality, *Cavalerie 2824* (P), *Ducloux 485* (K); 1914, *Maire s.n.* (G); 1921, *Maire s.n.* (P), *H. T. Tsia 58329* (KUN), *C. W. Wang 71653* (NAS).—XIZANG: Zayu, *Xizang Exped. 860* (HNWP)); *1216* (HNWP); Zhag'yab, *Qinghai-Xizang Exped. 12304* (PE); Qamdo, *Xizang Exped. 2722* (HNWP); Medog, *Ecol. Group, Xizang Exped. 11535* (PE), *P. S. Li & S. Z. Chang 2165* (PE), *3341* (PE), *Xizang Exped. 1676* (HNWP); Bomi, *Y. T. Chang & K. L. Lang 478* (PE), *T. S. Ying & D. Y. Hong 386* (PE), *412* (PE), *567* (PE), *1265* (PE); Mainling, *Qinghai-Xizang Exped. 75807* (PE); Nyingchi, *Y. T. Chang & K. Y. Lang 1067* (PE), *1131* (PE); Yigong, *T. S. Ying & D. Y. Hong 567* (PE); "Cyamde Chu," *Ludlow et al. 6304* (BM); Sog Xian, *D. D. Tao 10925* (PE); Conag, *British Museum Exped. 12399A* (BM), *Qingzang Exped. 751583* (PE), *751703* (PE); Lasa, *Y. T. Chang & J. Y. Lang 1593* (PE); Yadong, *K. H. Fu 901* (PE), *K. H. Fu et al. 1100* (PE), *Ludlow & Sherriff 10033* (BM), *Qinghai-Xizang Exped. 74-2136* (PE), *74-2395* (PE); 1975, *Qinghai-Xizang Exped. s.n.* (PE); Nyalam, *Xizang Med. Pl. Exped. 1086* (PE), *1537* (PE); Lulangpinzhan, *Y. T. Chang & K. Y. Lang 951* (PE); "Yar-gong," *Soulie 3978* (P); no locality, *Fougnia 446* (P).

Epilobium wallichianum is one of the most common species of the genus in the eastern Himalaya and south-central China. It is similar to *E. sikkimense* in geographical distribution and general morphology, but differs by possessing an underground stem devoid of scales or turions and a usually tetragonal stem with pronounced raised lines.

As noted by Raven (1962a), *E. wallichianum* is remarkably diverse morphologically, especially in density of vestiture, style length, flower size, leaf blade size and shape, and petiole length. None of these characters are consistently correlated with one another or with geography. Raven (1962a) recognized *E. wallichianum* subsp. *souliei* mainly by its more densely glandular inflorescence, generally larger flowers, and more open habit. Among the many additional collections that we have been able to examine, especially from southwestern China, we find that glandular hairs are quite common on "typical" specimens of *E. wallichianum,* and are not strongly correlated with flower size, habit, or other characters, nor with altitude or geography. For this reason, and in view of the overall variability found in this species, we no longer recognize *E. wallichianum* subsp. *souliei,* nor any other infraspecific taxa.

Our better collections and definition of *E. wallichianum* allow us to include in it two species that earlier seemed distinct. Raven (1962a) recognized *E. sykesii,* based on a single collection from central Nepal, by virtue of its broadly clavate stigma and cinereous inflorescence. Although the most common stigma shape in *E. wallichianum* is subcapitate, many collections from China have broadly clavate stigmas. The type collection of *E. sykesii* has unusually small leaves and denser pubescence on the inflorescence, but we agree with Hara (1979) that these characters no longer seem sufficient to warrant formal status for this collection.

Epilobium tanguticum is also now included in *E. wallichianum* on the basis of our examination of the type from LE. It is a large, robust specimen with large bracts and ovate leaves, and resembles a collection from eastern Sichuan (Jinfo Shan, *G. F. Li 64601,* PE). In other respects, it does not differ from most specimens of *E. wallichianum,* and it may simply represent a particularly robust collection from near the limits of the geographical range.

25. Epilobium laxum Royle, Illus. Bot. Himal. Mount. 211, t. 43, fig. 2. 1835.—TYPE: INDIA. Uttar Pradesh, Mussooree, *Royle s.n.* (holotype: LIV, not located).

Epilobium tetragonum var. *?amplectens* Bentham ex C. B. Clarke in J. D. Hooker, Fl. Brit. India 2: 587. 1879. [*Epilobium amplectens* Bentham in Wallich, Numer. List 216, n. 6330. 1832, nomen nudum.] *Epilobium amplectens* (Bentham ex C. B. Clarke) Haussknecht, Monogr. Epil. 208. 1884.—TYPE: INDIA. Uttar Pradesh, Kumaun, 1832, *Blinkworth* in *Wallich 6330* (holotype: K!; isotypes: BM! E! G!).

Epilobium duthiei Haussknecht, Monogr. Epil. 205, t. 9, fig. 54. 1884.—TYPE: Cultivated at the Royal Botanic Gardens, Kew, Jun 1880; seed from: INDIA. Uttar Pradesh, Tehri Garhwal, Kedar Kanta, 3050–3350 m, *Duthie s.n.* (holotype: K!; isotype; JE!).

Epilobium sadae H. Léveillé, Bull. Herb. Boiss., sér. 2, 7: 588. 1907.—TYPE: INDIA. Uttar Pradesh, Tehri Garhwal, Rudugaira Gad, 4250–4850 m, 20 Jul 1883, *Duthie 1046* (holotype: G!; isotype: BM!).

Epilobium subnivale Popov ex Pavlov, Wiss. Ber. Moskau. Staatsuniv. 2: 329. 1934.—TYPE: U.S.S.R. Kirgiziya ("Turkestania ross., prov. Syr-Darja"), Tian Shan range, Khr. Talasskiy-Alatau, on moist banks by small streams near source of River Topchak-su, 2000 m, fl. et fr., 24 Jul 1931, *Pavlov 735* (holotype: MW; isotype: LE!).

Common name: Dahua Liuyecai.

Erect perennial herb, more or less clumped and forming fleshy turions (sometimes elongated) just below ground level that leave brown scales at base of stem. Stems 10–70 cm tall, 1.5–4 mm in diameter, branched above, rarely throughout, or simple, strigillose all over in upper part, glabrous below except for raised strigillose

FIG. 50. *Epilobium laxum*. a. Habit, base with fleshy turions. b. Flower (one petal and three stamens removed); note hairs near base of style. c. Mid-cauline leaf. d. Axils from lower stem. e. Capsule and subtending bract. f. Seed (adaxial side). (Based on *C. J. Chen & Hoch 86235*.)

lines decurrent from margins of the petioles. Leaves opposite except on inflorescence, sessile on upper stem, attenuate to petioles 2–8 mm long in lower pairs; cauline blades 2–7 cm long, 1.2–2.6 cm wide, often longer than internodes they subtend except where much reduced on inflorescence, ovate or broadly lanceolate above to obovate below, subglabrous with strigillose margins and veins, apex subacuminate, base rounded, the margins sharply denticulate with 15–20 teeth on each side to subentire in basal pairs, the lateral veins distinct, 4–5 on each side of midrib. Inflorescence nodding to suberect, strigillose and usually glandular. Flowers suberect; buds 4–7 mm long; ovaries 1.8–2.1 cm long, strigillose with an admixture of glandular hairs; pedicels 0–3 mm long; floral tube 1–1.5 mm long, 1.8–3 mm in diameter, with a ring of hairs at the mouth within; sepals 4–7.5 mm long, 1–1.8 mm wide, sublanceolate, mixed strigillose and glandular; petals (7–) 10–16 mm long, 4–9 mm wide, bright rose-purple, broadly obovate to obcordate, the notch 1.5–3 mm deep; filaments of longer stamens 4.5–6.6 mm long, those of shorter ones 3.3–5 mm long; anthers 0.8–1.1 mm long, 0.4–0.5 mm in diameter, style 4–9.5 mm long, erect; stigma subcapitate, 1.9–2.5 mm long, 1.6–3 mm in diameter, subequal to longer stamens or often exserted beyond them. Capsules 3.5–7.5 cm long, 1–1.5 mm in diameter, sparsely strigillose, erect and usually appressed to stem, borne on pedicels 1–7 (–10) mm long. Seeds 1.2–1.5 mm long, 0.44–0.6 mm in diameter, narrowly obovoid, with short chalazal collar, brown, the surface finely papillose; coma 7–8 mm long, white, easily detaching. Chromosome number: n = 18. Fig. 50.

Phenology. Collected in flower and fruit from July to September.

Distribution (Fig. 20). Western Himalayan region from Kumaun (Uttar Pradesh, India) to Chitral in the Hindu Kush (NW Frontier Province, Pakistan), and north to the Tian Shan region; in China only in the western Tian Shan of Xinjiang; along rocky streams; 2750–4300 m (2500 m in Tian Shan).

ADDITIONAL SPECIMEN EXAMINED. **China.** XINJIANG: Tian Shan Mountains, Belo, Sailimu Lake, *C. J. Chen & Hoch 86-235* (MO, PE).

Although our collection of this species in 1986 is the first report from China, *E. laxum* has been recorded from the western Tian Shan region of the U.S.S.R. under the name *E. subnivale*. We have seen the isotype of the latter, which had been considered an endemic in the Tian Shan (Steinberg 1949), and it is the same as our collection from Xinjiang, and not obviously different from plants in the Himalayan region, as suggested by Raven (1962a). The distribution of the species is probably continuous throughout the region from the western Himalaya to the Tian Shan, the disjunction due to a lack of collections rather than a true absence of the species. *Epilobium laxum* is very distinctive by virtue of its turions, large flowers, ovate and acuminate leaves, short-pedicellate capsules, and relatively large seeds.

26. Epilobium sikkimense Haussknecht, Oesterr. Bot. Zeitschr. 29: 52. 1879.—TYPE: SIKKIM. Lachen, 4250 m, 14 Jul 1849, *Hooker* s.n. (lectotype, designated by Raven 1962a: K!; isolectotypes: BM! W!).

Epilobium sikkimense subsp. *ludlowianum* Raven, Bull. Brit. Mus (Nat. Hist.) Bot. 2: 373. 1962.—TYPE: BHUTAN. Singhi Dzong (approximately 27°55′N, 91°13′E), 2400 m, in *Abies* forest, 4 Aug 1949, *Ludlow, Sherriff & Hicks 21386* (holotype: BM!).

Epilobium trilectorum Raven, Bull. Brit. Mus. (Nat. Hist.) Bot. 2: 374. 1962.—TYPE: NEPAL. Lamjung Himal (approximately 28°28′N, 84°15′E), 4100 m, forming small clumps on open slopes, 11 Jul 1954, *Stainton, Sykes & Williams 6271* (holotype: BM!).

Epilobium soboliferum Raven, Bull. Brit. Mus. (Nat. Hist.) Bot. 2: 375. 1962.—TYPE: CHINA. Xizang (Tibet), Rong Chu, Tumbatse, Kongbo, (approximately 29°42′N, 94°47′E), 3800 m, on gravel bed of river, 1 Jul 1938, *Ludlow, Sherriff & Taylor 5075* (holotype: BM!).

Epilobium squamosum Raven, Bull. Brit. Mus. (Nat. Hist.) Bot. 2: 380. 1962.—TYPE: NEPAL. Chhoyang Khola, west of Num, Arun Valley (approximately 27°32′N, 87°14′E), 3500 m, on rocky hillside, 20 Jun 1956, *Stainton 726* (holotype: BM!).

Common name: Linpian Liuyecai (Li 1980).

Clumped erect perennial herb, forming thick fleshy soboles just at or below ground level that leave persistent brown coriaceous scales at the stem base, the scales obovate to spatulate, gradually elongated upwards, 4–20 mm long, 2–7 mm wide, apex rounded to subacute, the margins entire to remotely denticulate. Stems (5–) 10–25 (–60) cm tall, 1.5–6 mm in diameter, simple or sometimes sparsely branched, rarely well branched, glabrous except for 2 (or 4) raised strigillose lines decurrent from margins of the petioles, sometimes strigillose all over inflorescence with scattered glandular hairs. Leaves opposite, alternate only on the inflorescence, herbaceous or sometimes membranous, sessile and slightly clasping, the lower pairs with petioles 1–3 mm long; cauline blades (0.8–) 1.5–7.5 cm long, (0.6–) 1–3.7 cm wide, the basal pairs smaller and narrower, usually equal to or shorter than subtending internodes, ovate to somewhat elliptic or oblong-lanceolate, glabrous except for strigillose veins and margins, apex subobtuse to acute, base broadly cuneate or rounded, the margins serrulate with 10–35 teeth on each side, the lateral veins 4–5 (–6) on each side of midrib. Inflorescence mostly nodding, initially compact with bracts more than half as long as the ovaries they subtend, later more open. Flowers erect or nodding in bud; buds 5–8 mm long, 3.5–5 mm in diameter, oblong-ovoid, apiculate, strigillose and glandular; ovaries 1.5–3.5 cm long, strigillose and glandular; pedicels 0.5–0.8 cm long; floral tube 1.3–1.5 mm long, 1.8–3 mm in diameter, with a ring of villous pubescence at the mouth within; sepals 5.5–8 mm long, 1.4–1.8 mm wide, oblong-lanceolate, keeled; petals 7–14 mm long, 4.5–8 mm wide, pink to rose-purple, broadly obcordate or obovate, the notch 1.5–2.6 mm deep; filaments of longer stamens 3.8–4.4 mm long, those of shorter ones 2–2.5 mm long; anthers 0.8–1.3 mm long, 0.6–0.8 mm in diameter, oblong-ovate; style 3.5–5 mm long, erect, usually with a few spreading translucent hairs near base; stigma capitate, 1.2–2.5 mm long, 1.2–2.6 mm in diameter, the longer stamens shedding pollen onto stigma at anthesis. Capsules 5–9 cm long, 1.2–1.6 mm in diameter, erect, scattered strigillose and glandular, borne on pedicels 0.6–2 (–2.5) cm long. Seeds 1–1.3 mm long, 0.35–0.45 mm in diameter, narrowly obovoid, with short chalazal collar, grey-brown, the surface coarsely papillose; coma 6–8 mm long, dull white, detaching easily. Chromosome number: $n = 18$. Fig. 51.

Phenology. Collected in flower from July (rarely from June) to August, in fruit from August to September.

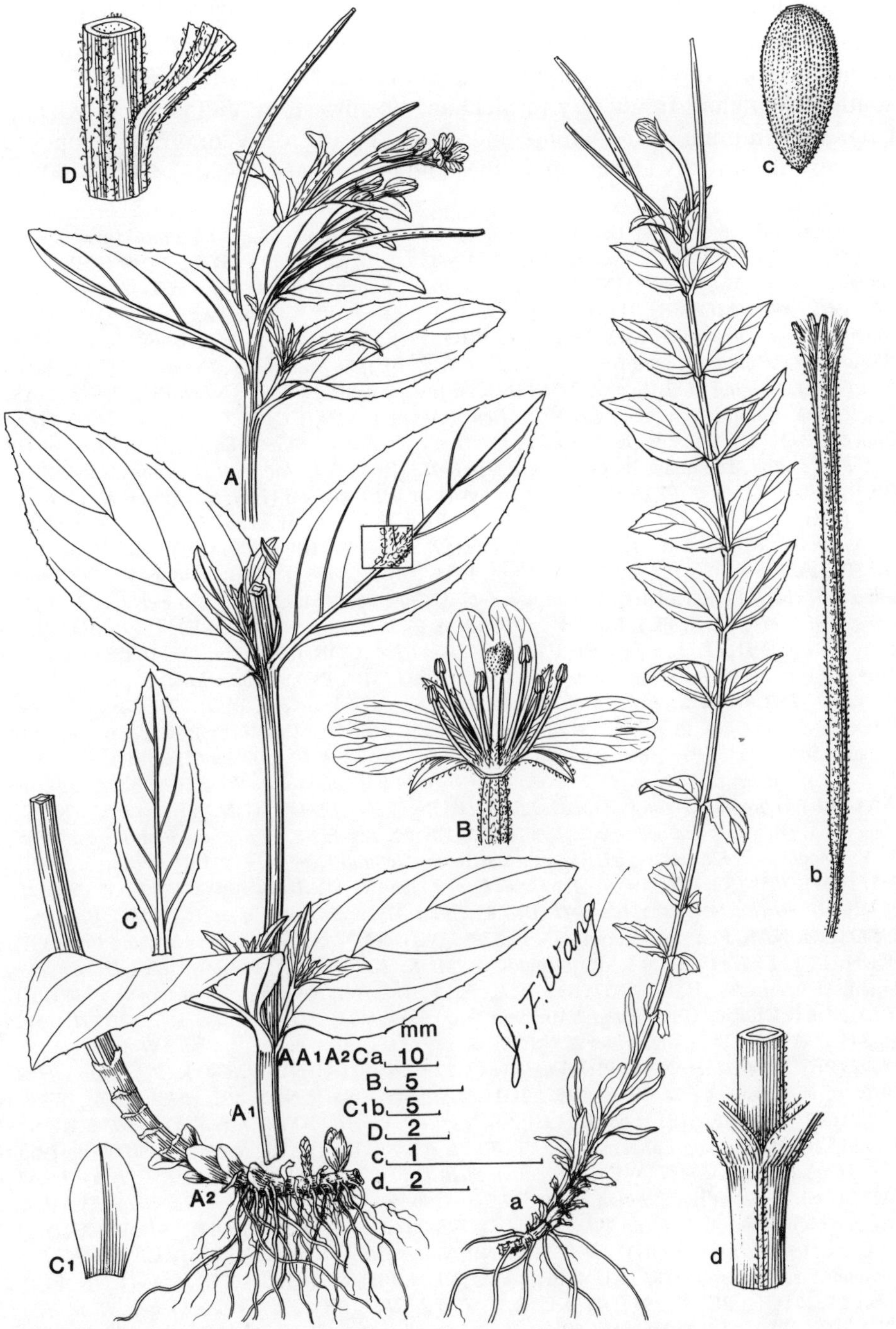

FIG. 51. *Epilobium sikkimense.* A. Inflorescence. A^{1}. Portion of mid-stem, with detail showing leaf midrib hairs. A^{2}. Base of stem, with persistent scales and fleshy soboles. B. Flower (one petal and two stamens removed): note hairs at mouth of floral tube and near base of style. C. Leaf from upper stem. C^{1}. Coriaceous scale from stem base. D. Axil from upper stem. a. Habit. b. Capsule. c. Seed. d. Axils from mid-stem. (A–D based on *Z. P. Soong 39029;* a–d based on *C. W. Wang 68351.*)

Distribution (Fig. 20). From central and southwestern China through the Himalayas in northern Burma, Bhutan, Sikkim, and Nepal to Uttar Pradesh in northern India; in China rarely in western Shaanxi (Taibai Shan), southeastern Gansu, and southern Qinghai, frequently in Sichuan, northwestern Yunnan, and Xizang (Tibet); high montane and alpine meadows, moist rocky or grassy slopes along streams, or on rocky glacial outwashes and gravel bars; (2400–) 3200–4700 m.

ADDITIONAL SPECIMENS EXAMINED. **China** SHAANXI: Taibai Shan, *Lanzhou Univ. Exped. 289* (LZU), *T. N. Liou & P. C. Tsoong 2751* (PE).—GANSU: "Xigu," *T. P. Wang 14869* (PE).—QINGHAI: Nanggen, *Y. C. Yang 1136* (HNWP, PE).—SICHUAN: Songpan, *W. L. Chen & C. J. Chen 8405* (PE), *8408* (PE), *8478* (MO, PE), *W. P. Fang 4106* (E, GH, K, P. PE); 1914, *Weigold s.n.* (W); Aba, *W. L. Chen & C. J. Chen 7878* (MO, PE), *7880* (MO, PE), *7881* (MO, PE); Heishui, *X. Li 73035* (PE); Wolung, Wenchuan, *Sichuan Inst. Biol. 8220* (CDBI, PE); Li Xian, *D. Z. Deng 2003* (SZ); Barkam, *P. X. Li* (PE), *Sichuan Med. Exped. 22363* (NAS); Jinchuan, *X. Li 77788* (NAS, PE), *77941* (NAS, PE), *77959* (PE); Xiaojin, *S. S. Chang & Y. X. Ren 5823* (PE), *6234* (CDBI); Guan Xian, *D. Z. Fu & Z. L. Zhai 87-2202* (PE); Baoxing, *K. L. Chu 2923* (PE), *Z. P. Soong 39029* (PE, SZ), *T. H. Tu 4430* (PE), *T. T. Yu 2223* (PE), Tianquan, *W. G. Hu & Z. He 10012* (PE), *X. L. Jian & J. H. Xiong 34264* (PE), *34589* (PE); Luding, *G. Y. Xu 23341* (CDBI); Hongya, *C. W. Yao 3993* (PE); Mt. Emei, *K. C. Kuan et al. 4008* (PE), *T. C. Lee 2867* (PE), *T. H. Tu 47* (PE); Ebian, *C. W. Yao 3055* (PE); Kangding, 1923, *Cunningham s.n.* (E), *K. C. Kuan et al. 373* (PE), *Z. S. Liu 994* (PE), *1063* (PE), *Mussot 157* (P), *Soulie 161* (P); 1929, *Stevens s.n.* (S); Zheduo Shan, 1953, *X. L. Jiang s.n.* (PE); Jiulong, *W. L. Chen & C. J. Chen 6402* (MO, PE); Yajiang, *W. L. Chen & C. J. Chen 6642* (MO, PE), *S. Jang et al. 2631* (PE), *K. Y. Lang et al. 2967* (KUN, PE); Muli, *W. L. Chen & C. J. Chen 5956* (MO, PE), *5979* (MO, PE), *Y. C. Yang 7329* (CDBI); *T. T. Yu 7616* (PE), *Q. S. Zhao 6324* (CDBI); Zamtang, *W. L. Chen & C. J. Chen 7368* (MO, PE); Garze, *W. L. Chen & C. J. Chen 6933* (MO, PE), *Sichuan Med. Exped. 15411* (NAS); Baiyu, *W. L. Chen & C. J. Chen 6878* (MO, PE); Serxu to Dainkog, *W. L. Chen & C. J. Chen 7096* (MO, PE), *7108* (MO, PE), *7368* (MO, PE); Litang, *S. Jiang 5328* (KUN, PE), *Kingdon-Ward 4380* (E); Batang, *Soulie 3637* (P); Mianning to Jiangkou, *W. L. Chen & C. J. Chen 6190* (MO, PE), *S. G. Wu 2278* (PE); Xichang, *W. L. & C. J. Chen 6123* (MO, PE); Yanyuang, *W. L. Chen & C. J. Chen 5607* (MO, PE), *Qinghai Xizang Exped. 12208* (KUN, PE), *12247* (KUN, PE), *12657* (KUN, PE); Daocheng, *Qinghai-Xizang Exped. 5893* (KUN, PE); Xiangcheng, *Hengduan Shan Exped. 3788* (PE), *K. Y. Lang et al. 2426* (KUN, PE); western Sichuan, *Sichuan Exped. 1435* (PE), *Wilson 3650* (BM, K), *3651* (BM), *3657* (K).—YUNNAN: Dali, *Dali Exped. 205* (YNU), *Delavay 985* (P), *K. M. Feng et al. 80-27* (KUN), *Forrest 4459* (BM, E), *6971* (BM, E), *Sino-American Bot. Exped. 1132* (MO, PE), *H. T. Tsai 53840* (GH, NAS, PE), *53948* (GH, NAS), *53957* (GH, KUN, NAS, PE), *H. C. Wang 1094* (PE), *2284* (KUN, PE), *Li Hen 4* (cult., MO); Jingdong, *M. K. Li 577* (PE); Heqing, 1885, *Delavay s.n.* (P); Lijiang, *Forrest 2491* (E, K), *5809* (BM, E, K, P, PE, UC, W), *Hengduan Shan Exped. 2138* (PE), *2389* (PE), *2456* (PE), *2508* (PE), *Handel-Mazzetti 3515* (S, W, WU), *Rock 4827* (GH, S), *4958* (E), *5426* (E), *9443* (E), *9894* (E, UC), *Schneider 2284* (GH, K), *3773* (G); Bijiang, *H. T. Tsai 58043* (GH, NAS, PE), *58092* (PE), *58244* (GH, NAS, PE); Fugong, *Q. Lin 791910* (KUN), *792034* (KUN); Weixi, *H. T. Tsai 59706* (GH, KUN), *C. W. Wang 64551* (GH, PE), *64651* (GH, NAS, PE), *64853* (PE), *67220* (PE), *67288* (GH, PE), *67635* (PE), *68351* (PE), *68471* (NAS, PE), *69741* (PE), *J. S. Yang 6987* (KUN), *T. T. Yu 8885* (PE); Nu Jiang-Lancang Jiang, *T. T. Yu 20808* (GH, E); Zhongdian, *K. M. Feng 1958* (GH, PE), *Handel-Mazzetti 4637* (WU), *Hengduan Shan Exped. 2800* (PE), *3226* (PE), *T. T. Yu 11864* (PE), *12233* (PE), *12394* (PE), *Zhongdian Exped. 63–3290* (KUN); Cang Shan, *Zhongdian Exped. 63-3945* (KUN); Gongshan, *X. F. Deng 791316* (KUN), *791420* (KUN), *791510* (KUN), *791559* (KUN), *Q. Ling & X. F. Deng 100588* (KUN), *790447* (KUN), *Nujiang Bot. Exped. 791508* (KUN), *792074* (KUN), *Qinghai-Xizang Exped. 8351* (KUN, PE), *8372* (KUN, PE), *8511* (KUN, PE), *8555* (KUN, PE), *C. W. Wang 67220* (GH, PE), *67288* (GH, KUN), *T. T. Yu 22167* (PE), *22732* (PE); Deqen, *K. M. Feng 5386* (PE), *6161* (PE), *6598* (PE), *6861* (PE), *Hengduan Shan Exped. 2248* (KUN, PE), *2793* (KUN, PE), *3839* (KUN, PE), *Qinghai-Xizang Exped. 2540* (KUN, PE), *2655* (KUN, PE), *2667* (KUN, PE), *2709* (KUN, PE), *2709* (KUN, PE), *2793* (KUN, PE), *2978* (PE), *C. W. Wang 64853* (GH, PE), *68749* (GH, NAS, PE), *68767* (GH, NAS, PE), *69070* (KUN, NAS, PE); *T. T. Yu 9057* (GH, PE), *9238* (PE), *9523* (PE), *9704* (KUN), *9773* (PE); "Yundshi Mtn," *Rock 23465* (E); N.-W. Yunnan, *Delavay 4* (P), *Mombeig 117* (E), *Schneider 3773* (GH), *H. T. Tsai 57722* (GH, PE), *S.-W. Union Univ. R.N.4069* (KUN), *11576* (KUN), *T. T. Yu 6276* (GH), *8885* (GH); no locality, *Soulie 2969* (P).—XIZANG:

Mainkung, *T. T. Yu 22167* (E, GH, KUN); *22482* (E, GH, PE); Zayu, 1984, *H. S. Kung et al. s.n.* (MO, PE); *Qinghai-Xizang Exped. 10166* (KUN, PE), *10181* (KUN, PE), *10475* (KUN, PE); Medog, *S. Z. Cheng & C. S. Li 126* (PE), *284* (PE), *Kingdon-Ward 754* (E), *J. Y. Zhuang 788* (PE); Baxoi, 1908, *Mombeig s.n.* (P), *Qinghai-Xizang Exped. 12391* (PE), *12436* (PE), *Soulie 2230* (P); Phung Chu, 1922, *Norton s.n.* (K); Bomi, *T. S. Ying & D. Y. Hong 386* (PE); Nyingchi, *Y. T. Chang & K. Y. Lang 1131* (PE); Mainling, *HWNP Xizang Exped. 1898* (HNWP), *Ludlow et al. 5140* (BM), *Xizang Med. Exped. 3785* (PE); Chickchar, *Kingdon-Ward 11902* (BM); Gongbo'gyamda, *Ludlow et al. 15446* (BM); Tama La, *Ludlow & Sherriff 2182* (BM): Gyaca, *C. Y. Wu et al. 751440* (PE); Cona, *B. Z. Go & W. Y. Wang 22995* (HNWP), *C. Y. Wu et al. 75-888* (PE), *757974* (PE); "Laktang," *Kingdon-Ward 3296* (E); Damxung, *Ludlow & Sherriff 11008* (BM); Lhasa, *Ludlow & Sherriff 9776* (BM); Khambut, *Gould 2357* (K); Zoxiao, *P. C. Tsoong 6145* (PE); Namling, *Xizang Med. Exped. 822* (PE); Rongxar, *Y. T. Chang & K. Y. Lang 705* (PE); Yadong, s.d., *Hobson s.n.* (K), *P. C. Tsoong 5770* (PE), *Waddell 32* (K), *C. Y. Wu 750293* (PE), *C. Y. Wu et al. 750361* (PE); Gyirong, *Qinghai-Xizang Exped. 7160* (PE), *Xizang Med. Exped. 551* (PE); Burang, *Qinghai-Xizang Exped. 76-8421* (PE); "Lonok," *Younghusband 196* (G, K).

Epilobium sikkimense occurs at high elevations in the Hengduan Shan of southwestern China westwards to the central Himalaya. This species is particularly polymorphic in its perennating structures, which vary from short fleshy soboles with compact scales similar to those of a turion to more elongated, lax, and even subfiliform stolons with scattered leaflike scales. Other variable characters include stature, density of leaves on stem, leaf shape and size, inflorescence nodding or erect, and flower size. Despite its variability, *E. sikkimense* has some consistent features that distinguish it from other taxa. The plants usually possess a tuft of coriaceous brown scales at the base of the stems, a feature that is present in relatively few species of the genus. This, combined with the mostly simple stems; fleshy soboles; 2 (or rarely 4) hairy lines decurrent from the margins of the petioles; leaves with a broadly cuneate base and often an obtuse apex; large flowers; capitate stigmas; and styles usually with glassy hairs on the lower part, serves to distinguish *E. sikkimense* from other species with which it might be confused. This species resembles *E. laxum,* a well-marked entity of the western Himalaya and Tian Shan region, which differs in its characteristic acuminate leaves and larger flowers and seeds.

After examining a large number of collections of *E. sikkimense* from throughout its geographical range, and especially from China, we have found it difficult to maintain several taxa that have been segregated from this variable species, and we have combined them in this treatment. Raven (1962a) recognized *E. sikkimense* subsp. *ludlowianum* as a taller, more robust plant than the typical form, but the material from southwestern China is extremely variable in stature and the variation is not correlated with other characters or geography. Raven (1962a) also recognized *E. trilectorum, E. soboliferum,* and *E. squamosum* as distinct species related to *E. sikkimense,* based largely on differences in the perennating structures, which are extremely variable in this and some other species of the genus (Keating et al. 1982). Those new taxa were based on very few collections (in two taxa, on a single sheet); more abundant recent collections and material from China not previously examined now suggests that they are better treated as part of a widespread, modally autogamous, and highly variable species. Hara (1979) also treated *E. trilectorum* as a synonym of *E. sikkimense,* but placed *E. squamosum* in synonymy with *E. williamsii;* he did not include *E. soboliferum* in his treatment, since it does not occur in Nepal.

Epilobium sikkimense grows more or less sympatrically with several other species and appears to form hybrids with them. For example, plants collected in Yunnan (without definite locality, *Soulie 2969,* P; pollen stainability 72%) and Sichuan (Kangding, *Z. S. Liu 1000,* PE) are probably hybrids between *E. sikkimense* and *E. wallichianum;* they have intermediate morphology, with scales at the base of the stems but leaf shape like that of *E. wallichianum*. Several collections might be of hybrid origin between *E. sikkimense* and *E. royleanum;* they all have scales on the underground stems: one from Xizang (Lhasa, *Ludlow & Sheriff 9766,* BM, with pollen stainability 71%) has stems with strigillose pubescence all over and lance-elliptic leaves with long petioles and serrulate margins; another from Yunnan (*H. T. Tsai 57722,* KUN; pollen stainability 45%) has a well-branched stem with hairy lines and pubescence all over; and a third collection from Yunnan (Dali, *H. T. Tsai 53948,* KUN; pollen stainability 0%) also has the mixed pattern of raised lines and pubescence all over.

An unusual collection from Xizang (Yadong, near Biwutang, *Qinghai-Xizang Exped. 74-2316,* PE) has fleshy underground runners 1–2 cm long, each terminating in a compact leafy rosette, and slender stems. It may be a distinct species from *E. sikkimense*.

27. Epilobium williamsii Raven, Bull. Brit. Mus. (Nat. Hist.) Bot. 2: 378. 1962.—TYPE: NEPAL. Taglung, south of Tukucha, Kali Gandaki (approximately 28°39′N, 83°38′E), 3800 m, open grass slopes, 15 Jul 1954, *Stainton, Sykes & Williams 1800* (holotype: BM!; isotype: E!).

Common name: Gaoshan Liuyecai (Li 1986).

Low-growing clumped perennial herb, forming fleshy elongated basal soboles 1.5–7 cm long with persistent brown coriaceous scales, the scales 3–7 mm long, 2–3.5 mm wide, broadly to narrowly obovate, obtuse or rounded, the margins entire. Stems 4–17 (–25) cm tall, 1–3 mm in diameter, erect or ascending, usually well branched from base, sometimes simple, strigillose above, mixed with glandular hairs on inflorescence, subglabrous to sparsely strigillose on lower part with more or less raised strigillose lines decurrent from margins of the petioles. Leaves opposite, alternate on the inflorescence, often pale green when dry, often densely arranged on stems and longer than the internodes they subtend, sessile or on petioles 1–2 mm long; cauline blades 0.7–2.2 cm long, 0.3–1 cm wide, ovate or sometimes elliptic-ovate, subcoriaceous, subglabrous except for sparsely strigillose veins, apex acute to subacuminate, base subrounded to broadly cuneate or subcordate, the margins more or less densely serrulate with 6–18 (–26) teeth on each side, the lateral veins 3–4 (–5) on each side of midrib. Inflorescence often nodding before anthesis, densely strigillose and glandular, bracts mostly less than half the length of subtending ovaries. Flowers slightly nodding or erect in bud; buds 3–4 mm long, 2–2.5 mm in diameter, ovoid to oblong-ovoid, apiculate, often purplish, sparsely strigillose and glandular; ovaries 1–2 cm long, densely glandular and strigillose; pedicels 0.3–0.6 cm long; floral tube 1–1.2 mm long, 1.5–2 mm in diameter, with a ring of spreading hairs at the mouth within; sepals 3–4.5 mm long, 1–1.2 mm wide, lance-oblong, keeled, strigillose and glandular; petals 5–6.5 mm long, 3–3.5 mm wide, rose-purple, obcordate, the notch 0.8–1 mm deep; filaments purplish, those of longer stamens 2–4 mm long, those of shorter ones 1.5–1.8 mm long; anthers 0.5–0.7 mm long, 0.35–0.45 mm in diameter, broadly oblong; style 2–4 mm long, glabrous, erect, purplish; stigma capitate, 0.7–1.3 mm long, 0.1–1.2 mm in diameter, at least the longer stamens shedding pollen onto stigma at anthesis. Capsules 3.5–5 (–6) cm long, sparsely strigillose, borne on pedicels 0.4–1 cm long. Seeds 0.9–1 (–1.2) mm long, 0.35–0.4 mm thick, narrowly obovoid, with very short chalazal collar, grey-brown, the surface finely papillose; coma 5–6 mm long, dull white, deciduous. Chromosome number unknown. Fig. 52.

Phenology. Collected in flower from July to August, in fruit from August to September.

Distribution (Fig. 34). From south-central China west throughout the Himalaya to Sikkim, Nepal, and northern India; in China distributed in eastern Qinghai, Sichuan, northern Yunnan, and southern Xizang (Tibet); in open alpine meadows, moist places by streams, or on gravel banks of glacial lakes in the high mountains; 3350–4900 m (in Qinghai from 2900 m).

ADDITIONAL SPECIMENS EXAMINED. **China.** QINGHAI: Tongren, *S. W. Liu & D. S. Lo 1392* (HNWP); Yushu, *Qinghai Exped. 423* (HNWP, PE).—SICHUAN: Tianquan, *Y. B. Yang 40037* (CDBI); Kangding, *W. L. Chen & C. J. Chen 6300* (PE), *6589* (PE), *C. S. Liu 1062* (PE), *K. Y. Lang et al. 871* (KUN, PE); Jiulong, *Sichuan Inst. Biology 21038* (CDBI); Muli, *W. L. Chen & C. J. Chen 6358* (MO, PE), *Handel-Mazzetti 7308* (S, WU), *Qinghai-Xizang Exped. 14239* (KUN, PE), *Q. S. Zhao et al. 6738* (CDBI); Garze, *W. L. Chen & C. J. Chen 6933* (MO, PE), *Y. W. Tsai 4274* (PE); Yajiang, *K. Y. Lang 2827* (KUN, PE); Batang, *Qinghai-Xizang Exped. 5232* (KUN, PE); Litang River divide, *Kingdon-Ward 4304* (E); Daocheng, *Qinghai-Xizang Exped. 5174* (KUN, PE); Xiangcheng, *Qinghai-Xizang Exped. 3788* (KUN, PE), *4886* (KUN, PE); Muli, *Qinghai-Xizang Exped. 14239* (KUN, PE).—YUNNAN: "Wo mengshan," *Y. P. Chang 577* (KUN); Qiaojia, *B. X. Sun 1081* (PE); Huize, *N.-E. Yunnan Exped. 258* (KUN), *276* (KUN), *335* (KUN); Dali, 1882, *Delavay s.n.* (P), *H. T. Tsai 53866* (A, KUN, PE); Bijiang, *Bijiang Exped. 1773* (KUN); Lushui, *Bijiang Exped. 1773* (KUN); Zhongdian, *Qinghai-Xizang Exped. 3228* (PE), *Zhongdian Exped. 1329* (KUN, PE, YNU), *63-3290* (KUN, PE); Weixi, *C. W. Wang 64560* (A, KUN, PE), *Qinghai-Xizang Exped. 2562* (KUN, PE), *2709* (KUN, PE), *T. T. Yu 9704* (PE); Deqen, *Kunming Bot. Inst. 24070* (KUN, PE), *Qinghai-Xizang Exped. 76-2709* (KUN, PE), *81-2562* (KUN, PE); Gongshan, *X. F. Deng 791533* (KUN), *K. M. Feng 7839* (PE), *7857* (KUN, PE), *C. W. Wang 67288* (PE).—XIZANG: Kalaree, "Zayu," *Lepcha 447* (E, K); Medog, *Chapman 88* (K), *P. S. Li et al. 756* (PE); Baxoi, *Qinghai-Xizang Exped. 12196* (PE); Denggen, *G. C. Xia 1715* (KUN); Songdo, Mainling, *Y. T. Zhang & K. Y. Lang 1397* (PE); Ngingchi, *P. S. Li et al. 6160* (PE); Nang Xian, *B. Z. Go & W. Y. Wang 22925* (HNWP), *Kingdon-Ward 11902* (BM); Cona, *B. Z. Go & W. Y. Wang 22878* (HNWP), *C. Y. Wu et al. 75-888* (KUN, PE), *75-1974* (KUN, PE); Reting, *Ludlow & Sherriff 8851a* (BM); Lhasa, *Y. T. Chang & K. Y. Lang 1593* (KUN, PE), *Richardson 193* (K); Yadong, *Qinghai Xizang Exped. 742593* (KUN, PE); Namling, *Xizang Med. Exped. 873* (PE); Qamdo to Bamda, *Y. T. Chang & K. Y. Lang 405* (PE); Xigaze, *Xizang Exped. 1724* (PE); Jilong, *Qinghai-Xizang Exped. 7160* (PE); Burang, *Qinghai N.-W. Plant Inst. Biol. 4013* (XJBI), *Qinghai-Xizang Exped. 76-8421* (KUN, PE); Rutog, *Qinghai-Xizang Exped. 76-9120* (PE); Kalaree, "Zayu," *Lepcha 447* (E, K); no locality, *King's collector 146* (BM, K, WU).

Epilobium williamsii is one of a complex of apparently closely related Himalayan species that includes at least *E. sikkimense,* with a very similar range, and *E. laxum,* from the western Himalaya and Tian Shan region (Raven 1962a; Hoch 1981). Among the Chinese populations, *E. williamsii* is distinguished from *E. sikkimense* by its dwarfed and well-branched habit, smaller flowers, and acute to short-acuminate leaves. The type of *E. williamsii* from Nepal has only very scattered glandular hairs on the inflorescence and in that respect resembles *E. chitralense* Raven of the western Himalaya; most populations, however, typically have

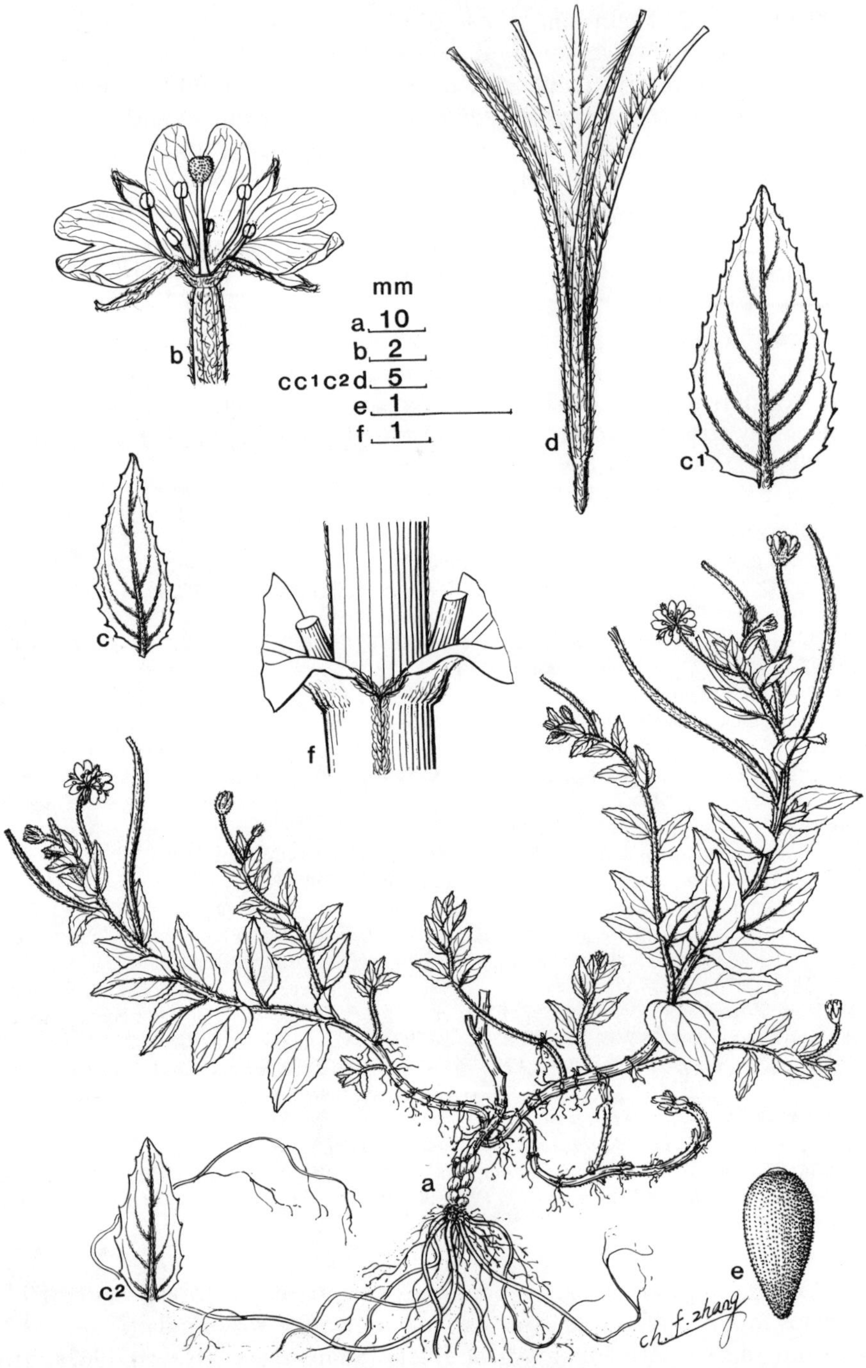

FIG. 52. *Epilobium williamsii.* a. Habit; note persistent scales at base of stem, inflorescences suberect before anthesis. b. Flower (one petal and two stamens removed). c. Leaf from inflorescence (bract). c[1]. Mid-cauline leaf. c[2]. Basal leaf. d. Capsule. e. Seed. f. Axils from mid-stem. (Based on *Xizang Exped. 1724.*)

dense glandular pubescence on the inflorescence. The relationship of *E. williamsii* and *E. chitralense* needs further investigation.

Three collections from northwestern Yunnan (Deqen, *Kunming Inst. Bot. 24070,* PE; Gongshan, *K. M. Feng 7857,* KUN, PE, and *C. W. Wang 67288,* PE) have nearly smooth (reticulate) rather than papillose seeds and unusually long pedicels (up to 2.7 cm long in *Feng 7857* and 3.5 cm in *Wang 67228*). They do not differ in other respects from most collections of *E. williamsii* in the region, and may represent a local pattern of differentiation. However, differences in seed surface sculpturing tend to be uniform in most species of *Epilobium* worldwide (Raven & Raven 1976; Hoch 1978; Solomon 1982), so this variation may indicate greater differentiation among these unusual populations.

Two collections from northeastern Yunnan (Huize, *N.-E. Yunnan Exped. 258* and *276,* both KUN) have tall but simple stems, with larger leaves than is usual for the species. In these aspects they are similar to *E. kermodei,* and may be of hybrid origin with that species. An unusual specimen from Xizang (Songdo, Mainling, 4170 m, *Y. T. Zhang & K. Y. Lang 1397,* PE) has marked petioles, long capsules (to 6 cm long), and seeds to 1.2 mm long. In these features it resembles *E. royleanum;* however, it also has well-marked decurrent lines of hairs on the stem and very finely papillose seeds, and seems best included in *E. williamsii.*

28. Epilobium subcoriaceum Haussknecht, Oesterr. Bot. Zeitschr. 29: 56. 1879.—TYPE: CHINA. Near border between Gansu and Qinghai, Datong River ("Tetung-gol"), 13–25 Jul 1872, *Przewaldski 268* (holotype: LE!).

Common name: Yagezhi Liuyecai.

Erect perennial herb, forming fleshy hypogeous basal turions that leave brown basal scales. Stems 15–45 cm tall, 2–4 mm in diameter, simple or branched, often reddish especially below, subglabrous except for 2 or 4 distinct hairy lines decurrent from margins of the petioles. Leaves opposite, alternate on the inflorescence, subsessile or with petioles 1–3 mm long; cauline blades 1.5–5.5 cm long, 0.5–1.5 cm wide, narrowly ovate to lanceolate, often subcoriaceous, subglabrous except for sparsely strigillose veins and margins, apex acute, base broadly cuneate, the margins serrulate with 13–22 teeth on each side, the lateral veins 4–5 on each side of midrib. Inflorescence somewhat nodding in bud, sparsely strigillose, with scattered glandular hairs; the bracts subtending the flowers much reduced, less than half the length of the ovaries. Flowers somewhat nodding in bud; buds 3–4.5 mm long, 2.5–3 mm in diameter, ovoid, apiculate, strigillose and glandular; ovaries 1.5–2.5 cm long, strigillose and glandular; pedicels 0.4–0.8 cm long; floral tube 1–1.2 mm long, 1.8–2.2 mm in diameter, a ring of white villous pubescence at the mouth within; sepals 3.5–6 mm long, 1–1.2 mm wide, lanceolate, keeled; petals 5.5–11 mm long, 3–5 mm wide, pink to rose-purple, obovate, the notch 1 mm deep; filaments of longer stamens 4–6 mm long, those of shorter ones 2–2.5 mm long; anthers 0.7–1 mm long, 0.4–0.5 mm in diameter, oblong; style 3–5.5 mm long, erect, glabrous or sometimes with scattered villous pubescence near the base; stigma capitate, 1.6–1.8 mm long, 1.5–1.8 mm in diameter, at least the longer stamens shedding pollen onto stigma at anthesis. Capsules 3–7 cm long, sparsely glandular and strigillose, borne on pedicels 0.4–1.2 cm long. Seeds 1.1–1.5 mm long, 0.35–0.45 mm in diameter,

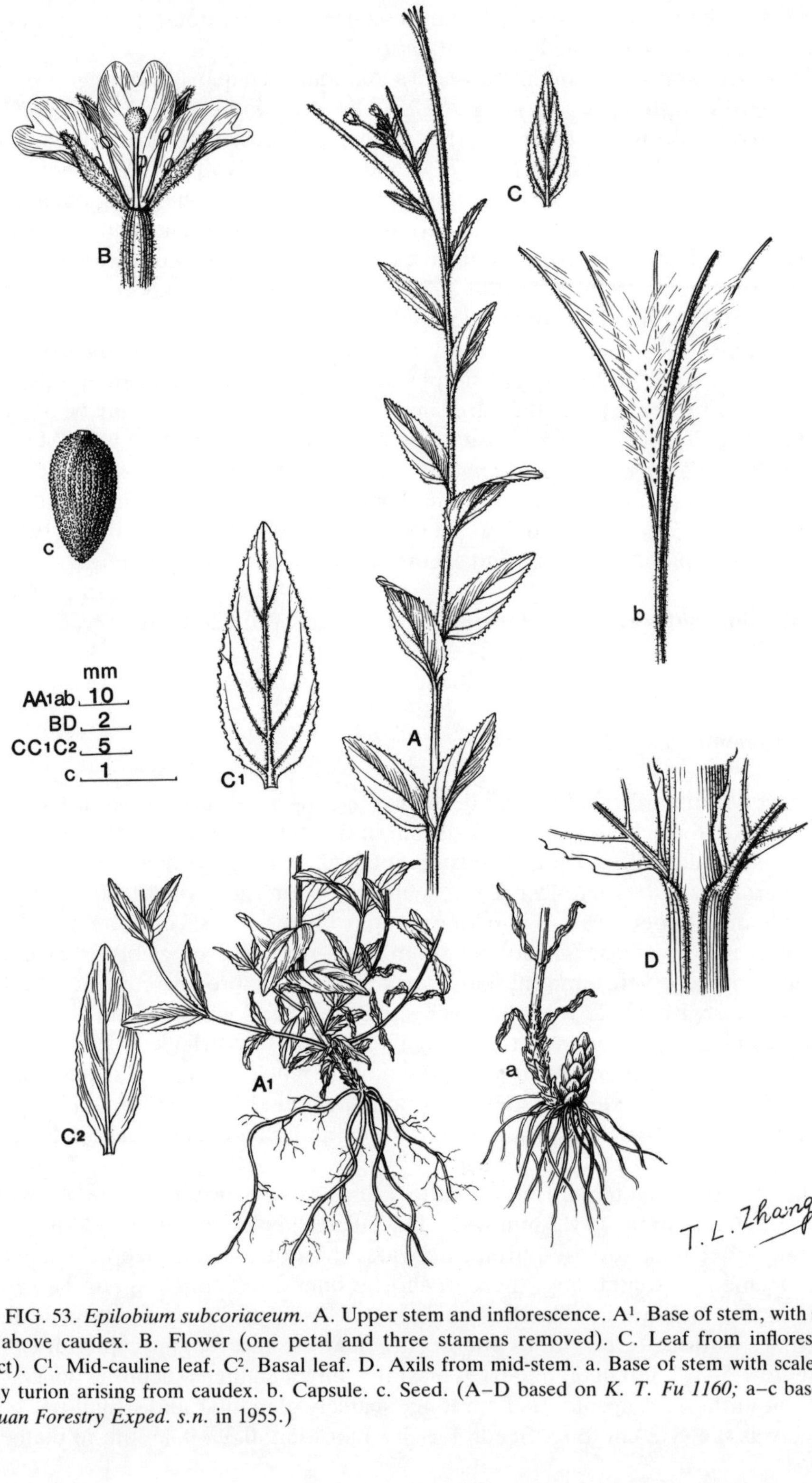

FIG. 53. *Epilobium subcoriaceum*. A. Upper stem and inflorescence. A^1. Base of stem, with scales just above caudex. B. Flower (one petal and three stamens removed). C. Leaf from inflorescence (bract). C^1. Mid-cauline leaf. C^2. Basal leaf. D. Axils from mid-stem. a. Base of stem with scales and fleshy turion arising from caudex. b. Capsule. c. Seed. (A–D based on *K. T. Fu 1160;* a–c based on *Sichuan Forestry Exped. s.n.* in 1955.)

oblanceolate or narrowly obovoid, slightly acuminate with a short pellucid chalazal collar, color unknown, the surface coarsely papillose; coma 6–7 mm long, dull white, readily detaching. Chromosome number unknown. Fig. 53.

Phenology. Collected in flower from July to August, in fruit from August to September.

Distribution (Fig. 42A). Endemic to China, in western Shaanxi, southeastern Gansu, eastern Qinghai, western Sichuan, northern Yunnan, and southeastern Xizang (Tibet); in moist places in the mountains, along streams, boggy areas, and weedy places; 2400–3700 m.

ADDITIONAL SPECIMENS EXAMINED. **China.** SHAANXI: Taibei Shan, *Lanzhou Univ. Exped. 289* (LZU); Gaoyang, *T. P. Wang 16200* (KUN, PE); no locality, *T. P. Wang 7664* (PE).—GANSU: "Gargannar," *R. C. Ching 909* (GH); Xiahe, *K. T. Fu 1160* (PE), *T. P. Wang 7368* (PE).—QINGHAI: Tongren, *S. W. Liu & D. S. Lo 2-1392* (HNWP).—SICHUAN: Nanping to Songpan, *W. L. Chen & C. J. Chen 8856* (MO, PE); Songpan, *W. L. Chen & C. J. Chen 8208* (PE), *8314* (MO, PE); Mt. Emei, *W. K. Hu 8257* (E), *C. W. Yao 3367* (PE); Kangding, *W. P. Fang 6058* (E), *Smith 10608* (S, UPS), *10747* (S); Muli, *W. L. Chen & C. J. Chen 5776* (MO, PE), 1955, *Sichuan Forestry Exped. s.n.* (PE), *Qinghai-Xizang Exped. 13028* (KUN, PE), *13050* (KUN, PE), *13741* (KUN, PE); Daocheng, *Qinghai-Xizang Exped. 5904* (KUN, PE), *T. T. Yu 14355* (BM, E, GH).—YUNNAN: Zhaotong, *Kunming Inst. Bot. 235* (KUN); Huize, *Dongchuan Exped. 63-174* (KUN); Dali, *Zhongdian Exped. 63-3807* (KUN), *H. T. Tsai 53948* (PE); Lanping, *H. T. Tsai 53788* (PE); Lushui, *Nujiang Exped. 1835* (KUN); Lijiang, *Schneider 3297* (GH), *Y. Z. Zhao 22070* (PE); Weixi, *C. W. Wang 67635* (PE); Deqen, *C. W. Wang 69741* (KUN, NAS, PE).—XIZANG: Jomda, *Xizang Exped. 2259* (HNWP); Zayu, *Xizang Exped. 1228* (HNWP), *1235* (HNWP); Nyingchi, *B. Z. Go 23300* (HNWP).

This species, which has been ignored for the most part since its original description in 1879, was based on a single collection. More recent collections from southwestern China allow us to verify and extend Haussknecht's concept. *Epilobium subcoriaceum* is similar to *E. sikkimense* in having scales resulting from fleshy soboles on the stem bases, and distinct raised pubescent lines decurrent from the margins of the petioles. It differs from *E. sikkimense* by its much reduced bracts and open inflorescence, leaf shape, and seed shape and surface. *Epilobium subcoriaceum* is also similar to *E. wallichianum* in habit and vestiture pattern, but differs from that species in having scales on the stem bases and short pedicels.

29. Epilobium gouldii Raven, Bull. Brit. Mus. (Nat. Hist.) Bot. 2: 371, pl. 35b. 1962.—TYPE: CHINA. Xizang (Tibet), Gyangze to Pagri ("Gautsa to Phari Drong"), north of Yadong, approximately 28°53′N, 89°33′E, 3650–4350 m, 13 Aug 1938, *Gould 1452* (holotype: K!).

Common name: Lingen Liuyecai (Li 1986).

Erect perennial herb, forming round fleshy basal turions that leave coriaceous scales 5–8 cm below ground level at base of stem. Stems 23–30 cm tall, 1.5–3 mm in diameter, strict, simple or sparingly branched, glabrous except for raised strigillose lines decurrent from margins of the petioles. Leaves opposite, alternate on the inflorescence, herbaceous, sessile; cauline blades 2–3 cm long, 1–1.4 cm wide, ovate, subglabrous except for sparsely strigillose veins and margins, apex acute to short-acuminate, base broadly rounded, the margins acute-serrulate with 16–28 teeth on each side, the lateral veins 4–5 on each side of midrib. Inflorescence

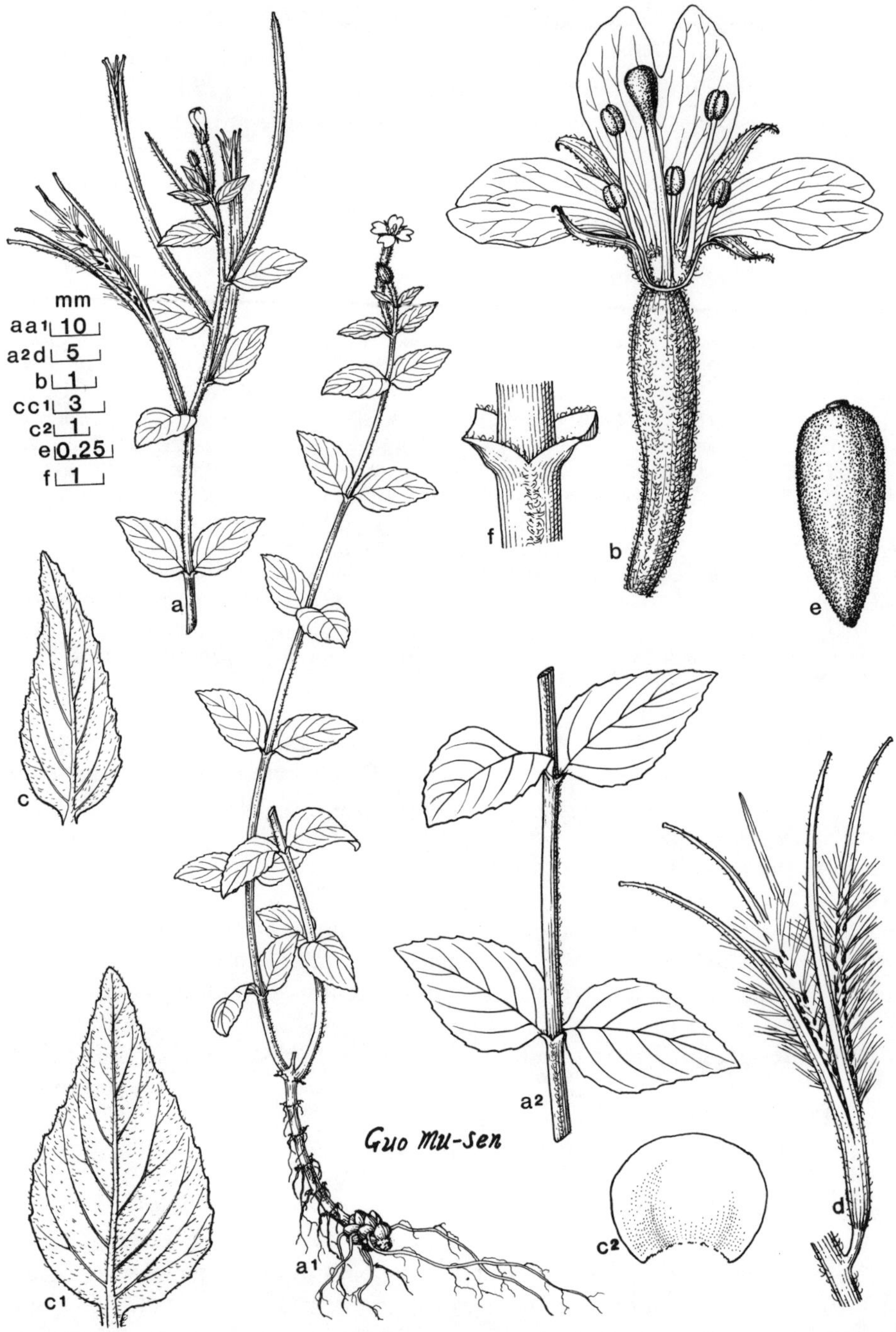

FIG. 54. *Epilobium gouldii*. a. Upper stem and inflorescence. a[1]. Plant (without main stem) with persistent scales at base. b. Flower (one petal and three stamens removed). c. Leaf from inflorescence (bract). c[1]. Mid-cauline leaf. c[2]. Coriaceous scale from base of stem. d. Capsule. e. Seed. f. Axils from mid-stem. (Based on *J. W. Zhang & G. X. Fu 1135*.)

slightly nodding before anthesis, strigillose and glandular, bracts mostly more than half the length of ovaries they subtend. Flowers slightly nodding in bud; buds 3–5 mm long, narrowly ovoid, apiculate; ovaries 1.2–1.8 cm long, strigillose and glandular; pedicels 3–6 mm long; floral tube 0.7–0.9 mm long, 1.8–2 mm in diameter, with a ring of villous pubescence at the mouth within; sepals 3.5–4.5 mm long, 0.7–1.2 mm wide, lance-oblong, apiculate, with scattered strigillose and glandular pubescence and a tuft of long spreading hairs at junctures of sepal bases; petals 4.5–5.5 mm long, 2.5–3.5 mm wide, rose-purple, obcordate, the notch 0.7–1 mm deep; filaments of longer stamens 2.5–3 mm long, those of shorter ones 1.5–2 mm long; anthers 0.5–0.6 mm long, 0.3–0.45 mm in diameter, broadly ovoid; style 2.5–3 mm long, erect, glabrous; stigma clavate to subcapitate, 1–1.2 mm long, 0.7–0.9 mm in diameter, at least the longer stamens shedding pollen onto stigma at anthesis. Capsules 4.5–6 cm long, borne on pedicels 0.6–1.5 cm long. Seeds 0.9–1 mm long, 0.35–0.36 mm wide, obovoid, with short chalazal collar, light brown, the surface papillose; coma 5–6 mm long, white, easily detached. Chromosome number unknown. Fig. 54.

Phenology. Collected in flower from July to August, never collected in fruit.

Distribution (Fig. 42B). Central Himalaya, including Sikkim and adjacent Tibet; in China only in southern Xizang; in alpine meadows; 3650–4350 m.

ADDITIONAL SPECIMENS EXAMINED. **China.** XIZANG: "Nathu La to Champitang," *Gould 409* (K); Yadong to Pagri, *J. W. Zhang & G. X. Fu 1135* (PE).

This rare, or at least little-collected, species seems most closely allied to *E. laxum* of the western Himalaya, and perhaps to *E. sikkimense,* but is distinguished from them by its fleshy turions and somewhat smaller seeds. A more precise delimitation of *E. gouldii* must await the collection of more complete specimens, however, especially some with mature seeds.

30. Epilobium fangii C. J. Chen, Hoch & Raven, sp. nov.—TYPE: CHINA. Sichuan, Mt. Omei, on the way from Jiulaodong to Caipengzi, ca. 2000 m, in rocky places, expedition sponsored by W. P. Fang, 11 May 1952, *Hsiung, Chang & Tsiang 30490* (holotype: PE!; photograph: MO!).

Common name: Chuanxi Liuyecai.

Herba perennis, turionibus hypogaeis oblongis carnosis et squamis brunneis praedita. Caulis simplex vel raro paulo ramosus, undique strigillosus, praeter lineas duo subelevatas indistinctas strigillosas e petiolorum marginibus decurrentes. Folia manifeste petiolata [petiolis 1–4 (–6) mm longis], elliptica vel elliptico-oblonga, 1.5–4 cm longa, 0.5–1.5 cm lata. Inflorescentia dense strigillosa et sparsim glandulosa, bracteis reductis. Petala subrosea vel roseo-purpurea, 6–7.5 mm longa; stigma capitatum. Semina 1.1–1.4 mm longa, 0.3–0.45 mm crassa, papillosa, rostro brevissimo pellucido instructa.

Erect perennial herb, forming fleshy elongated basal turions 2–3.5 cm underground that leave coriaceous brownish scales at base of stem, the scales 5–8 mm long, 2–7 mm wide, obovate to broadly elliptic. Stems 15–40 cm tall, 1.2–3 mm in diameter, simple or sometimes branched on the upper part, strigillose all over with scattered glandular hairs above and with 2 indistinct elevated lines of hairs decurrent from margins of petioles. Leaves opposite, the upper ones alternate, subcoriaceous, green or often purplish on abaxial surface, with conspicuous petioles 1–4 (–6) mm long; cauline blades 1.5–4 cm long, 0.5–1.5 cm wide, elliptic or elliptic-oblong, subglabrous except for scattered strigillose veins and margins, apex subobtuse or sometimes acute, base cuneate or broadly cuneate, the margins obscurely serrulate with 5–18 teeth on each side, lateral veins inconspicuous, 4–5 on each side of midrib. Inflorescence erect, densely strigillose with a few glandular hairs, bracts about half as long as the ovaries they subtend. Flowers erect; buds 3–4.5 mm long, 2–2.5 mm in diameter, oblong-ovoid, apiculate, strigillose and glandular; ovaries 1.5–3 cm long, glandular; pedicels 0.3–0.7 cm long; floral tube 0.6–1.1 mm long, 1.1–1.7 mm in diameter, with a ring of spreading hairs at the mouth within; sepals 4–5 mm long, 0.8–1.2 mm wide, oblong-lanceolate, keeled; petals 6–7.5 mm long, 2.8–3.5 mm wide, pink to rose-purple, narrowly obcordate, the notch 1–1.5 mm deep; filaments pinkish, those of longer set 2.8–3.5 mm long, those of shorter ones 2.4–2.6 mm long; anthers 0.7–1 mm long, 0.5–0.7 mm in diameter; style 3–4 mm long, erect, glabrous or rarely with a few spreading hairs near base; stigma capitate, 0.8–1.1 mm long, 1–1.2 mm in diameter, at least long stamens shedding pollen onto stigma at anthesis. Capsules 3–7 cm long, 1–1.8 mm in diameter, subglabrous or scattered strigillose, borne on pedicels 0.5–1.5 cm long. Seeds 1.1–1.4 mm long, 0.3–0.45 mm wide, narrowly obovoid, with a very short and inconspicuous pellucid chalazal collar, the surface finely papillose; coma 6–7 mm long, dingy white, detaching easily. Chromosome number: n = 18. Figs. 2b, 55.

Phenology. Collected in flower from May to July (rarely to August), in fruit from June to August (rarely to October).

Distribution (Fig. 42A). Endemic to China; mostly in western Sichuan and one station in northern Yunnan; in rather open places along stream banks, at bases of rock walls or scree slopes; (1100–) 1700–3500 m.

ADDITIONAL SPECIMENS EXAMINED. **China.** SICHUAN: Mt. Emei, *H. C. Chow 9875* (GH), *Y. H. Dao 50405* (SZ), *W. P. Fang et al. 20453* (PE), *K. C. Kuan et al. 505* (PE), *T. N. Liou 10301* (PE), *T. N. Liou & C. Wang 1208* (PE), *W. W. Ma 2602* (PE), *2611* (KUN), *Raven et al. 26101* (MO, PE), *26114* (MO, PE), *26132* (MO, PE), *26133* (PE), *26136* (MO, PE), *Sino-U.S.S.R. Bot. Exped. 980* (PE), *S. C. Sun & K. Chang 489* (GH), *T. H. Tu 47* (PE), *230* (PE), *S. X. Wang 451* (PE), *C. Y. Wu et al. 53* (PE), *76* (KUN, PE), *G. H. Yang 55395* (NAS, PE), *C. W. Yao 5112* (PE), *X. B. Zhang 363* (NAS), *G. S. Zhou 81240* (PE); Ebian, *C. W. Yao 3055* (PE), *T. T. Yu 832* (GH, PE); Hongya, *C. W. Yao 3942* (KUN, NAS); Leibo, *Z. T. Guan 8714* (PE), *Sichuan Econ. Pl. Exped. 975* (PE); Luding, *K. Y. Lang et al. 325* (KUN, PE); Baoxing, *S. S. Chang et al. 5464* (CDBI), *T. P. Soong 38485* (PE, SZ); Tianquan to Erlang Shan, *Y. T. Chang & K. Y. Lang 7* (PE), *K. L. Chu 2587* (BM, PE, W); Erlang Shan, *D. P. He 4406* (SZ), *H. C. Tsiang & C. H. Hsiung 34209* (PE), *34587* (PE), *Sichuan Econ. Pl. Exped. 831* (CDBI, PE); Daxianling, *Smith 10118* (S, UPS); Shimian, *W. L. Chen & C. J. Chen 6229* (PE); Xichang, *T. T. Yu 1274* (PE); Pege, *Sichuan Med. Pl. Exped. 25025* (NAS): Western Sichuan, *W. P. Fang 4542* (NAS), *7696* (NAS), *7697* (NAS).—YUNNAN: Qiaojia, *B. X. Sun 1081* (YNU).

Epilobium fangii, like *E. blinii* and several other species of the genus, occurs only in a restricted area of southwestern China, in this case in relatively inaccessible regions of western Sichuan and northern Yunnan. Although the species may be somewhat more widespread than the few collections indicate, *E. fangii* nevertheless appears to be a narrow endemic and apparently grows in rather specialized habitats, especially along the base of rock walls. It may be allied to *E. gouldii,* but

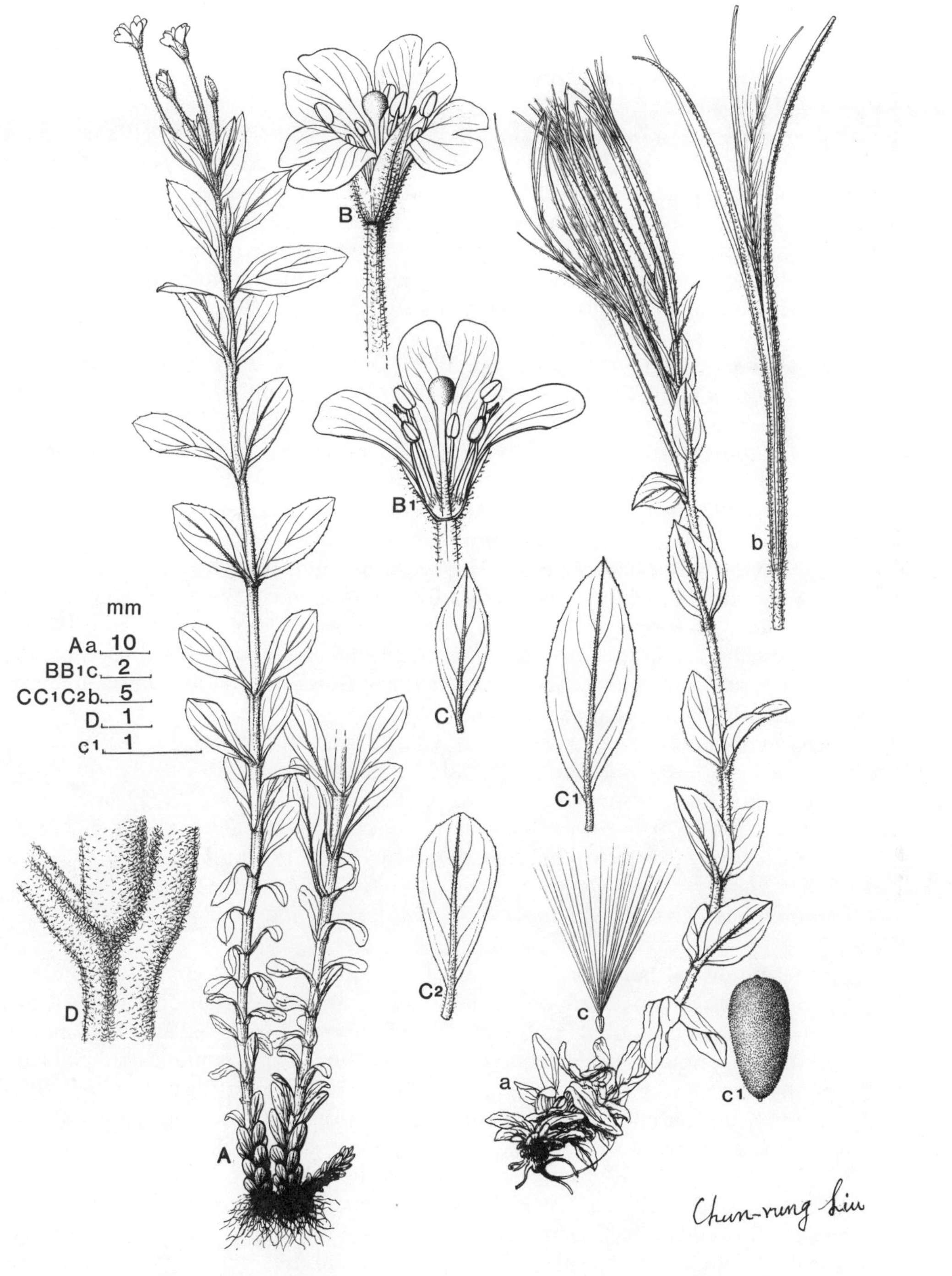

FIG. 55. *Epilobium fangii.* A. Habit; persistent scales and fleshy turion at base of stem. B. Flower. B^1. Flower (one petal and three stamens removed); note hairs at mouth of floral tube and near base of style. C. Leaf from inflorescence (bract). C^1. Mid-cauline leaf. C^2. Basal leaf. D. Axils from mid-stem. a. Habit; plant in fruit, with persistent scales at base of stem. b. Capsule. c. Seed with coma. c^1. Seed (coma removed). (A–D based on *C. H. Hsiung et al. 30490,* type; a–c^1 based on *C. W. Yao 5112.*)

differs from it in leaf shape and petiole length, flower size, and seed size. It may also be related to *E. taiwanianum,* an endemic from Taiwan, with which it shares similar turions, seeds, and general habit, but from which it differs in stem pubescence and leaf shape.

This species is named in honor of Prof. Fang Wen Pei (1899–1983), senior professor of Sichuan University (1937–1983), an acknowledged expert on Chinese Aceraceae, *Rhododendron,* and the Sichuan flora, and one of the major figures in modern Chinese botany. His personal collections (more than 20,000), and those gathered on expeditions organized and sponsored by him during more than fifty years of botanical activity, have made tremendous contributions to our knowledge and understanding of the flora of Sichuan.

31. Epilobium palustre Linnaeus, Sp. Pl. 1: 348. 1753. *Epilobium palustre* var. *typicum* C. B. Clarke in J. D. Hooker, Fl. Brit. India 2: 585. 1879.—TYPE: Presumably from Sweden, *Linnaeus* (holotype: LINN 486/6!).

Epilobium rhynchocarpum Boissier, Diagn. ser. 2, 2: 53. 1856. *Epilobium fischerianum* Pavlov, Bjull. Moskovsk. Obsc. Isp. Prir., Otd. Biol. n.s., 38: 105. 1929, nom. superfl.—TYPE: IRAN. "Northern Persia," near Lar at the foot of Mt. Demawend, *Kotschy 616* (holotype: G!; isotypes: BM! W!). Pavlov based his species on Haussknecht's informal "forma," which is not considered to be validly published unless validated by a subsequent author (Hoch & Raven 1990). Typification of Pavlov's epithet rests with Haussknecht's treatment; because Haussknecht included in his discussion the Kotschy collection that is the holotype of Boissier's validly published name, the Pavlov name is superfluous.

Epilobium palustre var. *majus* C. B. Clarke in J. D. Hooker, Fl. Brit. India 2: 585. 1879. No authentic material located; cf. Raven (1962a).

Epilobium palustre var. *minimum* C. B. Clarke in J. D. Hooker, Fl. Brit. India 2: 586. 1879.—TYPE: KASHMIR. Ladakh, Deosai, 3950 m, 31 Jul 1876, *Clarke 29816* (lectotype, designated by Raven 1962a: K!).

Common name: Zhaosheng Liuyecai (Li 1986).

Erect perennial herb, producing slender, filiform basal stolons 5–50 cm long with widely spaced small leaves, terminating in small fleshy turions, which leave brownish scales at base of plant. Stems (5–) 15–70 cm tall, 0.5–5.5 mm in diameter, simple or sometimes well branched, strigillose all over, sometimes subglabrous below midstem, lacking decurrent raised lines or with indistinct strigillose lines decurrent from margins of the petioles. Leaves opposite, alternate on inflorescence, sessile or rarely with petioles 1–3 mm long; cauline blades 1.2–7 cm long, 0.3–1.2 (–1.9) cm wide, sublinear to narrowly lanceolate or elliptic, sparsely strigillose on upper surface and on veins of lower surface, or rarely subglabrescent, apex acute or obtuse, base subrounded or cuneate, the margins entire to obscurely denticulate with 5–9 teeth on each side, occasionally revolute, the inconspicuous lateral veins 3–5 on each side of midrib. Inflorescence erect or slightly nodding in bud, densely strigillose, sometimes mixed with glandular hairs. Flowers suberect; buds 2–3 mm long, 1.8–2.2 mm in diameter, elliptic-ovoid, short-apiculate; ovaries 1.6–2.5 (–3) cm long, densely strigillose with a few glandular hairs; pedicels 0.8–1.5

cm long; floral tube 1–1.2 mm long, 1.3–2 mm in diameter, subglabrous or with a ring of spreading hairs at the mouth within; sepals 2.5–4.5 mm long, 1–1.2 mm wide, oblong-lanceolate, acute, densely strigillose and glandular; petals (3–) 5–7 (–9) mm long, 2–3 (–4.5) mm wide, white to pink or rose-purple, obcordate, the notch 0.8–1 mm deep; filaments of longer stamens 2–2.8 mm long, those of shorter ones 1.2–1.5 mm long; anthers 0.4–0.6 mm long, 0.2–0.4 mm wide, oblong; style 1.8–3.8 mm long, erect, glabrous; stigma clavate to subcylindric, 1–1.8 mm long, 0.4–0.7 mm in diameter, the longer stamens usually shedding pollen onto stigma at anthesis. Capsules 3–9 cm long, strigillose, borne on pedicels 1–5 cm long. Seeds (1.1–) 1.3–2.2 mm long, 0.38–0.55 mm in diameter, fusiform to narrowly obovoid, with conspicuous chalazal collar 0.08–0.3 mm long, brown, the surface very finely papillose; coma 6–9 mm long, dull white or occasionally tawny, not easily detaching. Chromosome number: n = 18. Figs. 2c, 6e, 7e, 56.

Phenology. Collected in flower from June to August, in fruit from August to September.

Distribution (Fig. 40). Circumboreal; in North America from Alaska and subarctic Canada south to California, Colorado, the Great Lakes region, and North Carolina; in southern Greenland; in northern and central Europe; in Asia throughout subarctic U.S.S.R. south to the Caucasus, the Caspian region, the Himalaya from Kashmir to Nepal and southern Xizang (Tibet), China, Japan, and the Soviet Far East; in China in Heilongjiang, Jilin, Liaoning, Nei Monggol, Hebei, Shanxi, Shaanxi, Gansu, Qinghai, Xinjiang, Sichuan, Yunnan, and Xizang (Tibet); in wet places, especially bogs, marshes, and along streams and rivers, and in meadows of subalpine areas; 200–4500 (–4950) m.

ADDITIONAL SPECIMENS EXAMINED. **China.** HEILONGJIANG: Ning'an, *Y. L. Zhang et al. 1207* (PE); Yichun, *T. H. Liou et al. 7776* (IFP, PE), *7872* (PE), *10184* (IFP, PE); Yilin, *Y. L. Zhang et al. 1974* (IFP); Dailing, *C. S. Sung 128* (PE), *197* (PE), *W. Wang et al. 922* (PE), *Z. Wang 4332* (IFP); Harbin, *Skvortzov & G. Z. Wang 1101* (IFP); Xiao Hinggan Ling, *W. Z. Fang 495* (KUN, NAS); Tieli, *N. E. China Exped. 495* (PE); Da Hinggan Ling, *Jilin Teacher's Univ. 657* (PE); Gen He, *D. Z. Fu 462* (PE), *T. N. Liou 8830* (NAS, PE); "Northern Manchuria," 1938, *Skvortzov s.n.* (GH); "Manchuria," 1896, *Chaffanjon s.n.* (P), *F. H. Chen 355* (NY, PE); no locality, *D. C. Zhao & Balarov 298* (IFP).—JILIN: Yanbian, 1959, *Yanbian Pl. Exped. s.n.* (IFP); Antu, *Yanbian Pl. Exped. 138* (PE), *502* (PE), *715* (PE), *P. Y. Fu et al. 1588* (IFP, PE); Shuangliao, 1918, *Y. Yabe s.n.* (NAS); Changbai Shan, *T. N. Liou et al. 1638* (IFP), *1785* (PE), *J. J. Chien 1031* (KUN, PE), *Yanbian Exped. 316* (PE), *W. Wang et al. 2189* (IFP, PE), *2363* (PE); Jiaohe, *Y. L. Zhang et al. 1011* (IFP, PE), *Y. L. Zhou et al. 2374* (NAS, PE), *P. Y. Fu 2321* (IFP, PE); Fusong, *Y. L. Zhang et al. 666* (IFP, PE), *T. N. Liou et al. 803* (IFP, PE), *S. G. Wang & S. P. Li 537* (IFP, PE); Dunhua, *Y. L. Zhang et al. 1087* (PE); near "O-muhsien," *H. W. Kung 1964* (PE), *2178* (NY); Changchun, 1910, *Yabe s.n.* (PE).—LIAONING: Benxi, *Z. Wang 1444* (IFP); "Sandaogas," 1917, *Yabe s.n.* (NAS).—NEI MONGGOL: "Weifuling," *N.E. Norm. Univ. 431* (NETU); Hailar, *Z. Wang 2114* (IFP, PE); Ergun Yougi, *Z. Wang 1399* (PE), *1723* (IFP); Hulun Buir Meng, *Y. Z. Liu 2* (NMU), *Nei Mong. Inst. For. 1174* (NMU), *G. Z. Wang 2784* (IFP); Jirem Meng, *S. Y. Jing 7* (NMU), *Y. C. Ma 23* (PE), *48* (HIMC, NMU); Ju Ud Meng, *Sino-German Exped. 8495* (PE), *Y. L. Zhang et al. 307* (IFP, PE); Xilin Gol Meng, *Nei Mong. Univ. 211* (PE); Duolun, *Duolun Pl. Exped. 4* (NMU); "Wulashan," *W. Y. Hsia 3109* (PE); Fengzheng, *Y. W. Tsui 1023* (PE); Ulanqab Meng, *X. Z. Zhang 134* (PE), *Nei Mong. Univ. 422* (HIMC, NMU); Hohhot, *Y. X. Hou 82* (NMU); Helan Shan, *Nei Mong. Univ. 166* (NMU); Hinggan Meng, *H. J. Meng 4* (NMU), *726* (NMU); Eer gu na qi, *C. Q. Lin et al. 1811* (IFP); E qi gu na, *Z. Wang 1974* (IFP); no locality, 1956, *X. Z. Lang s.n.* (PE).—HEBEI: Xiawutai Shan, *J. F. Li 11012* (NY, PE), *C. F. Li 11016* (NY, PE), *J. C. Liu 911* (UC), *T. P. Wang 698* (PE), *C. W. Wang 62078* (GH, PE); 1906, *Y. Yabe s.n.* (NAS); Zhuolu, *C. G. Yang 186* (KUN, NAS, PE), *1489* (PE); Zhangbei, *Hebei Exped. 8574* (PE); Laiyuan, *Hebei Univ. Agr. 42738* (PE), *42740* (PE); no locality, 1930, *E. Licent s.n.* (M).—SHANXI: Wutai Shan, *T. Tang 1111* (NY); Wutai, *K. C.*

FIG. 56. *Epilobium palustre*. a. Upper stem and inflorescence. a[1]. Base of stem; filiform stolons, bearing minute cataphylls, originate from caudex and lower axils. b. Flower. c. Leaf from inflorescence (bract). c[1]. Mid-cauline leaf. c[2]. Basal leaf. d. Capsule with subtending bract. e. Seed with persistent coma. e[1]. Seed (coma removed); note conspicuous chalazal collar. f. Axil from inflorescence. (Based on *C. G. Yang 1489*.)

Kuan & Y. L. Chen 2417 (PE), *J. W. Wang et al. 1* (PE); Ling Xian, *Huang He Exped. 1915* (PE).—SHAANXI: Suide, *Huang He Exped. 850* (PE).—GANSU: Tainzhu, *Y. Q. He 4151* (PE), 1967, *H. D. Zhao s.n.* (LZU); "Liangchow," *Trippner 343* (M); "Min Shan," *T. P. Wang 7598* (PE); Xiahe, *L. H. Zhou & H. T. Chang 1507* (HNWP), *Y. He 194* (LZU); Lintan, *Z. X. Peng 864* (LZU); border between Gansu and Qinghai, *Y. C. Wu 451* (NAS); Maqu, *G. Q. Yi 8* (LZU), *Maqu Exped. 56* (LZU); 1976, *Lanzhou Univ. s.n.* (LZU); near "Chuoni," *T. P. Wang 5345* (PE); Huangcheng, *Lanzhou Univ. Exped. 276* (LZU); "Labrang et Teotschou," *Fenzel 2635* (W); "Tsomousche, Schibsa et Labrang," *Fenzel 2555* (W).—QINGHAI: Huzhu, *B. Z. Guo 6591* (HNWP); Yongian, *X. M. Zhou & Q. Sha 79* (HNWP); Qingshizui, *G. F. Yang & R. L. Zhao 100* (HNWP); Haiyan, *P. C. Tsoong 8419* (PE); Menyuan, *Z. L. Ching 1302* (PE), *B. Z. Guo & W. I. Wang 12206* (HNWP), *Qinghai Exped. 1302* (HNWP), *1477* (HNWP), *P. C. Tsoong 8997* (KUN, PE); Datong, *J. M. Liu 6139a* (PE), *K. M. Liou 6576* (PE), *K. C. Kuan 77278* (PE); Gonghe, *B. Z. Guo & W. I. Wang 12043* (HNWP, NAS, PE); Tongren, *L. H. Zhou & L. N. Sun 944* (HNWP), *S. W. Liu & D. S. Lo 1195* (HNWP); Xinghai, *B. Z. Guo 6219* (HNWP); Henan, *L. H. Zhou & L. N. Sun 2333* (HNWP), *B. Z. Guo 9976* (HNWP); Qumarleb, *S. W. Liu & Y. F. Huang 869* (HNWP); Jigzhi, *Xizang Med. Pl. Exped. 380* (HNWP), *411* (HNWP), *553* (HNWP), *758* (HNWP), *Golog Exped. 574* (HNWP); Baima, *D. S. Lo et al. 80-1301* (PE), *80-1382* (PE); Chindu, *D. S. Lo et al. 80-0942* (PE); Yushu, *B. Z. Gong & W. I. Wang 8429* (HNWP), *Qinghai Exped. 247* (HNWP); Nangqen, *Y. C. Yang 1136* (HNWP), *1339* (HNWP, PE); Zadoi, *S. W. Liu 423* (HNWP); no locality, *Hopkinson 1256* (PE).—XINJIANG: Fuhai, 1964, *G. L. Chu et al. s.n.* (PE); Altay, *T. Y. Chou 652123* (HNWP, NAS, PE), *Xinjiang Exped. 1548* (PE); Xiaogonggou, NE of Altay, *C. J. Chen & Hoch 86081* (MO, PE); Jaishihanagai, NW of Altay, *C. J. Chen & Hoch 86132* (MO, PE), *Konta & K. Y. Lang 163* (PE); Qitai, *G. J. Liu et al. 6523* (XJBI); Tacheng, *K. C. Kuan 2838* (PE); Wenquan, *C. J. Chen & Hoch 86263* (MO, PE), *K. C. Kuan 4570* (PE); Urumqi, *Bayi Coll. Agr. 14470* (PE), *Konta & K. Y. Lang 86* (PE), *T. N. Liou 2855* (PE), *3032* (PE), *Xinjiang Exped. 5404* (XJBI); Dabancheng, *C. J. Chen & Hoch 86026* (MO, PE); Tian Shan, *T. Y. Chou et al. 651201* (HNWP, NAS, PE), *K. C. Kuan 1954* (PE); Bogda Shan, *T. N. Liou 3506* (PE); Fukang, *C. J. Chen & Hoch 86053*, *K. C. Kuan 4264* (PE); Toli, *K. C. Kuan 2446* (PE), *2685* (PE).—SICHUAN: Nanping, *W. L. Chen & C. J. Chen 8847* (MO, PE), *8854* (PE); Songpan, *W. L. Chen & C. J. Chen 8212* (MO, PE); *8243* (MO, PE); Hongya, *C. W. Yao 3942* (PE); Aba, *W. L. Chen & C. J. Chen 7909* (MO, PE); Barkam, *Sichuan Med. Pl. Exped. 22520* (NAS); Tanggor, *Sichuan Econ. Pl. Exped. 349* (CDBI); Zoige, *Sichuan Veg. Exped. 10298* (CDBI), *10402* (CDBI); Hongyuang, *S. K. Wu 2* (MO), *B. Z. Ni 354* (CDBI); Dawu, *Sichuan Med. Pl. Exped. 15627* (NAS); Luhuo, *W. L. Chen & C. J. Chen 7358* (MO, PE); Longriba, *X. Li 72152* (PE); N. W. Sichuan, *Smith 4272* (UPS); Kangding, *X. N. Jiang 36261* (PE), *H. S. Kung 6058* (PE); Sertar, *W. L. Chen & C. J. Chen 7492* (PE), *S. G. Li 6578* (CDBI, PE); Garze, *W. L. Chen & C. J. Chen 6934* (MO, PE), *Sichuan Med. Pl. Exped. 15414* (NAS); Dege, *Sichuan Exped. 7533* (CDBI, PE); Shiqiu, *Q. Q. Wang 7633* (CDBI); Daocheng, *Qinghai-Xizang Exped. 5584* (PE); Xiancheng, *Qinghai-Xizang Exped. 3710* (PE).—YUNNAN: Huize, *N. E. Yunnan Exped. 252* (KUN); Zhongdian, *Zhongdian Exped. 3691* (KUN).—XIZANG: Zayu, 1984, *H. S. Kung et al. s.n.* (MO, PE); Qamdo, *Qinghai-Xizang Exped. 12729* (PE); Cona, *Qinghai-Xizang Exped. 74-2746* (PE); Zetang, *C. Y. Wu et al. 75-782* (KUN, PE); Damzong, *Y. H. Tong et al. 024* (CISNR), *142* (CISNR), *143* (CISNR); Lhasa, 1904, *Walton s.n.* (CGE), *Richardson 185* (BM, RSA), *B. Z. Ge & W. Y. Wang 22227* (HNWP); Gyangze, 1905, *Walton s.n.* (B, K, UC); Sangang, *P. C. Tsoong 5722* (PE); Namling, 1972, *Xizang Med. Pl. Exped. s.n.* (PE); Ngamring, *Qinghai-Xizang Exped. 7321* (PE); Burang, *Qinghai-Xizang Exped. 76-8548* (KUN, PE); "Western Tibet," 1848, *Thomson s.n.* (K); Zanda, *Xizang Exped. 3803* (NWBI, XJBI); Rutog, *Xizang Exped. 3489* (HNWP, PE); Gyangze, 1925, s.c. (K); no locality, 1914, *Purdom s.n.* (K).

This species is widely distributed in boreal regions of North America (Munz 1965) and Eurasia (Raven 1968), and variable particularly in stature, leaf and flower size, and general vestiture (Haussknecht 1884). Autogamy is common, since most individuals have relatively small flowers and, like most species of *Epilobium,* are only infrequently visited by pollinators (Raven 1979). Perennation by turion-tipped filiform stolons, apparently an adaptation to the palustrine habitats favored by the species (Keating et al. 1982), is often quite vigorous and results in large if somewhat inconspicuous populations.

Plants of *E. palustre* in eastern Asia vary conspicuously in stature, leaf and seed size, and length of seed collar. Collections from high elevations on the Qinghai-

Xizang Plateau and the Hengduanshan range, like those from far northern areas in Asia and North America, are often short, unbranched plants with nodding inflorescence and small, entire, revolute leaves; these plants often have been segregated (var. *minimum;* also *E. oliganthum* Michaux in North America). However, these differences are primarily quantitative in nature, show a clinal pattern of variation, and largely disappear when the different morphological types are grown under uniform conditions (Hoch et al., unpubl.). Overall, there are no strong correlations between morphological variability and geographical or ecological range differences, and for that reason we do not formally recognize any infraspecific taxa in *E. palustre.*

32. Epilobium fastigiatoramosum Nakai, Bot. Mag. (Tokyo) 33: 9. 1919.—TYPE: KOREA. Koje Island ("Ko-ei"), c. 34°50′N, 128°40′E, in sandy places along streams, where it is quite plentiful, *Nakai 3280* (holotype: TI!).

Epilobium baicalense Popov, Bot. Mater. Gerb. Bot. Inst. Komarova Akad. Nauk SSSR 18: 6. 1957.—TYPE: U.S.S.R. Irkutskaya Oblast, Lake Baikal, western shore, Maloye More, on a promontory "Ujuga" not far from the village "Chala," 26 Aug 1951, *Popov & Bardunov s.n.* (holotype: NS!).

Common name: Duozhi Liuyecai (Liou et al. 1977).

Erect perennial herb, with short leafy soboles, usually lacking stolons, or rarely with short filiform, hypogeous stolons with scattered cataphylls. Stems 7–50 (–80) cm tall, 1.3–4 mm in diameter, simple to densely branched, strigillose all over, or often subglabrous in lower part. Leaves opposite, alternate on inflorescence, sessile or with petioles 1–2 mm long; cauline blades (1–) 2–7 cm long, (0.2–) 0.3–1.7 cm wide, lance-elliptic or sometimes lance-oblong, sparsely strigillose on adaxial surface and on veins of abaxial surface, apex acute, sometimes just the tip obtuse, base cuneate or subrounded, the margins subentire, the lateral veins 4–6 on each side of midrib. Inflorescence erect, densely strigillose with an admixture of glandular hairs. Flowers erect; buds 2–3 mm long, 1.5–2.5 mm in diameter, oblong-elliptic, apiculate; ovaries 1.2–2.5 cm long, 0.8–1.8 mm in diameter, densely strigillose and glandular; pedicels 0.4–1.3 cm long; floral tube 0.5–0.9 mm long, 0.8–1.5 mm in diameter, a sparse ring of hairs or subglabrous at the mouth within; sepals 2.5–3.3 mm long, 1–1.2 mm wide, narrowly ovate to lanceolate, slightly keeled; petals 3–4 (–4.7) mm long, 1.8–2.5 mm wide, white, obcordate, the notch 0.5–0.8 mm deep; filaments of longer stamens 1.5–2 mm long, those of shorter ones 1–1.5 mm long; anthers 0.4–0.6 mm long, 0.3–0.4 mm wide, broadly clavate; style 1.5–2.5 mm long, erect, glabrous; stigma subcapitate or sometimes clavate, (0.5–) 0.9–1.4 mm long, (0.3–) 0.5–0.9 mm in diameter, the longer or both sets of stamens shedding pollen onto stigma at anthesis. Capsules 1.7–7 cm long, strigillose, borne on pedicels 0.9–2.1 cm long. Seeds 0.9–1.3 mm long, 0.36–0.5 mm in diameter, narrow-obovoid or oblanceoloid, the chalazal collar very short and inconspicuous, brown, the surface very minutely papillose; coma 7–12 mm long, tawny, not detaching easily. Chromosome number: $n = 18$. Fig. 57.

Phenology. Collected in flower from July to August, in fruit from August to September.

FIG. 57. *Epilobium fastigiatoramosum*. A. Inflorescence. A¹. Portion of mid-stem. A². Base of stem. B. Flower (one petal and three stamens removed); stigma capitate. C. Leaf from inflorescence (bract). C¹. Mid-cauline leaf. C². Basal leaf. D. Capsule. E. Seed. F. Axil from inflorescence. a. Inflorescence. a¹. Portion of mid-stem. a². Base of stem with filiform stolons. b. Flower (one petal and three stamens removed); stigma broadly clavate. (A–F based on *T. P. Wang 3202;* a–b based on *T. Tang 1111.*)

Distribution (Fig. 32). Temperate eastern Asia, including Japan, Korea, the Ussuri to Dauria regions of the U.S.S.R., Mongolia, and northeastern to southwestern China; in China in Heilongjiang, Jilin, Liaoning, Nei Monggol, Hebei, Shanxi, Shandong, Shaanxi, Ningxia, Gansu, Qinghai, and Sichuan; in wet places, especially bogs, lake shores, wet meadows, and streambanks; 400–2000 (–3300) m.

Additional Specimens Examined. **China.** Heilongjiang: Raohe, *G. Z. Wang 596* (IFP); Dailing, *G. Z. Wang & C. H. Chang 4124* (IFP); Heilongjiang, *Heilongjiang Econ. Plant Exped. 626* (PE); Harbin, 1926, *Jettmar s.n.* (W); 1937, *Skvortzov s.n.* (GH); Lunliang, *W. Wang et al. 872* (PE); Hailun, 1933, *Suzuki-Tokio s.n.* (TAI); "Manchuria," 1928, *Licent s.n.* (BM); 1860, *Maximowicz s.n.* (GH); Hulin, *G. Z. Wang 386* (IFP); Shangzhi, *G. Z. Wang 1624* (IFP); Yilan, *Y. L. Zhang 1974* (IFP, PE); Ganhe, Da Hinggan Ling, *T. Z. Zhu 657* (KUN); Yichun, *T. N. Liou et al. 10184* (IFP).—Jilin: Antu, *T. N. Liou 4334* (IFP); Changchun, 1910, *Yabe s.n.* (NAS); Changbai Shan, *S. X. Li et al. 726* (PE), *W. Wang et al. 2189* (IFP, PE), *G. S. Zhou 50* (PE); Beicheng, *Beicheng Exped. 2939* (IFP, PE); Ehmu, *Suzuki-Tokio 9891* (TAI).—Liaoning: Xiuyan, *W. Wang et al. 1377* (IFP, PE), *1596* (IFP, NAS, PE); Qian Shan, *P. Y. Fu et al. 2475* (IFP, KUN, PE); Caohekuo, 1910, *Yabe s.n.* (NAS); Xifeng, *X. L. Cuo 92* (IFP), *C. X. Sun 146* (IFP), *X. C. Jiang 290* (IFP); Benxi, *C. Q. Lin 959* (IFP); Xinbin, *Xinbin Pl. Exped. 90* (IFP).—Nei Monggol: Xinghe, *Y. Z. Zhau 77-205* (NMU); Xuguit Qi, *G. Z. Wang 2831* (IFP); Jarud Qi, *Z. H. Zhang 1-830* (IFP); Manhan Shan, *Y. C. Ma 35* (NMU); Wuchingtai, *Beijing Univ. 192* (PE); "Eastern Mongolia," *David 1990* (P).—Hebei: Qinglong, *Chengdu Exped. 1622* (PE); Chongli, *S. Y. Huo 20483* (PE); Dongling, *H. T. Tsai 50355* (BP); Longguan, *Hebei Exped. 64575* (PE); Fuping, *K. M. Liou 3290* (NAS, PE), *3782* (NAS); Xinlong, *T. N. Liou 4697* (IFP); Beijing, *W. Y. Hsia 2210* (PE), *Lazaristes 2918* (P), *Raven & Raven 26099* (MO); 1961, *Sojak s.n.* (PR); Changping, *J. Zhang 2034* (KUN); Xiaowutai Shan, *Hebei Pl. Exped. 796* (PE), *2353* (PE), *H. W. Kung 1085* (PE), *Smith 136* (UPS), *138* (UPS), *140* (UPS), *Y. W. Tsui 2671* (NMU), *C. W. Wang 62460* (PE), *T. P. Wang 698* (PE), *Wu & Yang 37398* (PE); Zhangbei, *Hebei Exped. 8574* (PE); Daqin Shan, *Hebei Pl. Exped. 3735* (PE); Liayuan, *Hebei Univ. Agr. 42439* (PE); "Paita," *Licent 9769* (W).—Shanxi: Datong, *Shanxi Exped. 1347* (PE); Wutai, *K. C. Kuan & Y. L. Chen 2158* (PE), *2417* (PE), *T. Tang 1111* (NAS, PE), *K. C. Kuan & Y. L. Chen 2417* (PE), *Serre 2386* (W); Xin Xian, *T. P. Wang 3202* (PE), *Huanghe Exped. 2566* (PE); Zhaocheng, *Smith 6886* (UPS), *7305* (UPS), *7420:2* (UPS); Pu Xian, *X. Y. Liu 21045* (PE); no locality, *Licent 2386* (PE), *9769* (PE); "Ha-ha-ye," *Licent 11036* (PE, W).—Shandong: Tai Shan, *T. Y. Chou 7358* (NAS, PE), *7533* (NAS, PE); "Eastern China," 1910, *Balfour s.n.* (E).—Shaanxi: Jingbian, *Huang He Exped. 8718* (PE); Taibai Shan, *T. N. Liou & P. C. Tsoong 2736* (PE).—Ningxia: Helan Shan, *Ningxia Med. Pl. Exped. A185* (PE); Longde, *T. P. Wang 13685* (HNWP, PE).—Gansu: Zhenyuan, *T. P. Wang 17272* (PE); Tianzhu, *Y. Q. He 4536* (PE), *4563* (PE).—Qinghai: Zadoi, *S. W. Liu 423* (HNWP, PE); Datong, *B. W. Li 72-7* (HNWP); Huangzhong, *H. P. Tsui et al. 80-46* (PE); Nangqen, *Xizang Medic. Exped. 1180* (PE); Tongren, *T. P. Wang 6215* (PE); no locality, *Licent 4847* (W).—Sichuan: 40 km N of Songpan, *W. L. Chen & C. J. Chen 8217* (MO, PE); Kangding, *H. S. Kung 6058* (MO); Zamtang, *W. L. Chen & C. J. Chen 7508* (MO, PE); Xiangcheng, *Hengduan Shan Exped. 3710* (PE); Daocheng, *H. Li 1039* (MO).

Epilobium fastigiatoramosum, similar to *E. palustre* in overall appearance and habit, has subentire leaves, strigillose pubescence throughout, and sometimes filiform stolons. It differs from *E. palustre* in having smaller flowers, a more or less subcapitate stigma, smaller seeds with an inconspicuous chalazal collar, and elliptic leaves apically acute/subobtuse. Although both species perennate by filiform stolons, this feature is apparently poorly developed and often absent in *E. fastigiatoramosum.* Because so few collections show this feature clearly, it is not even certain that *E. fastigiatoramosum* forms the compact terminal turions that are so characteristic of *E. palustre.* The distribution of *E. fastigiatoramosum* is exclusively in eastern Asia, whereas that of *E. palustre* is circumboreal.

33. Epilobium minutiflorum Haussknecht, Oesterr. Bot. Zeitschr. 29: 55. 1879. 1884. *Epilobium tetragonum* β. *minutiflorum* (Haussknecht) Boissier, Fl. Orient. Suppl. 240. 1888.—TYPE: TURKEY. Gaziantep ("Aintab"), banks of the River Sajur Suyi ("Sadschur"), 26 Jun 1865, *Haussknecht 812* (lectotype, designated by Raven 1962a: JE!).

Epilobium modestum Haussknecht, Oesterr. Bot. Zeitschr. 29: 55. 1879.—TYPE: PAKISTAN. Ladakh, Nubra Valley, 4 Aug 1848, *Thomson s.n.* (lectotype, designated by Raven 1962a: K!).

Epilobium propinquum Haussknecht, Monogr. Epil. 213. 1884, nom. nov. *Epilobium decipiens* Haussknecht, Oesterr. Bot. Zeitschr. 29: 57. 1879, non *Epilobium decipiens* F. Schultz, 1861.—TYPE: CHINA. Beijing, Gubeikou ("Ku-pei-ku"), 1868–1871, *Wawra 877* (lectotype, here designated: W!; isolectotype: JE!).

Common name: Xizi Liuyecai (Li 1986).

Erect perennial herb, producing short fleshy basal soboles or leafy rosettes. Stems 15–100 cm tall, 2–8 mm in diameter, well branched, rarely simple, densely strigillose all over especially on upper stem, often subglabrous below, sometimes with 2 (rarely 4) inconspicuous lines decurrent from margins of petioles. Leaves opposite, alternate only on inflorescence, subsessile above, with petioles 1–6 mm long in basal pairs; cauline blades 2–7 cm long, 0.4–1.7 cm wide, oblong-lanceolate to narrowly ovate, glabrous except for strigillose margins and veins, apex subobtuse or acute, base cuneate or subrounded, the margins serrulate with 20–41 teeth on each side, the prominent lateral veins 4–7 on each side of midrib. Inflorescence suberect before anthesis, white-canescent with scattered glandular hairs. Flowers erect; buds 2–3 mm long, 1.5–2.5 mm in diameter, globose to ovoid, apiculate; ovaries 1.5–4 cm long, densely white-canescent with a few glandular hairs; pedicels 0.4–1.5 cm long; floral tube 0.8–1.2 mm long, 1.2–1.4 mm in diameter, a ring of spreading hairs at the mouth within; sepals 2.4–4 mm long, 0.7–1.2 mm wide, oblong-lanceolate, acute, sometimes keeled; petals 3–4.3 (–5) mm long, 1.2–1.5 (–2) mm wide, white to rarely pink or rose, oblong to obovate, the notch 0.5–1 mm deep; filaments of longer stamens 1.5–2.5 mm long, those of shorter ones 1–2 mm long; anthers 0.4–0.8 mm long, 0.2–0.3 mm in diameter, oblong-lanceolate; style 1.4–2.5 mm long, erect, glabrous; stigma clavate or occasionally subcapitate, 0.7–1.5 mm long, 0.8–1.1 mm in diameter, the stamens shedding pollen onto stigma at anthesis. Capsules 3–8 cm long, strigillose or rarely glabrescent, borne on pedicels 0.5–2 cm long. Seeds 0.8–1.2 mm long, 0.38–0.48 mm in diameter, narrowly obovoid, with a conspicuous pellucid chalazal collar, brown, the surface minutely papillose; coma 5–7 mm long, white, detaching readily. Chromosome number: n = 18. Fig. 58.

Phenology. Collected in flower from June to August, in fruit from July to October.

Distribution (Fig. 22). Widespread across central and southern Asia, from Korea, northeastern to northwestern China, Mongolia, the Tian Shan region, central Asia, and the Himalaya west of Uttar Pradesh, to central Anatolia in Turkey; in China in Jilin, Liaoning, Nei Monggol, Hebei, Shanxi, Shaanxi, Ningxia, Gansu,

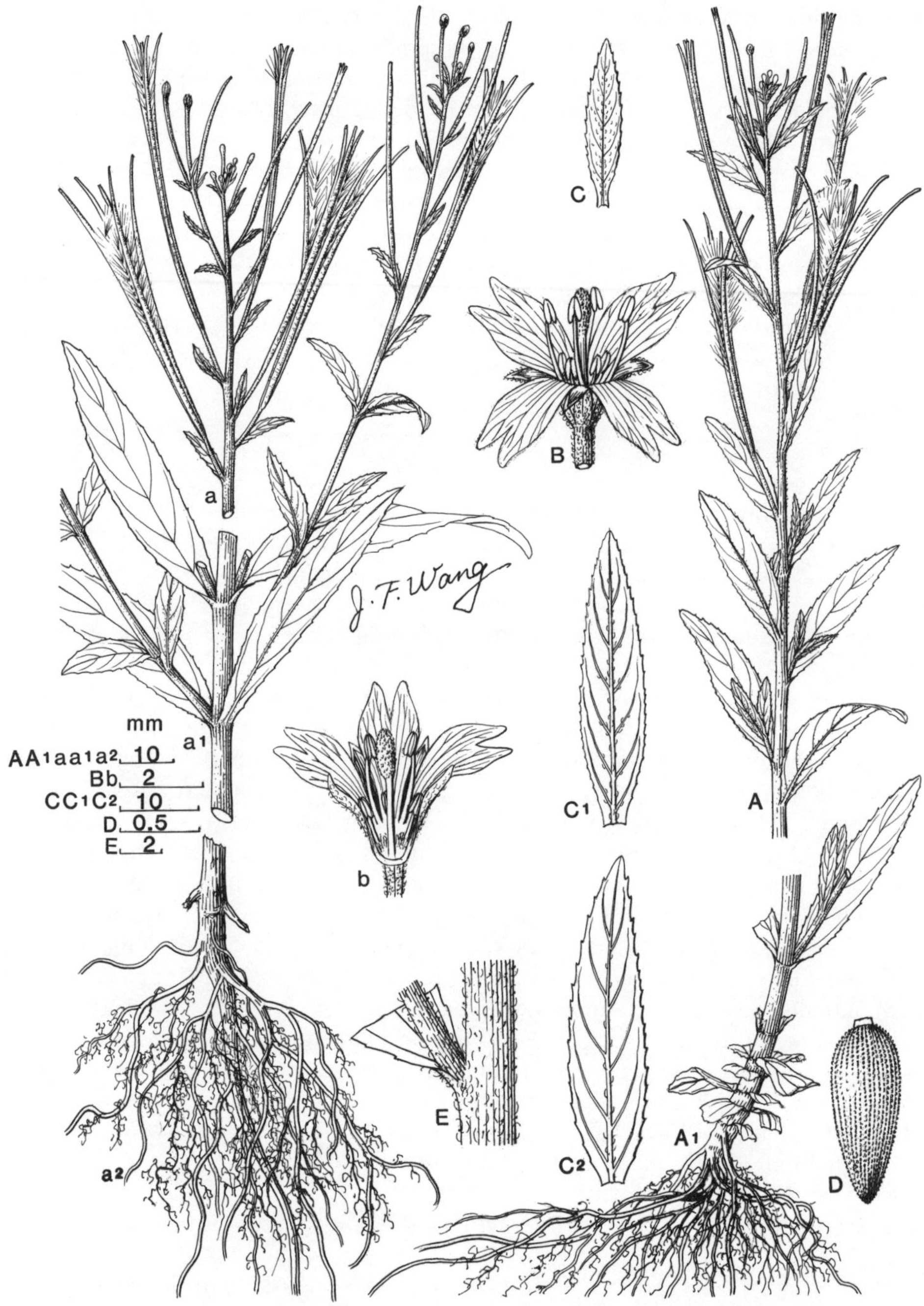

FIG. 58. *Epilobium minutiflorum.* A. Upper stem and inflorescence. A[1]. Base of stem. B. Flower. C. Leaf from inflorescence (bract). C[1]. Mid-cauline leaf. C[2]. Basal leaf. D. Seed. E. Axil from inflorescence. a. Inflorescence. a[1]. Portion of mid-stem. a[2]. Base of plant. b. Flower (one petal and three stamens removed). (A–E based on *Beijing University 64–344;* a–b based on *Xinjiang Exped. 4746.*)

Xinjiang, and far western Xizang (Tibet); in generally hot, dry regions, locally growing in wet, often disturbed places by streams, bogs, or river banks; 500–1800 m.

ADDITIONAL SPECIMENS EXAMINED. **China.** JILIN: Wangqing, *P. Y. Fu 952* (PE); Changbai Shan, *J. J. Qian 760* (KUN, PE).—LIAONING: Jianping, *C. S. Wang et al. 2793* (IFP, PE).—NEI MONGGOL: Ulanqab Meng, *X. Z. Zhang 134* (PE); Ih Ju Meng, *Nei Mong. Exped. 189* (NMU), *Z. Z. Cheng 3* (NMU); Daqing Shan, *T. P. Wang 2466* (PE), *Nei Mong. Univ. 101* (NMU), *Y. C. Ma & Q. R. Wu 65-165* (NMU), *Q. R. Wu 77* (NMU); Tengelialqi, *Z. Y. Gong 1063* (NMU); "Ortus Qi," *Beijing Univ. 64-344* (PE); "Chimisto Gol," *Y. L. Keng 637* (NAS); "Gehol," *David 1975* (JE, P); "Liutaok'ow," 1927, *T. H. Ting s.n.* (PE); "Yangch'angtzukow," 1927, *T. H. Ting s.n.* (PE).—HEBEI: Dongling, *Y. Liu 11920* (PE), *Yu 354* (PE); Beijing, *Z. Y. Cao 126* (PE), *J. C. Liu 1147* (NY, UC), *Raven & Raven 26100* (MO), 1961, *Sojak s.n.* (PR), *C. W. Wang 60004* (PE), *J. W. Wang 169* (HIMC), 1903–1935, *Chanet & Serre s.n.* (P); Xiawutai Shan, *H. W. Kung 1085* (NY, PE), *Y. Liu 10655* (PE), *Smith 971* (UPS), *C. Y. Yang & Z. G. Li 36241* (PE); Neiqiu, *Y. Liu 13168* (NAS); Baihe, *Licent 328* (BM, P); Laiyuan, *Hebei Univ. Agr. 3903* (PE); Xingtai, *X. Y. Liu 603* (PE); no locality, s.d., *s.c. 687* (PE).—SHANXI: Wutaishan, *W. Y. Hsia 1316* (PE), *K. C. Kuan & Y. L. Chen 2049* (PE); Linwu, *J. M. Liu 1602* (PE); Xia Xian, *Smith 6226* (UPS); Jincheng, *Smith 7582* (UPS); Huoxien, *T. P. Wang 972* (PE).—SHAANXI: Huayin, *W. Y. Hsia 3622* (PE); Taibai Shan, *Biol. Dept. Lanzhou Univ. 719* (LZU).—NINGXIA: Helan Shan, *Y. Q. He 7278* (PE), *Huanghe Exped. 7993* (PE), *Ningxia Exped. 338* (PE), *Y. Y. Pai 157* (PE).—GANSU: Tianshui, *G. C. Geng 597* (LZU); Wen Xien, *Z. Y. Chang 6229* (HNWP), *Z. Y. Zhang 15161* (PE); Hexi, *Gansu Exped. 2068* (LZU).—XINJIANG: Turpan, *Xinjiang Exped. 329* (PE, XJBI), *2127* (PE, XJBI), *C. J. Chen & Hoch 86170* (MO, PE); Toksun, *A. R. Li & J. N. Zhu 7285* (KUN, PE), *Xinjiang Exped. 176* (PE), *7285* (PE, XJBI); Qitai, *Xinjiang Exped. 6414* (XJBI); Dabanchang, *C. J. Chen & Hoch 86027* (MO, PE); Urumqi, *Bayi Coll. Agr. 15506* (PE), *R. C. Ching 147* (PE), *T. N. Liou 2855* (PE), *Xinjiang Exped. 4746* (XJBI), *650070* (LZU), *G. L. Chu & W. Y. Xu 5431* (NAS, PE), *Aimait 650079* (NAS), *Sdik & Kurbanjiang 65117* (HNWP), 1973, *G. L. Zhang et al. s.n.* (LZU); Tianchi, *C. J. Chen & Hoch 86054* (MO, PE); Miquan, *First Aug. Coll. Agr. 50* (NAS); Shawan, *K. C. Kuan 1316* (PE, XJBI), *1589* (PE); Tacheng, *C. W. Chang et al. 4518* (XJBI), *A. R. Li & J. N. Zhu 10680* (PE, XJBI), *10878* (KUN, PE, XJBI); Altay Shan, *R. C. Ching 2929* (PE), *G. L. Chu & W. Y. Xiu 6854* (PE); Qinghe, Altay, *R. C. Ching 1666* (KUN), *T. Y. Chou 652051* (HNWP, NAS, PE, XJBI), *G. J. Liu et al. 680* (XJBI), *Xinjiang Exped. 1580* (PE), *1666* (PE, XJBI), *1901* (HNWP); Habahe, *C. W. Chang et al. 4363* (XJBI); Toli, *K. C. Kuan 2424* (PE); Shanshan, *A. R. Li & J. N. Zhu 6688* (PE), *Xinjiang Exped. 1960* (XJBI); Wenquan, *C. J. Chen & Hoch 86262* (MO, PE); Huocheng, *Z. W. Zhang et al. 3347* (XJBI), *A. W. Zhang 4043* (XJBI); Ili, *K. C. Kuan 3078* (PE, XJBI); Gongliu, *Z. X. Peng 5137* (LZU), *Q. R. Wang 2664* (PE); Shache, *L. R. Tsui 1177* (XJBI); Kashi, *Konta & K. Y. Lang 524* (PE); no locality, *G. J. Liu et al. 680* (XJBI).—XIZANG: Western Himalaya, s.d., *Hooker & Thomson s.n.* (P); s.d., *Thomson s.n.* (K, P, W).

Raven (1962a) pointed out that there were "no consistent and correlated differences" between *E. minutiflorum* and *E. modestum* in terms of petiole length, serrulation of the leaf margin, or differences in the seeds, and he merged the two into one species. Although Haussknecht also described *E. modestum* as having glabrescent stems, buds, and capsules, he informally discussed *E. minutiflorum* f. *glabrescens* as well. Our examination of additional material, especially from China, has reinforced Raven's earlier conclusion that two species cannot be maintained.

Haussknecht (1884) also described a third species, *E. propinquum,* which he distinguished by its capitate-subclavate stigmas, compared to the clavate or narrowly clavate stigmas in the other two species. He also noted that *E. propinquum* occurred in China and Mongolia, whereas the other two species were found from Turkey ("Syria") to Ladakh. The analysis of many collections, including numerous cultivated strains, shows stigma shape to vary from narrowly to broadly clavate, but without any marked discontinuity, and without correlation with any other variable character or with geography. The range of this widespread species is continuous. Although plants from the northeastern part of the range of this species tend to have subglabrous stems, this character is not correlated with stigma shape or any other differences.

One very distinctive collection from the Altay region (Xinjiang, Altay Bridge Forestry Center to Halong, *G. J. Liu et al. 680,* XJBI) has unusually large flowers (petals 5.5–6.5 mm long) and seeds (to 1.2 mm long), weak but definite raised lines on the stem, and markedly subcapitate stigmas, but in other characteristics is typical of *E. minutiflorum.* It may be of hybrid origin (the other parent possibly being *E. roseum*), but there is no reduction in seed set, and pollen fertility is 86%.

Epilobium minutiflorum has been confused with *E. palustre* in some Chinese taxonomic papers (Wu 1979), since its clavate stigma and pubescent stems are similar to those of *E. palustre,* and the two species often grow together at lower and middle elevations in China. *Epilobium minutiflorum* is quite distinct from *E. palustre* by virtue of its smaller flowers and seeds, white-canescent ovaries, broader serrulate leaves, and basal rosettes rather than filiform stolons.

34. Epilobium ciliatum Rafinesque, Med. Repos. II. 5: 361. 1808. *Epilobium ciliatum* subsp. *ciliatum.*—TYPE: U.S.A. Northern Pennsylvania, *Rafinesque s.n.* (not located).

Epilobium maximowiczii Haussknecht, Oesterr. Bot. Zeitschr. 29: 57. 1879.—TYPE: JAPAN. Hokkaido, 1861, *Maximowicz s.n.* (holotype: LE; isotype: W!).

Epilobium punctatum H. Léveillé, Bull. Acad. Geogr. Bot. 11: 316. 1902.—TYPE: JAPAN. Hokkaido, Aomori, Aug 1901, *Kinashi 7 ("919")* (lectotype, here designated: KYO!).

Epilobium kurilense Nakai, Bot. Mag. (Tokyo) 22: 83. 1908. *Epilobium glandulosum* var. *kurilense* (Nakai) H. Hara, J. Jap. Bot. 18: 238. 1942.—TYPE: U.S.S.R. Kurile Islands, Aug 1903, *Yendo s.n.* (holotype: TI!).

Epilobium glandulosum var. *asiaticum* H. Hara, J. Jap. Bot. 18: 241. 1942.—TYPE: U.S.S.R. Sakhalin, Toyohara, 6 Aug 1923 (5 Aug on label), *Sawada s.n.* (holotype: TI!).

Common name: Dongbei Liuyecai (Liou et al. 1958).

Erect perennial herb, forming compact leafy rosettes or rarely fleshy hypogeous turions that leave brown basal scales. Stem (10–) 25–90 (–150) cm tall, 2–5 (–8) mm in diameter, more or less branched, especially above, strigillose and glandular-pubescent all over on upper parts, sparsely pubescent or glabrescent or glabrescent below, with raised strigillose lines decurrent from margins of the petioles. Leaves opposite, alternate on the inflorescence, herbaceous, sessile, or in lower pairs with petioles 1–3 mm long; cauline blades 2.5–6 (–7) cm long , 0.6–1.5 (–2) cm wide, lanceolate to narrowly ovate, sparsely strigillose on veins and margins, older leaves often glabrescent, apex acute to subacuminate, base rounded or rarely subcordate, the margins serrulate with (7–) 15–25 (–32) teeth on each side, the conspicuous lateral veins 5–8 on each side of midrib. Inflorescence erect, strigillose and glandular, open, the bracts very small, narrowly lanceolate to sublinear. Flowers erect; buds 2–2.5 mm long, 1.5–2 mm in diameter, broadly ovoid to oblong-ovoid, apiculate; ovaries 0.8–2.8 cm long, densely strigillose with an admixture of glandular hairs; pedicels 0.2–0.5 cm long; floral tube 0.8–1.2 mm long, 1.2–1.8 mm in diameter, with a sparse ring of spreading hairs at the mouth within; sepals 2.4–3.5 mm long, 0.7–1 mm wide, lance-oblong, keeled, strigillose and glandular; petals 3.5–5 (–7) mm long, 2–3 mm wide, pink to rose-purple, occasionally white, oblong-obovate, the notch 0.7–1.2 mm deep; filaments of longer stamens 2.5–3.2 mm long, those of shorter ones 1.5–2.5 mm long; anthers 0.5–0.6 mm long, 0.25–0.35 mm in diameter, broadly ovoid; style 1.8–3.2 mm long, erect, glabrous; stigma clavate to cylindrical, 0.8–2 mm long, 0.6–0.8 in diameter, at least the longer stamens shedding pollen onto stigma at anthesis. Capsules 4.5–7 cm long, erect, sparsely strigillose and glandular, borne on pedicels 0.5–0.8 (–1.4) cm long. Seeds 0.8–1.2 mm long, 0.25–0.4 mm wide, narrowly obovoid or oblong-elliptic, with a fairly conspicuous chalazal collar 0.08–0.1 mm long, brown, the surface with conspicuous longitudinal ridges of flattened, fused papillae; coma 6–8 mm long, dull white, readily detaching. Chromosome number: n = 18. Figs. 5c, 7c, 8j–l, 59.

Phenology. Collected in flower from July to August (rarely to September), in fruit from August to October.

Distribution (Fig. 34). North America from Alaska, subarctic Canada, and Greenland south throughout boreal and montane western North America and Mexico to Guatemala, and throughout boreal eastern North America to Oklahoma, Iowa, Illinois, Ohio, Tennessee, and North Carolina; South America in southern Chile and Argentina; Far East region of the U.S.S.R., Japan, Korea, and northeastern China; also naturalized in Europe and the European part of the U.S.S.R., and in Australasia; in China only in northern Heilongjiang and eastern Jilin; usually in moist places, especially where disturbed, such as grassy slopes, seeps, stream banks, and gravel beds of rivers, sometimes also by hot springs; (700–) 1230–2100 m.

Additional Specimens Examined. **China.** Heilongjiang: Boli, *Y. L. Chang 2113* (IFP, PE); Yichun, *T. N. Liou et al 10193* (IFP); Dailing, *T. N. Liou et al. 5984* (IFP), *6035* (IFP), *W. Wang et al. 872* (IFP), *C. Wang 4444* (IFP); Ning'an, *P. Y. Fu et al. 3273* (IFP).—Jilin: Yanbian, *Yanbian Exped. 121* (IFP); Wangqing, *P. Y. Fu 333* (IFP), *P. Y. Fu et al. 952* (IFP), *1012* (IFP), *C. S. Wang et al. 1413* (IFP, KUN); Antu, *J. D. Li et al. 129* (NETU), *T. N. Liou 3886* (IFP), *3891* (IFP), *4094* (IFP), *4103* (IFP), *T. N. Liou et al. 4188* (IFP), *C. S. Wang 603* (IFP); Changbai Shan, *S. X. Li et al. 1027* (IFP), *T. N. Liou et al. 1516* (KUN, NAS), *1638* (IFP, KUN), *J. J. Qian 462* (PE), *Tonghua Exped. 74* (PE), *W. Wang et al. 2188* (IFP), *2393* (PE), *2401* (IFP), *Y. T. Zhao et al. 543* (NETU), *4188* (IFP), *19723* (NETU); Manjiang, *N. E. China Teacher's Univ. 331* (PE).

As we currently delimit the species, *E. ciliatum* is distinguished from all but one other species in the genus by its very distinctive, ridged seed surface. It also is characterized to a lesser degree by its mostly lanceolate, deeply veined leaves; well-branched, open inflorescences; and leafy basal rosettes (Hoch 1978, unpubl.; Solomon 1982). This species, one of the most aggressive in the genus, is capable of rapidly colonizing a wide range of habitats; this is accomplished by a combination of modal autogamy, unusually high seed production, and vigorous vegetative growth (Raven & Raven 1976; Hoch 1978).

In North America, we recognize three subspecies that have discrete but partly overlapping eco-geographic ranges, and that sometimes intergrade morphologically (Hoch 1978). Most collections of this species from China have the branched stems,

FIG. 59. *Epilobium ciliatum* subsp. *ciliatum*. a. Inflorescence. a^1. Portion of mid-stem. a^2. Base of stem, with leafy rosette at caudex. b. Flower (one petal and two stamens removed). c. Leaf from inflorescence (bract). c^1. Mid-cauline leaf. c^2. Basal leaf. d. Capsule. e. Seed; note pattern of longitudinal lines (ridges) on surface. f. Axil from inflorescence. f^1. Axils from mid-stem. (Based on *S. X. Li et al. 1027*.)

leafy rosettes, lanceolate leaves, and small flowers typical of subsp. *ciliatum* from North America, although like the North American populations they vary in flower color, leaf shape, and branching pattern (Hoch 1978). A collection from Jilin (Changbai Shan, *T. N. Liou et al. 1638*), with a simple stem, ovate leaves (up to 7 cm long, 3 cm wide), and a compact leafy inflorescence, tends towards subsp. *glandulosum* (Lehmann) Hoch & Raven, a taxon found in cooler and/or less disturbed habitats than subsp. *ciliatum* in North America and extreme northeast Asia. However, that collection has smaller flowers than are typical for subsp. *glandulosum,* and overall seems best treated as an immature or shaded specimen of subsp. *ciliatum.* One plant on the PE sheet of this *Liou et al.* collection has narrowly lanceolate, acuminate leaves and shrivelled capsules, and may be a hybrid between *E. ciliatum* and *E. palustre.*

Epilobium ciliatum subsp. *ciliatum* occurs widely in northeastern Asia, North America, and southern South America, and has been introduced into Europe and Australasia. This is the first report for this taxon in China. Although the earliest collections from China were made in the early 1950's, the taxon may be native, as it may be in other areas of eastern Asia bordering the north Pacific. The Chinese collections are from middle elevations in the eastern Xiao Hinggan Ling range in Heilongjiang and the Changbai Shanmai range in Jilin.

Y. L. Chang, in Liou et al. (1977), erroneously determined most collections of *E. ciliatum* subsp. *ciliatum* as *E. tenue* Komarov and *E. cylindrostigma* Komarov. We have seen the types of these taxa, and consider *E. tenue* to be a synonym of *E. amurense* subsp. *amurense,* and *E. cylindrostigma* to be a synonym of *E. amurense* subsp. *cephalostigma. Epilobium ciliatum* has general similarities to these taxa and hence might be confused with them, but is clearly distinguished by the numerous features mentioned above.

35. Epilobium clarkeanum Haussknecht, Monogr. Epil. 220. 1884.—TYPE: SIKKIM. Lachen, 3650 m, 16 Jul 1849, *Hooker s.n.* (lectotype, designated by Raven 1962a: K!; isolectotype: JE!).

Common name: Yazhi Liuyecai.

Loosely clumped perennial herb, forming short soboles that leave scales on the basal 2–4 cm of the stem at or below ground, the scales 4–7 mm long, 2–3 mm wide, narrow-obovate to spatulate, obtuse, coriaceous, brown. Stems 10–20 cm tall, 0.8–2 mm in diameter, simple or sometimes branched at the base, ascending, sparsely strigillose all over upper part, glabrescent below, with inconspicuous strigillose lines decurrent from margins of the petioles. Leaves opposite below inflorescence, subsessile or with petioles 1–2 mm long in basal pairs; cauline blades 1.1–1.6 cm long, 0.5–0.8 cm wide, ovate-elliptic to subobovate below, glabrous except for strigillose veins and margins, apex obtuse to subacute, base broadly cuneate or subrounded, the margins denticulate with 3–8 teeth on each side, the lateral veins inconspicuous, 3–4 on each side of midrib. Inflorescence nodding, sparsely strigillose, open, with upper internodes 1–2 times longer than the subtending leaves. Flowers slightly nodding; buds 3–4 mm long, broadly elliptic, ovaries 1.2–1.5 cm long, strigillose; pedicels 0.3–0.6 cm long; floral tube 0.8–1 mm long, 1.6–1.8 mm in diameter, a ring of white spreading hairs at the mouth within; sepals 2.5–3 mm

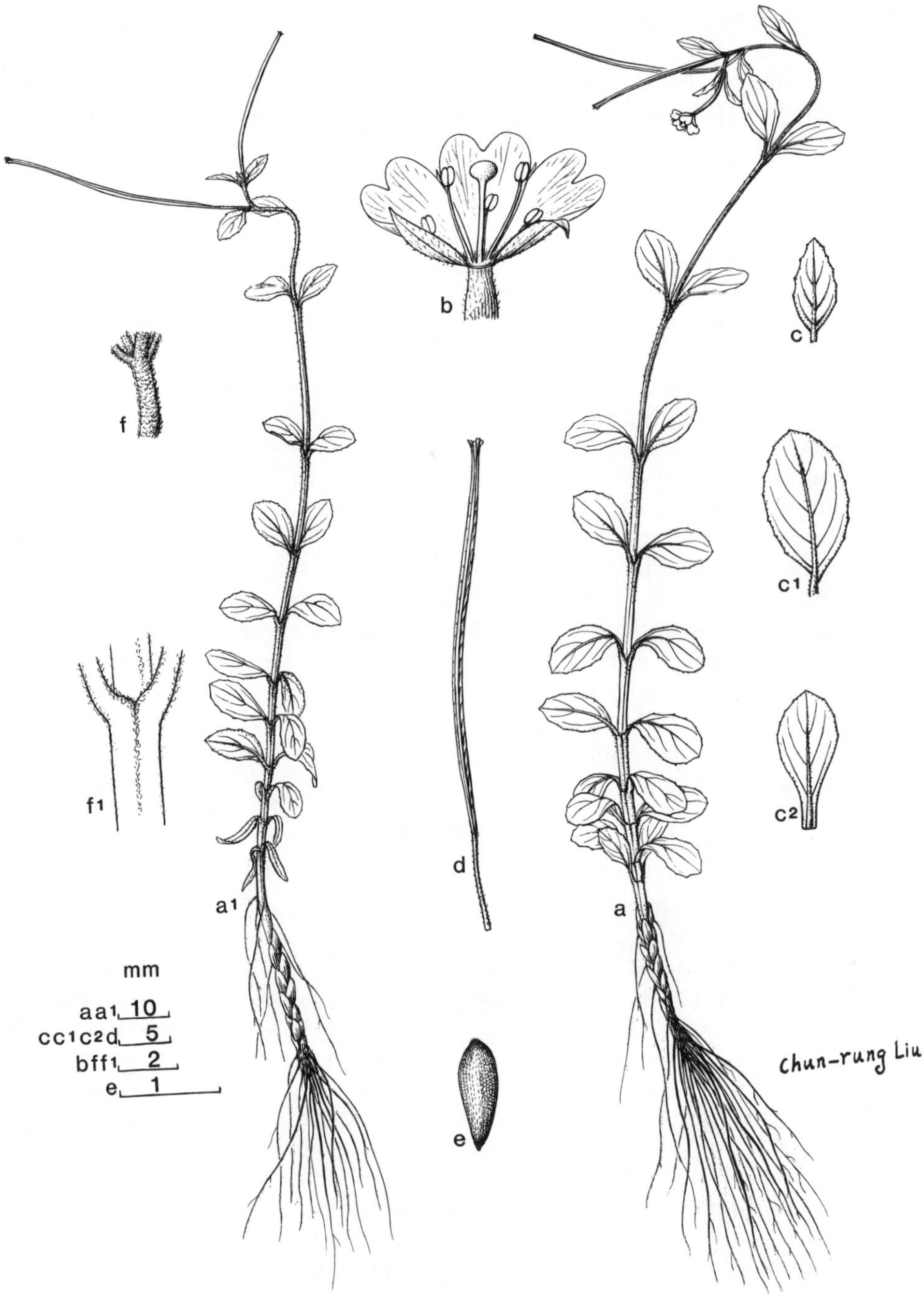

FIG. 60. *Epilobium clarkeanum.* a–a^1. Habit; note nodding inflorescences and scales at base of stems. b. Flower (one petal and three stamens removed). c. Leaf from inflorescence (bract). c^1. Mid-cauline leaf. c^2. Basal leaf. d. Immature capsule. e. Seed; note reticulate surface. f. Axils from inflorescence, showing hairs all over. f^1. Axil from mid-stem, showing hairs restricted to margins of petioles and lines on stems. (Based on *Delavay 5145.*)

long, 1–1.2 mm wide, lanceolate, keeled; petals 5–6 mm long, 3–3.5 mm wide, white, obcordate, the notch 1 mm deep; filaments of longer stamens 2.5–3 mm long, those of shorter ones 1.8–2 mm long; anthers 0.5–0.6 mm long, 0.35–0.4 mm in diameter, ovate-oblong; style 3–3.2 mm long, erect, glabrous; stigma capitate, 1.2 mm long, 1.3 mm in diameter, the stamens shedding pollen onto stigma at anthesis. Capsules 3.5–4 cm long, sparsely strigillose, borne on pedicels 0.6–0.7 cm long. Seeds 0.8–0.9 mm long, 0.3–0.35 mm in diameter, narrowly obovoid, with a short chalazal collar, light brown, the surface reticulate; coma 4.5–5 mm long, white, detaching rather easily. Chromosome number unknown. Fig. 60.

Phenology. Collected in flower from May (possibly from April) to July, in fruit from July to August.

Distribution (Fig. 42A). Southwestern China, northern Burma, and Sikkim; in China only in western Yunnan; in wet, boggy places in the mountains, ca. 360–4500 m (data from Burmese and Sikkimese collections only).

ADDITIONAL SPECIMENS EXAMINED. **China.** YUNNAN: Qiaojia, (Long-Ki), 1893, *Delavay s.n.* (P), *Delavay 5147* (P); Gongshan, *C. W. Wang 66859* (PE).

This species is one of the rarest in the genus in southern Asia; only three collections, all from Yunnan, are known from China, and only three additional collections are known (Haussknecht 1884; Raven 1962a). We are reporting *E. clarkeanum* from China for the first time; the Chinese specimens agree well with the type and the other two collections. The rarity of this species may be more apparent than real, since the paucity of specimens may also be due to its small size and relative inaccessibility in high alpine habitats.

This species is strikingly similar to *E. pengii* from Taiwan: the two have similar habits and perennation structures, stem pubescence, leaf shape and type, flower size and color, and seed surface. They clearly differ only in leaf size, capsule and pedicel length, and seed size, all qualitative characters. Both are highly restricted in distribution, in areas separated by more than 1500 km. In view of their few but clear dissimilarities and the large range disjunction, we recognize both as distinct species. As noted by Raven (1962a), *E. clarkeanum* (and by extension, *E. pengii*) may be related to the Alpinae group of Haussknecht, but final determination of its status and affinities must await additional collections and study.

36. Epilobium pengii C. J. Chen, Hoch & Raven, sp. nov.—TYPE: CHINA. Taiwan: Taizhong Xian, Hsuehshan (Snow Mt.), 24°24′N, 121°17′E, on trail to summit in partially shaded meadow by small waterfall, 3260 m, 7 Aug 1986, *Hoch & C.-I Peng 3381* (holotype: MO!; isotypes: HAST! PE!).

Common name: Wangzi Liuyecai.

Herba perennis caespitosa ascendens, sobolibus foliosis et squamis brunneis praedita. Caules simplices, undique strigillosi, et lineis pubescentibus et petiolorum marginibus decurrentibus. Inflorescentia nutans, dense strigillosa. Petala alba, postea rosea, 5–6.5 mm longa; stigma capitatum. Capsulae 4.5–5 cm longae, pedicelli 1.5–2 cm longi. Semina 1–1.1 mm longa, reticulata.

Perennial ascending herb, loosely clustered, arising from short leafy or fleshy soboles that leave scattered scales and very small leaves at stem base. Stems 7–25 cm tall, 1–2.5 mm in diameter, simple or very sparsely branched in large plants, subterete, strigillose on upper stem to subglabrous below, with distinct raised pubescent lines decurrent from margins of the petioles. Leaves opposite below inflorescence, crowded on lower stem to well spaced above, subsessile above midstem, with petioles 1–2 mm below; cauline blades 1.5–2.5 cm long, 0.8–1.2 cm wide, ovate to lanceolate above to broadly elliptic below, subglabrous except for strigillose margins and veins, especially on the abaxial midrib, apex acute above to obtuse below, based rounded above to cuneate below, the margins irregularly denticulate, 7–12 teeth on each side, the veins 3–5 on each side of midrib. Inflorescence nodding, densely strigillose to subcanescent, eglandular. Flowers erect; buds 3–4 mm long, broadly elliptic, apiculate; ovaries 1.2–2 cm long, strigillose; pedicels 6–10 mm long; floral tube 0.9–1.2 mm long, 1.4–1.5 mm in diameter, with ring of hairs at mouth within; sepals 4.5–5.5 mm long, 1–1.5 mm wide, lanceolate, strigillose mixed with a few glandular hairs; petals 5–6.5 mm long, 3–3.5 mm wide, white fading to pink, narrowly obovate, the notch 1.2–1.4 mm deep; filaments pinkish, those of longer set 3–3.8 mm long, those of shorter set 1.5–2.7 mm long; anthers 0.5–0.7 mm long, 0.5–0.6 mm in diameter, subrectangular; style 2.8–3 mm long, erect; stigma capitate, 1–1.3 cm long, 0.9–1.8 cm in diameter, at least the longer stamens shedding pollen onto stigma at anthesis. Capsules 4.5–5 cm long, sparsely strigillose, borne on pedicels 1.5–2 cm long. Seeds 1–1.1 mm long, 0.35–0.4 mm in diameter, narrowly obovoid, with short chalazal neck, light brown, the surface reticulate; coma 5–7 mm long, dingy, subpersistent. Chromosome number unknown. Fig. 61.

Phenology. Collected in flower from July to September, in fruit from September to October.

Distribution (Fig. 12). Endemic to Taiwan, China; in high montane to subalpine moist areas, in sheltered places along streams or in moist seeps in subalpine *Juniperus* stands; 3100–3700 m.

ADDITIONAL SPECIMENS EXAMINED. **China.** TAIWAN: Xinzhu Xian, Zhijiayangdashan (Sikayotaizan), *Masamune 1232* (TAI); Yilan Xian, Nanhutashan, *C.-I Peng 5795* (HAST, MO); Taizhong Xian, Hsuehshan, *C.-I Peng 7985* (HAST, MO, PE), (Mt. Silvia), 1936, *Mori s.n.* (TAI), Nantou Xian, Yushan, *Hoch 3269* (MO, PE); Hohuanshan, *C.-I Peng 8295* (HAST, MO); Hualian Xian, Mt. Tyuosenzan, *Fukuyama & Suzuki-Tokio 15091* (TAI).

Epilobium pengii, which was collected only three times prior to 1983 and appears to be uncommon in the high mountains of Taiwan, nevertheless occurs over a fairly wide area. Because it is clearly distinguished from other species in Taiwan by its reticulate seeds, loosely clumped ascending habit, and strigillose pubescence all over the stem including on raised lines, the affinities of this species do not appear to be with those taxa. As discussed under *E. clarkeanum, E. pengii* is morphologically similar to that species, although the two species are separated by some 1500 km. In general aspect and habit, *E. pengii* also resembles *E. lactiflorum* Haussknecht, a member of the circumboreal Alpinae group that reaches central Honshu in Japan (as *E. shiroumense* Matsumura & Nakai), but is not known from China. *Epilobium pengii* differs from *E. lactiflorum* by its capitate stigma, sparsely strigillose decurrent stem lines, broader leaves, and shorter pedicels. These two species may well be

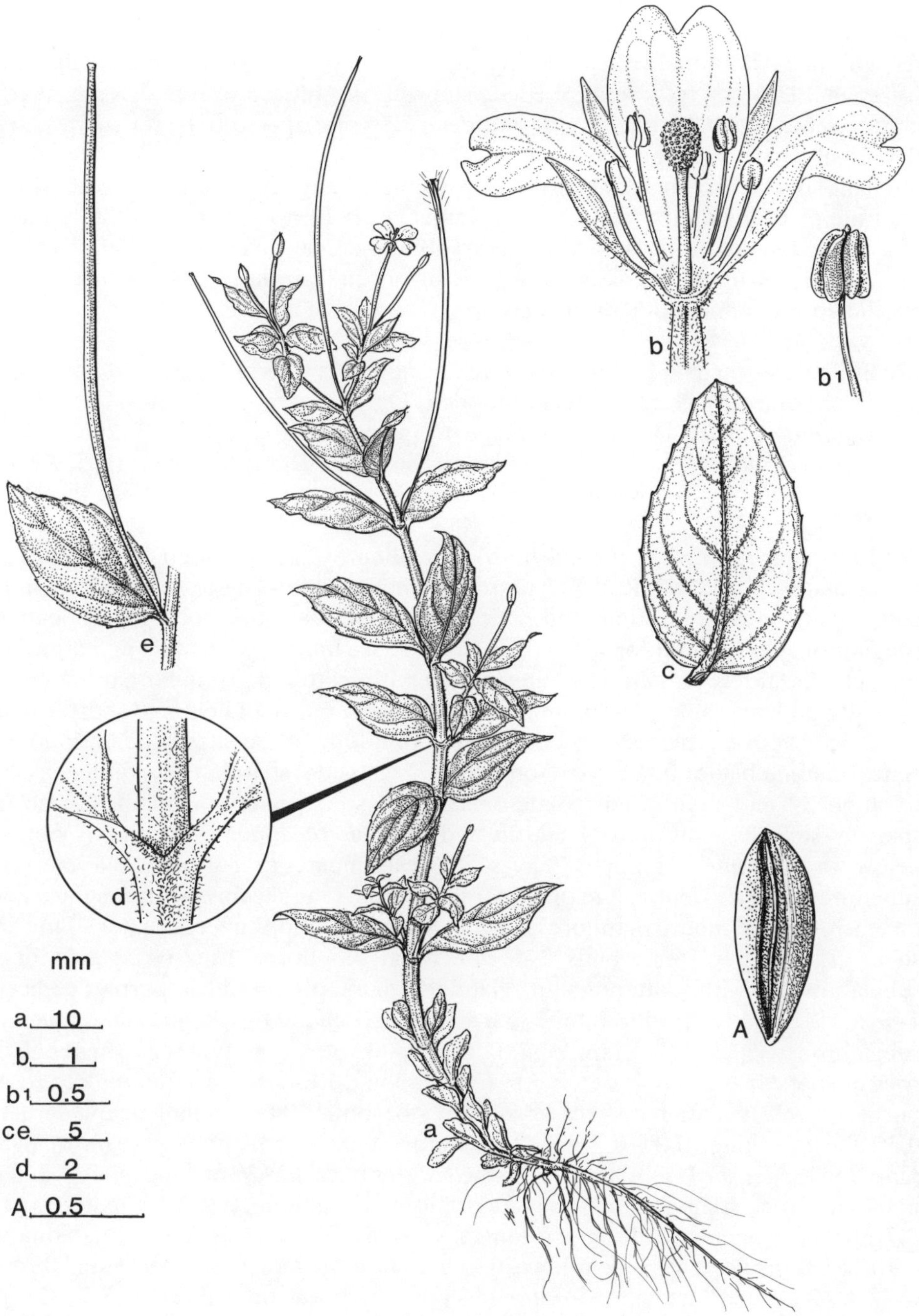

FIG. 61. *Epilobium pengii.* a. Habit. b. Flower (one petal and three stamens removed). b[1]. Detail of stamen, showing longitudinally dehiscent anther. c. Mid-cauline leaf. d. Mid-cauline axils. e. Immature capsule and subtending bract. A. Seed (adaxial view), surface reticulate. (a–e based on *Hoch & C.-I Peng 3381,* type; A based on *K. Mori s.n.* in 1936.)

closely related, but these differences and the wide geographical separation between the two taxa suggest that the distinctive populations on Taiwan warrant specific recognition. Additional collections are much needed, since only a single collection (*Mori* in 1936) has fully mature seeds showing the reticulate surface (the nearly mature seeds of *Hoch & Peng 3381* and *Peng 5795* also appear to be reticulate), and that highly diagnostic character needs to be verified.

This species is named in honor of Dr. Ching-I Peng, Curator of the herbarium, Institute of Botany, Academia Sinica, Taipei (HAST), whose careful observations and collections alerted us to the remarkable endemic diversity of *Epilobium* in Taiwan. He planned and assisted with fieldwork in Taiwan in 1986, which greatly clarified our understanding of this group.

37. Epilobium anagallidifolium Lamarck, Encycl. 2: 376. 1786.—TYPE: FRANCE. "J'ai trouve cette plante au Monde-d'Or" (holotype: P-LAM, microfiche!).
Epilobium dielsii H. Léveillé, Repert. Spec. Nov. Regni Veg. 3: 20. 1906.—TYPE: JAPAN. Norikusa, 2500 m, in rocky places, rare, 28 Aug 1905, *Faurie 6672* (holotype: not located; isotype: K!).

Low, mat-forming perennial herb, spreading by thin leafy epigeous soboles. Stems many, 3–20 cm tall, 0.5–1 mm in diameter, ascending to suberect, sigmoidally bent, sometimes thin and sparsely leaved, or more robust and leafier, subglabrous throughout, or with sparsely strigillose lines decurrent from margins of the petioles, rarely strigillose on upper stem with scattered glandular hairs. Leaves opposite on lower stem, alternate on inflorescence, crowded below to shorter than internodes above, attenuate to petioles 1–6 mm long, often subsessile on inflorescence; cauline blades 0.8–2.5 cm long, 0.25–1 cm wide, spatulate to oblong basally to elliptic at mid-stem to lanceolate or narrowly so in upper pairs, subglabrous or sparsely strigillose on abaxial midrib and margins of upper leaves, apex obtuse below to subacute on upper leaves, base attenuate to cuneate, the margins subentire to barely denticulate on upper leaves, the lateral veins inconspicuous, 2–4 on each side of midrib. Inflorescence nodding in bud, suberect later. Flowers suberect, few per stem; buds 2–5 mm long, elliptic; ovaries 6–20 mm long, subglabrous or with scattered short glandular hairs, often reddish purple; pedicels 1–6 (–15) mm long; floral tube 0.6–1.2 mm long, 0.8–1.8 mm in diameter, subglabrous; sepals 1.5–5 mm long, 0.6–1.5 mm wide, green or reddish, subglabrous or sparsely glandular; petals 2.5–6.5 mm long, 1.6–3.5 mm wide, pink to rose-purple, rarely white, narrowly obcordate, the notch 0.6–1.2 mm deep; anthers 0.35–0.6 mm long, 0.25–0.4 mm wide, subrectangular; filaments cream to light pink, the longer set 1.4–3.2 mm long, the shorter ones 0.7–2 mm long; style 1.2–2.5 mm long, erect; stigma broadly clavate or rarely subcapitate, 0.9–1.5 mm long, 0.4–0.7 mm in diameter, at least the longer stamens shedding pollen onto stigma at anthesis. Capsules 1.7–3.6 cm long, 0.1–0.2 cm in diameter, slender, subglabrous or with scattered short hairs, often reddish purple, borne on pedicels 1–3.5 (–5) cm long. Seeds 0.7–1.4 mm long, 0.3–0.45 mm wide, narrowly obovoid, with inconspicuous chalazal collar 0.02–0.1 mm long, light brown, the surface reticulate or rarely low-papillose; coma 1.5–4 mm long, dull white, very persistent. Chromosome number: n = 18. Fig. 62.

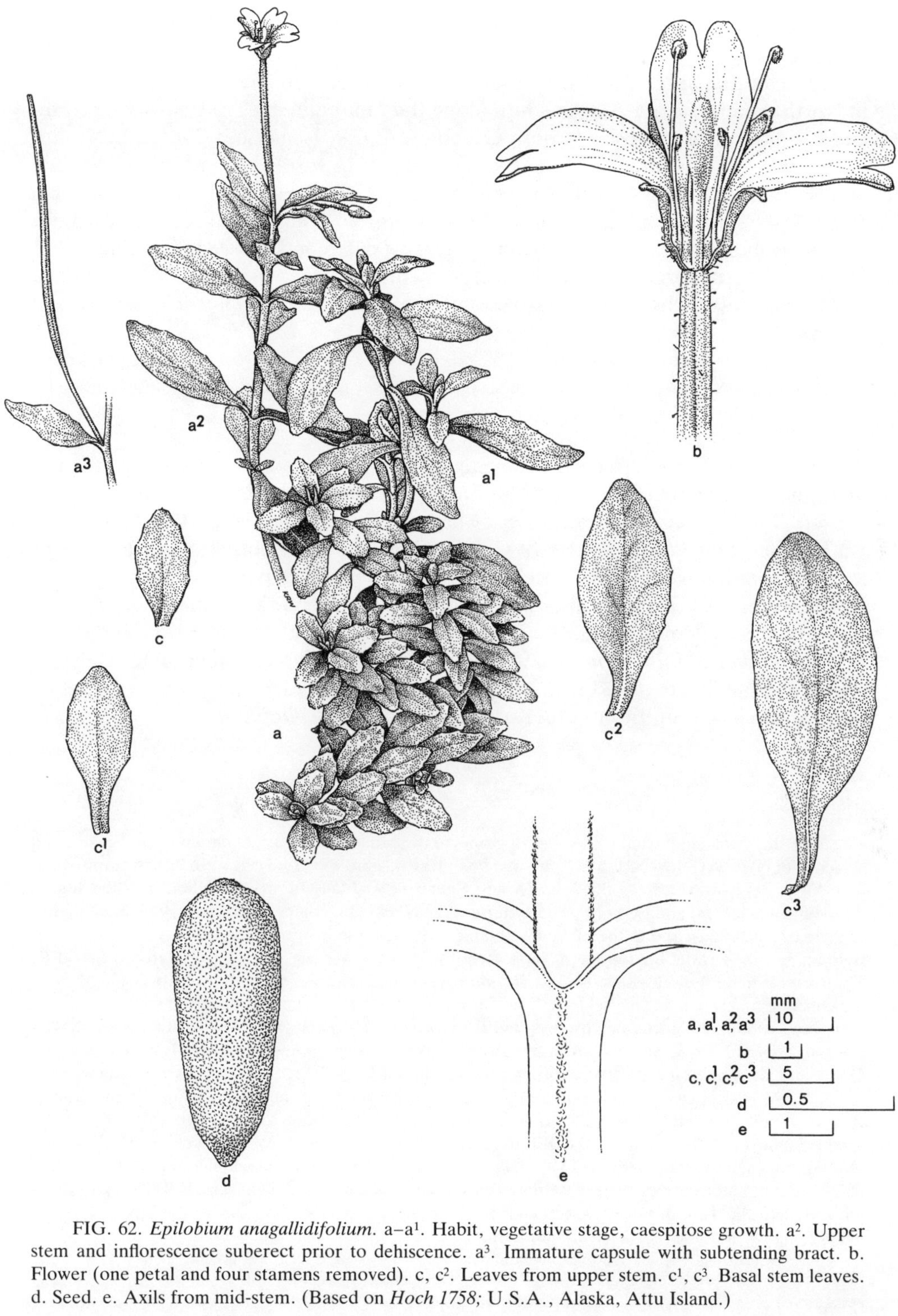

FIG. 62. *Epilobium anagallidifolium.* a–a¹. Habit, vegetative stage, caespitose growth. a². Upper stem and inflorescence suberect prior to dehiscence. a³. Immature capsule with subtending bract. b. Flower (one petal and four stamens removed). c, c². Leaves from upper stem. c¹, c³. Basal stem leaves. d. Seed. e. Axils from mid-stem. (Based on *Hoch 1758;* U.S.A., Alaska, Attu Island.)

Phenology. Collected in flower from July to August, in fruit from August to September.

Distribution (Fig. 47). Circumboreal in montane, alpine, and subarctic areas; in North America from Alaska south along the mountain axes to California, Utah, and Colorado, east across northern Canada to Baffin Island and Greenland, south to Québec, New Hampshire, and Newfoundland; in Eurasia from central Europe and Scandinavia across northern Asia to Siberia, the Kamtchatka Peninsula, and Japan (as *E. dielsii*); in China only known from the Altay Mountains of northern Xinjiang; moist rock-slides and stony places near streams or seeps in montane to alpine areas; scattered and uncommon from 1300–1500 m in Altay (from near sea level in the north to 4000 m in its southern distribution in North America and Europe).

Additional Specimens Examined. **China.** Xinjiang: Altay Mountains, ca. 20 km NE of Altay city, 1400 m, *C. J. Chen & Hoch 86080* (MO, PE).

This widespread, discontinuous-circumboreal species is reported here for the first time from China on the basis of a single collection from the Altay Mountains in August, 1986. It may be expected elsewhere in that region, and perhaps also in the Da Hinggan range of northeastern China, where the cool-temperate mixed conifer forest vegetation is also found. Relatively few collections have been made at higher elevations of these regions, and these plants are small and rather inconspicuous, so that the species may be more common than the single collection indicates. *Epilobium anagallidifolium* is a member of the distinctive circumboreal "Alpinae" group of Haussknecht, and with the possible exception of *E. clarkeanum,* it seems to have no close affinities to other Chinese species of *Epilobium.*

ACKNOWLEDGMENTS

This study was supported by a series of grants to the Missouri Botanical Garden from the National Science Foundation (DEB–8214879, BSR–8518906, BSR–8906848; Peter H. Raven, Principal Investigator), by grants to Raven from the John D. and Catherine T. MacArthur Foundation, by the Chinese Academy of Sciences, and by a grant to Hoch from the National Geographic Society. We extend thanks to these organizations, and to the Missouri Botanical Garden, the Institute of Botany, Academia Sinica, Beijing, and their staffs for long-term support of this project. We are especially grateful to David E. Boufford for helpful discussions, manuscript improvements, and valuable collections; to Ching-I Peng for enabling assistance during field work in Taiwan, useful discussions regarding the species endemic there, and comments on the manuscript; and to A. K. Skvortsov for valuable collections from Soviet Central Asia and for comments on the manuscript. We are also grateful to Kunio Iwatsuki, Hideaki Ohba, and K. Amano for assistance during a visit to herbaria in Tokyo, and to Hiroshige Koyama for assistance in Kyoto; to Qian Ying-qian (Y. C. Chien), Lu An-ming, and the staff of the Institute of Botany, Academia Sinica, Beijing, for their assistance during this study; to Li Shugang, Wen Longxing, Zhang Liyun, and Xie Shiao of the Xinjiang Institute of Biology, Pedology, and Desert Research, Urümqi, for their assistance with field work in Xinjiang; to Ching-san Chen and the staff of the Institute of Botany, Academia Sinica, Taipei, of their support and assistance with field work in Taiwan; to Yuen-po Yang and the Taiwan Forestry Research Institute for logistical support; and to Shih-wen Yeh and Yueh-Fong Chen of the Yushan National Park in Taiwan for assistance and collecting permits.

We gratefully acknowledge the excellent illustrations provided by the following artists: J. Y. Feng (Figs. 24, 28, 44, 49), M. S. Guo (Figs. 26, 39, 41, 48, 54), C. Z. Ji (Figs. 14, 16, 33, 43, 56), C. R. Liu (Figs. 21, 38, 45, 55, 60), J. Myers (Figs. 35, 36, 50, 61), J. F. Wang (Figs. 18, 30, 51, 58), K. R. West (Fig. 62), C. H. Wu (Figs. 11, 13, 19, 27), C. F. Zhang (Figs. 9, 31, 52, 57), and T. L. Zhang (Figs. 23,

37, 46, 53, 59). We thank the Horticulture staff at the Missouri Botanical Garden, especially Petra Malesevich and Kathy Pickett, for their excellent care of the living collections, and Gloria Hoch for typing the manuscript.

We also thank the directors and curators of the following herbaria for making their collections available for our study: A, AA, B, BH, BM, BP, BR, BRNM, C, CAL, CAN, CAS, CDBI, CGE, CISNR, CM, COLO, CSFU, DAO, DD, DS, E, F, FJSI, FJTU, G, GH, GOET, GZ, GZAC, GZBI, GZMI, H, HAL, HAST, HBG, HIB, HIIM, HIMC, HK, HNWP, HSFS, HTC, HZBG, IBG, IBK, IBSC, IFP, JE, K, KUN, KYO, L, LBG, LD, LE, LINN, LIV, LZU, M, MAK, MANCH, MHA, MICH, MO, MPU, MSC, NAS, NCU, ND, NETU, NMU, NS, NTUF, NU, NWBI, NWIP, NWP, NY, O, OXF, P, PE, PENN, PH, POM, PR, RSA, S, SHIN, SMU, SZ, TAI, TAIF, TI, TNS, UC, UPS, US, W, WIS, WTU, WU, WUK, XJBI, XZ, YNU, YUN, Z. If we have failed to mention anyone who has helped in any way, we offer our apologies and sincere thanks.

LITERATURE CITED

Alexander, M. P. 1969. Differential staining of aborted and non-aborted pollen. Stain Technol. 44: 117–122.

Benham, B. R. 1969. Insect visitors to *Chamaenerion angustifolium* and their behavior in relation to pollination. Entomologist 102: 221–228.

Böcher, T. W. 1962. A cytological and morphological study of the two-species hybrid *Chamaenerion angustifolium* × *C. latifolium*. Bot. Tidsskr. 58: 1–34.

Boufford, D. E. 1982. The systematics and evolution of *Circaea* (Onagraceae). Ann. Missouri Bot. Gard. 69: 804–994.

Boufford, D. E., and S. A. Spongberg. 1983. Eastern Asian-Eastern North American phytogeographical relationships—A history from the time of Linnaeus to the twentieth century. Ann. Missouri Bot. Gard. 70: 423–439.

Britton, N. L., and A. Brown. 1913. *An illustrated flora of the northern United States, Canada and the British possessions*, ed. 1. New York: Charles Scribner's Sons.

Chamberlain, D. F., and P. H. Raven. 1972. Onagraceae. In *Flora of Turkey and East Aegean Islands*, ed. P. H. Davis, 4: 183–196. Edinburgh: University Press.

Chen, C. J., P. C. Hoch, and P. H. Raven. 1988. Variation patterns and systematics of the widespread species *Epilobium angustifolium* L. (Onagraceae) in China. Acta Phytotax. Sin. 26: 81–95.

Cheng, Z. 1983. A comparative study of the vegetation in Hubei Province, China, and in the Carolinas of the United States. Ann. Missouri Bot. Gard. 70: 571–575.

Flint, F. P. 1980. Ecology and distribution of diploid and tetraploid *Epilobium angustifolium* (fireweed) in the Beartooth Mountains of Wyoming and Montana. Ph.D. dissertation. Ann Arbor: University Microfilms.

Forbes, F. B., and W. B. Hemsley. 1887. Index Florae Sinensis. J. Linn. Soc. Bot. 23: 307–309.

Handel-Mazzetti, H. 1929. Oenotheraceae. In *Symbolae Sinicae* 7(3): 600–605. Wien: Julius Springer.

Hara, H. 1942. Preliminary study on Japanese *Epilobium*. J. Jap. Bot. 18: 173–186.

———. 1965. *Epilobium*. In *Flora of Japan*, ed. J. Ohwi, pp. 656–658. Washington: Smithsonian Institution Press.

———. 1966. *The Flora of Eastern Himalaya*, pp. 223–224. Tokyo: University of Tokyo Press.

———. 1971. *The Flora of Eastern Himalaya*, Second Report, pp. 89–90. Tokyo: University of Tokyo Press.

———. 1979. *Epilobium*. In *An enumeration of the flowering plants of Nepal*, ed. H. Hara and H. J. Williams, 2: 174–175. London: British Museum (Natural History).

Haussknecht, C. 1884. *Monographie der Gattung Epilobium*. Jena.

Hoch, P. C. 1978. The systematics and evolution of the *Epilobium ciliatum* complex in North America (Onagraceae). Ph.D. dissertation. Ann Arbor: University Microfilms.

———. 1981. *Epilobium*. In *Flora of Pakistan*, ed. E. Nasir and S. I. Ali, 139: 8–35.

———. 1991. *Epilobium*. In *Flora of Bhutan*, by A. Grierson and D. G. Long, 2(1): 309–322. Edinburgh: Royal Botanic Garden.

Hoch, P. C., and P. H. Raven. 1990. Perspectives on "Monographie der Gattung *Epilobium*" of Carl Haussknecht. Haussknechtia 5: 21–28.

Holm, T. 1922. Contributions to the morphology, synonymy, and geographical distribution of arctic plants. Report Canad. Arctic Exped. 1913–1918, 5(B): 1–139. Ohawa: F. A. Acland.
Holub, J. 1972. Taxonomic and nomenclatural remarks on *Chamaenerion* auct. Folia Geobot. Phytotax. 7: 81–90.
Hou, H. Y. 1983. Vegetation of China with reference to its geographical distribution. Ann. Missouri Bot. Gard. 70: 509–549.
Howes, F. N. 1979. *Plants and beekeeping*. London: Faber and Faber.
Hsü, J. 1983. Late Cretaceous and Cenozoic vegetation in China emphasizing their connections with North America. Ann. Missouri Bot. Gard. 70: 490–508.
Hu, S. Y. 1980. The *Metasequoia* flora and its phytogeographic significance. J. Arnold Arbor. 61: 41–94.
Institute of Botany, Academia Sinica. 1972. Onagraceae. In *Iconographia Cormophytorum Sinicorum* 2: 1018–1021. Beijing: Science Press.
Izco, J. 1982. *Epilobium paniculatum* nueva adventicia para Europa. Candollea 38: 309–315.
Jiangsu Institute of Botany. 1982. Onagraceae. In *Flora Jiangsu* 2: 547–548. Nanjing: Jiangsu Scientific and Technical Press.
Keating, R., P. C. Hoch, and P. H. Raven. 1982. Perennation in *Epilobium* (Onagraceae) and its relation to classification and ecology. Syst. Bot. 7: 379–404.
Kitagawa, M. 1939. Onagraceae. In *Lineamenta Florae Manshuricae*, pp. 328–329.
Knuth, P. 1908. *Handbook of flower pollination*, Vol. 2. Based on Hermann Müller's work *The fertilization of flowers by insects*. Transl. by J. R. Ainsworth Davis. Oxford: Clarendon Press.
Komarov, V. L. 1901. Species novae Florae Asiae Orientalis. Acta Hort. Petrop. 18: 432–433.
———. 1905. *Flora Manshuriae*. Trudy Imp. S.-Peterburgsk. Bot. Sada 25: 88–95.
Lauener, L. A. 1972. Catalogue of the names published by Hector Léveillé. Notes Royal Bot. Gard. Edinburgh 31: 397–417.
Lee, S. G., and W. Y Chang. 1953. *Chinese geology*. Shanghai: Zhengfeng Press.
Lewis, H., and M. E. Lewis. 1955. The genus *Clarkia*. Univ. Calif. Publ. Bot. 20: 241–392.
Li, H. 1980. Onagraceae. In *Enumeration of the vascular plants of Xizang (Tibet)*, Editors Group of Flora Xizang, pp. 222–223. Beijing: National Commission of Science and Technology.
———. 1986. Onagraceae. In *Flora Xizangica*, ed. C. Y. Wu, 3: 355–367. Beijing: Science Press.
Li, S. Z. 1596. *Ben Cao Gang Mu* (Compendium of Materia Medica) 5: 123. 1963 edition. Beijing: Zhonghuashuju.
Linnaeus, C. 1753. *Species plantarum* 1: 347–348. Stockholm.
Liou, T. N. et al. 1955. *Illustrated Flora of ligneous plants of Northeast China*. Beijing: Science Press.
———. 1958. Onagraceae. In *Claves plantarum chinae boreali-orientalis*, pp. 235–238. Beijing: Science Press.
———. 1977. Onagraceae. In *Flora plantarum herbacearum Chinae boreali-orientalis* 6: 144–155. Beijing: Science Press.
Löve, A., and D. Löve. 1961. Chromosome numbers of central and northwest European plant species. Opera Bot. 5: 1–581.
Maximowicz, C. J. 1859. Primitiae florae amurensis. Mém. Acad. Imp. Sci. St.-Pétersbourg Divers Savans 9: 104–105.
Michaelis, P. 1954. Cytoplasmic inheritance in *Epilobium* and its theoretical significance. Adv. Genet. 6: 287–401.
Mosquin, T. 1963a. The chromosome races of *Epilobium angustifolium* L. Yearb. Amer. Philos. Soc. 345–346.
———. 1963b. The chromosomal races of *Epilobium angustifolium* L., and their taxonomic rank. Amer. J. Bot. 50: 636.
———. 1966. A new taxonomy for *Epilobium angustifolium* L. (Onagraceae). Brittonia 18: 167–188.
Moss, E. H. 1936. The ecology of *Epilobium angustifolium* with particular reference to rings of periderm in the wood. Amer. J. Bot. 23: 114–120.
Mulligan, G. A. 1957. Chromosome numbers in Canadian weeds I. Canad. J. Bot. 35: 779–787.
Munz, P. A. 1965. Onagraceae. N. Amer. Fl. II. 5: 1–278.
Northwest Institute of Botany, Academia Sinica. 1981. *Epilobium*. In *Flora Tsingling*, 1: 353–356. Beijing: Science Press.
Pavlov, N. 1934. Bemerkung über die neuen und wenig bekannten Arten der Gattung *Epilobium* in Mittel-Asien. Wissenschaftl. Berichte der Moskauer Staatsuniv. 2: 325–330.

Pellett, F. C. 1976. *American honey plants*, 5th ed. Hamilton, Illinois: Dadant and Sons.

Raven, P. H. 1962a. The genus *Epilobium* in the Himalayan region. Bull. Brit. Mus. (Nat. Hist.) Bot. 2: 325–382.

———. 1962b. The genus *Epilobium* in Turkey. Notes Roy. Bot. Gard. Edinburgh 24: 183–203.

———. 1964. Onagraceae. In *Flora Iranica*, ed. K. H. Rechinger, 7: 1–19. Akademische Druck- u. Verlagsanstalt.

———. 1967a. A revision of the African species of *Epilobium*. Bothalia 9: 309–333.

———. 1967b. The genus *Epilobium* in Malesia. Blumea 15: 269–282.

———. 1968. *Epilobium*. In *Flora Europaea*, ed. T. G. Tutin, 2: 308–311.

———. 1976. Generic and sectional delimitation in Onagraceae, Tribe Epilobieae. Ann. Missouri Bot. Gard. 63: 326–340.

———. 1977. Onagraceae. In *Flora Taiwan*, eds. H. L. Li et al., 3: 879–899.

———. 1979. A survey of reproductive biology in Onagraceae. New Zealand J. Bot. 17: 575–593.

———. 1980. Hybridization and the nature of species in higher plants. Canad. Bot. Assoc. Bull. Suppl. 13(1): 3–10.

———. 1988. Onagraceae as a model of plant evolution. In *Plant evolutionary biology*, ed. L. D. Gottlieb and S. K. Jain, pp. 85–107. New York: Chapman and Hall.

Raven, P. H., and T. E. Raven. 1976. The genus *Epilobium* (Onagraceae) in Australasia: a systematic and evolutionary study. New Zealand Dept. Sci. Industr. Res. Bull. 216: 1–321.

Schmitz, U. 1988. Dwarfism and male sterility in interspecific hybrids of *Epilobium*. 1. Expression of plastid genes and structure of the plastome. Theor. Appl. Genet. 75: 350–356.

Schreber, J. C. D. von. 1771. *Spicilegium florae lipsicae*, 146–147. Leipzig: Bibliopolio Dykiano.

Seavey, S. R., and K. S. Bawa. 1986. Late-acting self-incompatibility in angiosperms. Bot. Rev. 52: 195–219.

Seavey, S. R., R. E. Magill, and P. H. Raven. 1977. Evolution of seed size, shape, and surface architecture in the tribe Epilobieae (Onagraceae). Ann. Missouri Bot. Gard. 64: 18–47.

Seavey, S. R., and P. H. Raven. 1977a. Chromosomal evolution in *Epilobium* sect. *Epilobium* (Onagraceae). Plant Syst. Evol. 127: 107–119.

———. 1977b. Chromosomal evolution in *Epilobium* sect. *Epilobium* (Onagraceae), II. Plant Syst. Evol. 128: 195–200.

———. 1977c. Chromosomal differentiation and the sources of the South American species of *Epilobium* (Onagraceae). J. Biogeogr. 4: 55–59.

———. 1978. Chromosomal evolution in *Epilobium* sect. *Epilobium* (Onagraceae). III. Plant Syst. Evol. 130: 79–83.

Seavey, S. R., and J. C. Solomon. 1984. Cytogeography of the South American species of *Epilobium* (Onagraceae). Syst. Bot. 9: 17–21.

Shea, M. M., and M. Watson. 1989. Patterns of leaf and flower removal: their effect on fruit growth in *Chamaenerion angustifolium* (fireweed). Amer. J. Bot. 76: 884–890.

Shen, C. F., and D. E. Boufford. 1988. *Fagus hayatae* (Fagaceae)—A remarkable new example of disjunction between Taiwan and central China. J. Jap. Bot. 63: 96–101.

Skvarla, J. J., P. H. Raven, W. F. Chissoe, and M. Sharp. 1978. An ultrastructural study of viscin threads in Onagraceae pollen. Pollen et Spores 22: 5–143.

Skvarla, J. J., P. H. Raven, and J. Praglowski. 1975. The evolution of pollen tetrads in Onagraceae. Amer. J. Bot. 62: 6–35.

———. 1976. Ultrastructural survey of Onagraceae pollen. In *The Evolutionary significance of the exine*, ed. I. K. Ferguson and J. Muller, pp. 447–479. London: Linnean Society of London.

Small, E. 1968. The systematics of autopolyploidy in *Epilobium latifolium* L. (Onagraceae). Brittonia 20: 169–181.

Solomon, J. C. 1982. The systematics and evolution of *Epilobium* (Onagraceae) in South America. Ann. Missouri Bot. Gard. 69: 239–335.

Sprengel, C. K. 1793. *Das entdeckte Geheimnis der Natur.* Berlin: Friedrich Vieweg der Ältere.

Stace, C. A. (ed.). 1975. *Hybridization and the flora of the British Isles*. London: Academic Press.

Stearn, W. T. 1938. Ligneous plants described by H. Léveillé from eastern Asia, Index to notes on. J. Royal Hort. Soc. 63: 196.

Stebbins, G. L. 1950. *Variation and evolution in plants*. New York: Columbia University Press.

Steinberg, E. J. 1949. Onagraceae. In *Flora of the U.S.S.R.*, ed. B. K. Shishkin and E. G. Bobrov, 15: 565–637. Moskva-Leningrad: Izdater'stvo Akademii Nauk SSSR.

Stuckey, R. L. 1970. Distributional history of *Epilobium hirsutum* (Great Hairy Willow-herb) in North America. Rhodora 72: 164–181.

Tsoong, P. C., and H. P. Yang. 1979. *Flora republicae popularis sinicae* 67: 6, 55. Beijing: Science Press.

van Andel, J. 1975. A study on the population dynamics of the perennial plant species *Chamaenerion angustifolium* (L.) Scop. Oecologia 19: 329–337.

van Andel, J., W. Bos, and W. Ernst. 1978. An experimental study on two populations of *Chamaenerion angustifolium* (L.) Scop. (=*Epilobium angustifolium* L.) occurring on contrasting soils, with particular reference to the response to bicarbonate. New Phytol. 81: 763–772.

Vodalazskij, L. E. 1979. On variety in the adventitious buds and shoots of the root-sucker forming herbaceous perennial *Chamaenerion angustifolium* (L.) Scop. (Onagraceae). Bot. Zhürn. SSSR 64: 734–739. (in Russian).

———. 1982. The development of structure in the ontogenesis of *Chamaenerion angustifolium* and *Chamaenerion angustissimum* (Onagraceae). Bot. Zhürn. SSSR 67: 33–47. (in Russian).

Wallich, N. 1828–1849. A numerical list of dried specimens of plants . . . Manuscript lithographed and distributed by the author. (Portion with *Epilobium* published in 1832.)

Wang, W. T. 1980. Classification specierum sinicarum *Elatostematis* (Urticaceae). Bull. Bot. Lab. Northeast Forest. Inst. 7: 6, 55.

Weber, A. 1982. Contributions to the morphology and systematics of Klugieae and Loxonieae (Gesneriaceae) IX. The genus *Whytockia*. Notes Roy. Bot. Gard. Edinburgh 40: 359–367.

Weberling, F. 1988. The architecture of inflorescence in the Myrtales. Ann. Missouri Bot. Gard. 75: 226–310.

Wu, C. J. 1979. Oenotheraceae. In *Flora intramongolia*, ed. H. C. Fu and Y. C. Ma, 4: 129–135.

Wu, C. Y. 1979. The regionalization of Chinese flora. Acta Bot. Yunnan 1: 1–21.

———. 1984. *Epilobium*. In *Index florae yunnanensis*, 1: 310–311. Beijing: Science Press.

Wu, C. Y., and X. Y. Hou. 1980. *The vegetation of China*, pp. 19–27, 82–86. Beijing: Science Press.

Wu, C. Y. & and H. W. Li. 1977. *Rubiteucris*. In *Flora republicae popularis sinicae*, 65(2): 20–22. Beijing: Science Press.

Wu, C. Y., and H. S. Wang. 1983. In *Phytogeography section, natural geography of China*, ed. Editorial Committee of Natural Geography of China, Academia Sinica, 1: 1–129. Beijing: Science Press.

Ying, S. S. 1975. Alpine plants in Taiwan. Quart. J. Chin. Forest. 8: 121.

APPENDIX I

Chromosome numbers of species of *Epilobium* in China. Vouchers are deposited at MO. All were cultivated at the Missouri Botanical Garden. All counts are somatic unless otherwise noted.

Epilobium amurense subsp. *amurense:* 2n = 36, *Hoch M3016* (seed from CHINA. SICHUAN: Erlang Shan, Tianquan, 1981, *Li Hen s.n.*).

Epilobium amurense subsp. *cephalostigma:* 2n = 36, *Hoch M3023* (seed from JAPAN. MIYAGI PREFECTURE: Sendai, Miyagi-gun, Futakuchi, *H. Okazaki s.n.*).

Epilobium angustifolium subsp. *angustifolium:* 2n = 36, cultivated also at Institute of Botany, Beijing [seed from CHINA. JILIN: Changbai Shan, Aug 1982, *S. X. Li et al. s.n.* NEI MONGGOL: Hinggan Meng, *H. J. Meng 75,* NMU (MO as *Hoch M2959*). QINGHAI: Qilian Shan, *S. W. Liou 8101* (MO as *Hoch M2623*). XINJIANG: Altay, *Z. W. Zhang et al. 4443,* XJBI, PE. SICHUAN: Batang, 31 Aug 1981, *H. Li s.n.* (MO as *Hoch M2728*); Kangding, *W. L. Chen & C. J. Chen 6305,* PE; Aba, *W. L. Chen & C. J. Chen 7977,* PE; Minya Konka, *Andrews 12* (MO as *Hoch M2943*), *Andrews 16* (MO as *Hoch M2945*); Xiangcheng, Aug 1981, *M. Z. Wang s.n.* (MO as *Hoch M2628*); no locality, *S. K. Wu 1* (MO as *Hoch M2655*), *S. K. Wu 7* (MO as *Hoch M2657*). YUNNAN: no locality, *S. K. Wu 6* (MO as *Hoch M2656*)].

Epilobium angustifolium subsp. *circumvagum:* 2n = 72, cultivated also at Institute of Botany, Beijing [seed from CHINA. JILIN: Changbai Shan, *Shenyang Inst. Forest. & Pedol. s.n.* (MO as *Hoch M3179*). NEI MONGGOL: Hulun Buir Meng, Aug 1981, *Y. Z. Liu s.n.,* NMU, PE. GANSU: Wen Xian, *W. L. Chen & C. J. Chen 8807,* PE. HUBEI: Shennongjia, *Shennongjia Bot. Exped. 22844,* PE. SICHUAN: Nanping, *W. L. Chen & C. J. Chen 8450,* PE; Aba, *W. L. Chen & C. J. Chen 7981,* PE].

Epilobium blinii: 2n = 36, *Hoch M3340* (seed from CHINA. YUNNAN: Dali, *Sino-American 1984 Bot. Exped. 1019,* KUN, MO, PE).

Epilobium brevifolium subsp. *brevifolium:* 2n = 36, *Hoch M2013* (seed from NEPAL. Syabron, 1975, *Fukuda s.n.*).

Epilobium brevifolium subsp. *trichoneurum:* n = 18, *Seavey M1854* (seed from CHINA. YUNNAN: *Yunnan Inst. of Bot. s.n.);* and 2n = 36, *Hoch M2994* (seed from CHINA. YUNNAN: Dali Xian, Tsang Shan, *Li Hen 5*).

Epilobium conspersum: 2n = 36, *Hoch M2832* (seed from NEPAL. Eswa Valley, *P. R. Shakya 7272*).

Epilobium cylindricum: 2n = 36, *Hoch M3084* (seed from INDIA. Garhwal Himalaya, Pandukeshwar, *Khullar P-4*).

Epilobium fangii: n = 18, 2n = 36, *Hoch M2981* (seed from CHINA. SICHUAN: Omei Shan, Xian Feng Temple, *Raven et al. 26101,* MO).

Epilobium fastigiatoramosum: 2n = 36, *Hoch M2954* (seed from CHINA. LIAONING: Kuandian, Sep 1982, *S. X. Li s.n.*).

Epilobium kermodei: 2n = 36, *Hoch M2502* (seed from CHINA. SICHUAN: Mt. Emei, *Raven et al. 26110,* MO, PE).

Epilobium laxum: n = 18, *Seavey M997* (seed from INDIA. KASHMIR: Zanskar, Sankoo, *Kak 782,* MO).

Epilobium nankotaizanense: 2n = 36, *M1314* (seed from CHINA. TAIWAN: Yilan Xian, Nanhutashan, *Jeng 2023*).

Epilobium pannosum: 2n = 36, *Hoch M2996* (seed from CHINA. YUNNAN: Kunming, 1981, *Li Hen s.n.*).

Epilobium platystigmatosum: 2n = 36, *Hoch M2625* (seed from CHINA. SICHUAN: Luding Xian, *Y. C. Tang 01*); and n = 18, *Seavey M591* (seed from CHINA. Taiwan, Hualian Xian, *C. Chang* in 1971).

Epilobium roseum subsp. *subsessile:* 2n = 36, *Hoch M2824* (seed from AFGHANISTAN. Prov. Herat, Koh-i-Zyarat, *Podlech & Jarmal 29233,* M).

Epilobium sikkimense: 2n = 36, *Hoch M2685* (seed from CHINA. YUNNAN: Dali Xian, Tsangshan, *Li Hen 4*).

Epilobium sinense: 2n = 36, *Hoch M2688* (seed from CHINA. HUBEI: Shennongjia, *1980 Sino-American Exped. 599,* GH, KUN, PE).

Epilobium tianschanicum: 2n = 36, *Hoch M3240* (seed from U.S.S.R. KIRGHIZ SSR: N. Tien-Shan, Transily Alatau, Kaindy River, 1983, *Baitem s.n.*).

Epilobium wallichianum: 2n = 36, *Hoch M2069* (seed from BHUTAN. Between Nyakchung La and Sakden, *Dunbar 100*).

APPENDIX 2

Occurrence of species of *Epilobium* in China by province. The numbers in parentheses correspond to the numbers assigned to the provinces in Fig. 1.

Anhui (14): *E. amurense* subsp. *cephalostigma, E. brevifolium* subsp. *trichoneurum, E. hirsutum, E. pyrricholophum.*

Fujian (18): *E. amurense* subsp. *cephalostigma, E. brevifolium* subsp. *trichoneurum, E. pyrricholophum.*

Gansu (11): *E. amurense* subsp. *amurense, E. amurense* subsp. *cephalostigma, E. angustifolium* subsp. *angustifolium, E. angustifolium* subsp. *circumvagum, E. brevifolium* subsp. *trichoneurum, E. cylindricum, E. fastigiatoramosum, E. hirsutum, E. minutiflorum, E. palustre, E. parviflorum, E. platystigmatosum, E. royleanum, E. sikkimense, E. sinense, E. subcoriaceum, E. wallichianum.*

Guangdong (20): *E. amurense* subsp. *cephalostigma, E. brevifolium* subsp. *trichoneurum, E. hirsutum, E. pyrricholophum.*

Guangxi (22): *E. amurense* subsp. *amurense, E. amurense* subsp. *cephalostigma, E. brevifolium* subsp. *trichoneurum, E. kermodei, E. platystigmatosum, E. pyrricholophum.*

Guizhou (26): *E. amurense* subsp. *amurense, E. amurense* subsp. *cephalostigma, E. angustifolium* subsp. *circumvagum, E. blinii, E. brevifolium* subsp. *trichoneurum, E. cylindricum, E. hirsutum, E. kermodei, E. pannosum, E. parviflorum, E. pyrricholophum, E. royleanum, E. sinense, E. wallichianum.*

Hainan (21): no *Epilobium.*

Hebei (5): *E. amurense* subsp. *amurense, E. amurense* subsp. *cephalostigma, E. angustifolium* subsp. *angustifolium, E. angustifolium* subsp. *circumvagum, E. fastigiatoramosum, E. hirsutum, E. minutiflorum, E. palustre, E. parviflorum, E. platystigmatosum.*

Heilongjiang (1): *E. amurense* subsp. *cephalostigma, E. angustifolium* subsp. *angustifolium, E. angustifolium* subsp. *circumvagum, E. ciliatum, E. fastigiatoramosum, E. palustre.*

Henan (8): *E. amurense* subsp. *amurense, E. amurense* subsp. *cephalostigma, E. angustifolium* subsp. *circumvagum, E. brevifolium* subsp. *trichoneurum, E. hirsutum, E. parviflorum, E. platystigmatosum, E. pyrricholophum, E. royleanum, E. sinense.*

Hubei (24): *E. amurense* subsp. *amurense, E. amurense* subsp. *cephalostigma, E. angustifolium* subsp. *circumvagum, E. brevifolium* subsp. *trichoneurum, E. cylindricum, E. hirsutum, E. kermodei, E. parviflorum, E. platystigmatosum, E. pyrricholophum, E. royleanum, E. sinense, E. wallichianum.*

Hunan (23): *E. amurense* subsp. *amurense, E. amurense* subsp. *cephalostigma, E. brevifolium* subsp. *trichoneurum, E. hirsutum, E. kermodei, E. parviflorum, E. pyrricholophum, E. sinense.*

Jiangsu (15): *E. hirsutum, E. pyrricholophum.*

Jiangxi (17): *E. amurense* subsp. *cephalostigma, E. angustifolium* subsp. *circumvagum, E. brevifolium* subsp. *trichoneurum, E. hirsutum, E. pyrricholophum.*

Jilin (2): *E. amurense* subsp. *amurense, E. amurense* subsp. *cephalostigma, E. angustifolium* subsp. *angustifolium, E. angustifolium* subsp. *circumvagum, E. ciliatum, E. fastigiatoramosum, E. hirsutum, E. minutiflorum, E. palustre.*

Liaoning (3): *E. amurense* subsp. *cephalostigma, E. angustifolium* subsp. *circumvagum, E. fastigiatoramosum, E. hirsutum, E. minutiflorum, E. palustre.*

Nei Monggol (4): *E. amurense* subsp. *amurense, E. angustifolium* subsp. *angustifolium, E. angustifolium* subsp. *circumvagum, E. fastigiatoramosum, E. hirsutum, E. minutiflorum, E. palustre, E. parviflorum.*

Ningxia (10): *E. angustifolium* subsp. *angustifolium, E. angustifolium* subsp. *circumvagum, E. fastigiatoramosum, E. hirsutum, E. minutiflorum.*

Qinghai (12): *E. amurense* subsp. *amurense, E. angustifolium* subsp. *angustifolium, E. angustifolium* subsp. *circumvagum, E. conspersum, E. fastigiatoramosum, E. latifolium, E. palustre, E. platystigmatosum, E. royleanum, E. sikkimense, E. subcoriaceum, E. williamsii.*

Shaanxi (9): *E. amurense* subsp. *amurense, E. amurense* subsp. *cephalostigma, E. angustifolium* subsp. *circumvagum, E. brevifolium* subsp. *trichoneurum, E. conspersum, E. fastigiatoramosum, E. hirsutum, E. minutiflorum, E. palustre, E. parviflorum, E. platystigmatosum, E. royleanum, E. pyrricholophum, E. sikkimense, E. sinense, E. subcoriaceum.*

Shandong (7): *E. amurense* subsp. *amurense, E. amurense* subsp. *cephalostigma, E. angustifolium* subsp. *circumvagum, E. fastigiatoramosum, E. hirsutum, E. parviflorum, E. pyrricholophum.*

Shanxi (6): *E. amurense* subsp. *amurense, E. angustifolium* subsp. *angustifolium, E. angustifolium* subsp. *circumvagum, E. fastigiatoramosum, E. hirsutum, E. minutiflorum, E. parviflorum, E. palustre.*

Sichuan (25): *E. amurense* subsp. *amurense, E. amurense* subsp. *cephalostigma, E. angustifolium* subsp. *angustifolium, E. angustifolium* subsp. *circumvagum, E. blinii, E. brevifolium* subsp. *trichoneurum, E. conspersum, E. cylindricum, E. fangii, E. fastigiatoramosum, E. hirsutum, E. kermodei, E. kingdonii, E. palustre, E. pannosum, E. parviflorum, E. platystigmatosum, E. pyrricholophum, E. royleanum, E. sikkimense, E. sinense, E. subcoriaceum, E. tibetanum, E. wallichianum, E. williamsii.*

Taiwan (19): *E. amurense* subsp. *amurense, E. brevifolium* subsp. *trichoneurum, E. hohuanense, E. nankotaizanense, E. pengii, E. platystigmatosum, E. taiwanianum.*

Xinjiang (13): *E. anagallidifolium, E. angustifolium* subsp. *angustifolium, E. hirsutum, E. latifolium, E. laxum, E. minutiflorum, E. palustre, E. parviflorum, E. roseum* subsp. *roseum, E. roseum* subsp. *subsessile, E. royleanum, E. tianschanicum.*

Xizang (28): *E. amurense* subsp. *amurense, E. angustifolium* subsp. *angustifolium, E. angustifolium* subsp. *circumvagum, E. brevifolium* subsp. *brevifolium, E. brevifolium* subsp. *trichoneurum, E. conspersum, E. cylindricum, E. gouldii, E. hirsutum, E. kingdonii, E. latifolium, E. minutiflorum, E. palustre, E. parviflorum, E. royleanum, E. sikkimense, E. speciosum, E. subcoriaceum, E. tibetanum, E. wallichianum, E. williamsii.*

Yunnan (27): *E. amurense* subsp. *amurense, E. amurense* subsp. *cephalostigma, E. angustifolium* subsp. *angustifolium, E. angustifolium* subsp. *circumvagum, E. brevifolium* subsp. *brevifolium, E. brevifolium* subsp. *trichoneurum, E. clarkeanum, E. conspersum, E. cylindricum, E. fangii,*

E. hirsutum, E. kermodei, E. kingdonii, E. latifolium, E. palustre, E. pannosum, E. parviflorum, E. platystigmatosum, E. royleanum, E. sikkimense, E. sinense, E. subcoriaceum, E. tibetanum, E. wallichianum, E. williamsii.

Zhejiang (16): *E. amurense* subsp. *cephalostigma, E. brevifolium* subsp. *trichoneurum, E. hirsutum, E. pyrricholophum.*

APPENDIX 3

Literature reports of species of *Epilobium* occurring in China.

1. Epilobium latifolium Linnaeus, Sp. Pl. 1: 347. 1753; C. B. Clarke in J. D. Hooker, Fl. Brit. India 2: 583. 1879; Haussknecht, Monogr. Epil. 190. 1884; H. Léveillé, Icon. Gen. Epil. t. 225. 1911; Raven, Bull. Brit. Mus. (Nat. Hist.) Bot. 2: 348. 1962; et in K. H. Rechinger, Fl. Iran. 7: 5. 1964; et in Tutin et al., Fl. Europ. 2: 309. 1968; E. Small, Brittonia 20: 169–181. 1968; R. Stewart, Annot. Catal. Vasc. Pl. W. Pakistan & Kashmir 506. 1972; Hara in Hara & H. J. Williams, Enum. Fl. Pl. Nepal 175. 1979; Hoch in Nasir & Ali, Fl. Pakistan 139: 13. 1981. *Chamaenerion latifolium* (Linnaeus) Fries & Lange, Fl. Dan. xvii, 49: 7. 1877; Steinberg in Shishkin & Bobrov, Fl. URSS 15: 626. 1949; Wendelbo, Nytt. Mag. Bot. 1: 46. 1952; Edit. Group Fl. Xizang, Enum. Vasc. Pl. Xizang (Tibet) 223. 1980, pro parte; C. Y. Wu, Index Fl. Yunnan. 1: 310. 1984; H. Li in C. Y. Wu, Fl. Xizang. 3: 365. 1986, pro parte. *Chamerion latifolium* (Linnaeus) Holub, Folia Geobot. Phytotax. 7: 86. 1972.

Epilobium changaicum Grubov, Bot. Mater. Gerb. Bot. Inst. Komarova Akad. Nauk SSSR 17: 20. 1955.

Epilobium kesamitsui Yamazaki, J. Jap. Bot. 65: 141. 1990.

2. Epilobium speciosum Decaisne in Jacquemont, Voy. India IV, Bot.: 57, t. 69. 1844. *Epilobium latifolium* subsp. *speciosum* (Decaisne) Raven, Bull. Brit. Mus. (Nat. Hist.) Bot. 2: 349. 1962; R. Stewart, Annot. Cat. Vasc. Pl. W. Pakistan & Kashmir 506. 1972; Hara in Hara & Williams, Enum. Fl. Pl. Nepal 2: 175. 1979; Hoch in Nasir & Ali, Fl. Pakistan 139: 15. 1981. *Chamerion speciosum* (Decaisne) Holub, Folia Geobot. Phytotax. 7: 86. 1972.

Chamaenerion latifolium sensu H. Li in C. Y. Wu, Fl. Xizang. 3: 365. 1986, pro parte, quoad pl. Burang; non (Linnaeus) Fries & Lange.

3. Epilobium conspersum Haussknecht, Oesterr. Bot. Zeitschr. 29: 51. Feb 1879; et Monogr. Epil. 190, t. 6, Fig. 4b. 1884; H. Léveillé, Icon. Gen. Epil. t. 57. 1910; Raven, Bull. Brit. Mus. (Nat. Hist.) Bot. 2: 350. 1962; Hara in Hara & Williams, Enum. Fl. Pl. Nepal 2: 125. 1979. *Epilobium reticulatum* C. B. Clarke in J. D. Hooker, Fl. Brit. India 2: 583. May 1879, nom. superfl.; W. Smith & Cave, Rec. Bot. Surv. India 4: 198. 1911; W. Smith, loc. cit. 4: 373. 1913. *Chamaenerion reticulatum* (C. B. Clarke) Kitamura in Kihara, Fauna & Fl. Nepal Himal. 1: 185. 1953. *Chamaenerion conspersum* (Haussknecht) Kitamura, Results Kyoto Univ. Sci. Exped. Karak. & Hinduk. 1955, 8: 110. 1966; Edit. Group Fl. Xizang, Enum. Vasc. Pl. Xizang 223. 1980; C. Y. Wu, Index Fl. Yunnan. 1: 310. 1984; H. Li in C. Y. Wu, Fl. Xizang. 3: 364. 1986. *Chamerion conspersum* (Haussknecht) Holub, Folia Geobot. Phytotax. 7: 86. 1972.

Chamaenerion angustifolium sensu Inst. Bot. Bor.-Occ. Acad. Sin., Fl. Tsinling 1: 355. 1981, pro minima parte; non (Linnaeus) Scopoli.

4. Epilobium angustifolium Linnaeus, Sp. Pl. 1: 347. 1753; Maximowicz, Prim. Fl. Amur. 104. 1859; C. B. Clarke in J. D. Hooker, Fl. Brit. India 2: 582. 1879; Aitchison, J. Linn. Soc. Lond., Bot. 18: 60. 1880; Hance, J. Bot. 21: 321. 1883; Haussknecht, Monogr. Epil. 37. 1884; Forbes & Hemsley, J. Linn. Soc. Bot. 23: 307. 1887; Diels, Bot. Jahrb. Syst. 29: 484. 1900; Nakai, Bot. Mag. (Tokyo) 22: 75. 1908; et Fl. Kor. 1: 241. 1909; Hara in Ohwi, Fl. Japan 820. 1953, (Engl. ed.) 657. 1965; Raven, Bull. Brit. Mus. (Nat. Hist.) Bot. 2: 346. 1962; et Notes Royal Bot. Gard. Edinburgh 24: 186. 1962; et in K. H. Rechinger, Fl. Iran. 7: 5. 1964; et in Tutin et al., Fl. Europ. 2: 309. 1968; Mosquin, Brittonia 18: 167. 1966; Chamberlain & Raven in Davis, Fl.

Turkey 4: 185. 1972; R. Stewart, Ann. Cat. Vasc. Pl. W. Pakistan & Kashmir 505. 1972; Hara in Hara & Williams, Enum. Fl. Pl. Nepal 2: 174. 1979; Hoch in Nasir & Ali, Fl. Pakistan 139: 11. 1981; C. J. Chen, Hoch & Raven, Acta Phytotax. Sin. 26: 88. 1988. *Chamaenerion angustifolium* (Linnaeus) Scopoli, Fl. Carniol., ed. 2, 1: 271. 1772; Komarov, Trudy Imp. S.-Peterburgsk. Bot. Sada 25: 88. 1905; et Fl. Penin. Kamtsch. 2: 317. 1929; Handel-Mazzetti, Symb. Sin. 7: 602. 1933; Steinberg in Shishkin & Bobrov, Fl. URSS 15: 622. 1949; Wendelbo, Nytt. Mag. Bot. 1: 46. 1952; T. N. Liou et al., Clav. Pl. Chin. Bor.-Orient. 235. 1959; Dept. Biol. Beijing Norm. Univ., Fl. Beijing 2: 668. 1964; Q. H. Xing in S. Y. He et al., ibid. rev. ed. 1: 626. 1984; Inst. Bot. Acad. Sin., Icon. Corm. Sin. 2: 1021. 1972; Liaoning Inst. Flor. Ped., Fl. Pl. Herb. Chin. Bor.-Orient. 6: 155. 1977; C. J. Wu in H. C. Fu et al., Fl. Intramongol. 4: 129. 1979; Edit. Group Fl. Xizang, Enum. Vasc. Pl. Xizang (Tibet) 223. 1980; Inst. Bot. Bor.-Occ. Acad. Sin., Fl. Tsinling 1: 355. 1981; V. I. Grubov, Key Vasc. Pl. Mongol. 185. 1982; C. Y. Wu, Index Fl. Yunnan. 1: 310. 1984; H. Li in C. Y. Wu, Fl. Xizang. 3: 364. 1986. *Chamerion angustifolium* (Linnaeus) Holub, Folia Geobot. Phytotax. 7: 86. 1972.

4a. Epilobium angustifolium subsp. **angustifolium.**

Epilobium spicatum Lamarck, Fl. Franc. 3: 482. 1778; Boissier, Fl. Orient. 2: 745. 1872; H. Léveillé, Icon. Gen. Epil. t. 226. 1911. *Epilobium neriifolium* H. Léveillé, Monde Pl. 6: 125. 1896, nom. superfl.; et Bull. Acad. Int. Geogr. Bot. 9: 210. 1900.

4b. Epilobium angustifolium subsp. **circumvagum** Mosquin, Brittonia 18: 167. 1966; Czerepanov, Addit. Corr. Fl. URSS 350. 1973; Hara in Hara & Williams, Enum. Fl. Pl. Nepal 2: 125. 1979; C. J. Chen, Hoch & Raven, Acta Phytotax. Sin. 26: 89. 1988. *Chamaenerion angustifolium* var. *platyphyllum* Daniels, Fl. Boulder Colo. 176. 1911. *Chamaenerion angustifolium* subsp. *circumvagum* (Mosquin) Moldenke, Phytologia 27: 289. 1973.

5. Epilobium hirsutum Linnaeus, Sp. Pl. 1: 347. 1753; Boissier, Fl. Orient. 2: 746. 1872; C. B. Clarke in J. D. Hooker, Fl. Brit. India 2: 583. 1879; Haussknecht, Monogr. Epil. 53. 1884; Forbes & Hemsley, J. Linn. Soc. Bot. 23: 307. 1887; Nakai, Fl. Kor. 1: 242. 1909; H. Léveillé, Icon. Gen. Epil. tt. 230, 231. 1911; Handel-Mazzetti, Symb. Sin. 7: 600. 1933; Steinberg in Shishkin & Bobrov, Fl. URSS 15: 578. 1949; Hara, Enum. Sperm. Jap. 3: 367. 1954; T. N. Liou et al., Clav. Pl. Chin. Bor.-Orient. 236. 1959; et Fl. Pl. Herb. Chin. Bor.-Orient. 6: t. 57, fig. 2–3. 1977; Raven, Bull. Brit. Mus. (Nat. Hist.) Bot 2: 352. 1962; et Notes Royal Bot. Gard. Edinburgh 24: 189. 1962; et in K. H. Rechinger, Fl. Iran. 7: 7. 1964; et Bothalia 9: 309. 1967; et in Tutin et al., Fl. Europ. 2: 309. 1968; Dept. Biol. Beijing Norm. Univ., Fl. Beijing 2: 667. 1964; Q. H. Xing in S. Y. He et al., ibid. rev. ed. 1: 625. 1984; Chamberlain & Raven in Davis, Fl. Turkey 4: 187. 1972; R. Stewart in Nasir & Ali, Annot. Cat. Vasc. Pl. W. Pakistan & Kashmir 506. 1972; Inst. Bot. Acad. Sin., Icon. Corm. Sin. 2: 102. 1972; C. J. Wu in H. C. Fu et al., Fl. Intramongol. 4: 133. 1979; Inst. Bot. Bor.-Occ. Acad. Sin., Fl. Tsinling 1: 353. 1981; Hoch in Nasir & Ali, Fl. Pakistan 139: 16. 1981; Jiangsu Inst. Bot., Fl. Jiangsu 2: 547. 1982; Grubov, Man. Vasc. Pl. Mongol. 184. 1982; C. Y. Wu, Index Fl. Yunnan. 1: 312. 1984. *Chamaenerion hirsutum* (Linnaeus) Scopoli, Fl. Carniol. ed. 2, 270. 1772.

Epilobium villosum Thunberg, Prodr. 75. 1794. *E. hirsutum* var. *villosum* (Thunberg) H. Hara, J. Jap. Bot. 18: 178 1942; Hara in Ohwi, Fl. Japan. 820. 1953; (Engl. ed.) 656. 1965.

Epilobium tomentosum Ventenat, Descr. Pl. Nouv. Jard. Cels. t. 90. 1802. *Epilobium hirsutum* var. *tomentosum* (Ventenat) Boissier, Fl. Orient. 2: 746. 1872. *Epilobium velutinum* Nevski, Trudy Bot. Inst. Akad. Nauk SSSR, Ser. 1, Fl. Sist. Vyss. Rast. 4: 312. 1937, nom. superfl.

Epilobium hirsutum var. *sericeum* Bentham ex C. B. Clarke in J. D. Hooker, Fl. Brit. India 2: 584. 1879; Aitchison, J. Linn. Soc., Bot. 18: 60. 1880. *E. sericeum* Bentham in Wallich, Numer. List 216, n. 6325. 1832, nom. nud.; Royle, Illustr. Himal. 1: 211. 1839.

Epilobium hirsutum var. *laetum* Wallich ex C. B. Clarke in J. D. Hooker, Fl. Brit. India 2: 584. 1879. *E. laetum* Wallich, Numer. List 216, n. 6329. 1832, nom. nud., quoad B.

Epilobium nassirelmulcii Stapf, Denkschr. Kaiserl. Akad. Wiss. Wien Math.-Naturw. Kl. 51: 325. 1886.

6. Epilobium parviflorum Schreber, Spicil. Fl. Lips. 146, 155. 1771; Boissier, Fl. Orient. 2: 746. 1872; C. B. Clarke in J. D. Hooker, Fl. Brit. India 2: 584. 1879; Haussknecht, Monogr. Epil. 66. 1884;

H. Léveillé, Icon. Gen. Epil. t. 232. 1911; Handel-Mazzetti, Symb. Sin. 7: 601. 1933; Steinberg in Shishkin & Bobrov, Fl. URSS 15: 580. 1949; Raven, Bull. Brit. Mus. (Nat. Hist.) Bot. 2: 353. 1962; et in K. H. Rechinger, Fl. Iran. 7: 7. 1964; et Bothalia 9: 315. 1967; et in Tutin et al., Fl. Europ. 2: 309. 1968; Chamberlain & Raven in Davis, Fl. Turkey 4: 187. 1972; Inst. Bot. Acad. Sin., Icon. Corm. Sin. 2: 1019. 1972; R. Stewart, Annot. Cat. Vasc. Pl. W. Pakistan & Kashmir 507. 1972; C. J. Wu in H. C. Fu et al., Fl. Intramongol. 4: 133. 1979; Hara in Hara & Williams, Enum. Fl. Pl. Nepal 2: 175. 1979; Inst. Bot. Bor.-Occid. Acad. Sin., Fl. Tsinling 1: 354. 1981; Hoch in Nasir & Ali, Fl. Pakistan 139: 17. 1981; Bartholomew et al., J. Arnold Arb. 64: 74. 1983.

Epilobium parviflorum var. *vestitum* Bentham in C. B. Clarke in J. D. Hooker, Fl. Brit. India 2: 584. 1879. *E. vestitum* Bentham in Wallich, Numer. List 216, n. 6327. 1832, nom. nud.

7. **Epilobium blinii** H. Léveillé, Repert. Spec. Nov. Regni Veg. 7: 338. 1909; et Icon. Gen. Epil. t. 80. 1910; et Cat. Pl. Yunnan 185. 1916; C. Y. Wu, Index Fl. Yunnan. 1: 311. 1984.

Epilobium forrestii Diels, Notes Roy. Bot. Gard. Edinburgh 5: 254. 1912; Handel-Mazzetti, Symb. Sin. 7: 601. 1933.

8. **Epilobium nankotaizanense** Yamamoto, Suppl. Icon. Pl. Form. 2: 29, pl. 2. 1926; S. S. Ying, Quart. J. Chin. Forest. 8: 122. 1975; et Alpine Pl. Taiwan Color 1: 59. 1975; Raven in H. L. Li et al., Fl. Taiwan 3: 888. 1977; K. S. Hsu et al., Rare Threat. Pl. Taiwan 58. 1980.

9. **Epilobium roseum** Schreber, Spicil. Fl. Lips. 155. 1771. Boissier, Fl. Orient. 2: 749. 1872; Haussknecht, Monogr. Epil. 124. 1884; Steinberg in Shishkin & Bobrov, Fl. URSS 15: 588. 1949; Raven, Notes Roy. Bot. Gard. Edinburgh 24: 194. 1962; et in K. H. Rechinger, Fl. Iran. 7: 10. 1964; et in Tutin et al., Fl. Europ. 2: 310. 1968; Chamberlain & Raven in Davis, Fl. Turkey 4: 192. 1972.

9a. **Epilobium roseum** subsp. **roseum** Raven, Notes Royal Bot. Gard. Edinburgh 24: 194. 1962; et in K. H. Rechinger, Fl. Iran. 7: 10. 1964; Chamberlain & Raven in Davis, Fl. Turkey 4: 192. 1972; Czerepanov, Addit. Corr. Fl. URSS 350. 1973.

9b. **Epilobium roseum** subsp. **subsessile** (Boissier) Raven, Notes Roy. Bot. Gard. Edinburgh 24: 194. 1962; Raven in K. H. Rechinger, Fl. Iran. 7: 10. 1964; Chamberlain & Raven in Davis, Fl. Turkey 4: 192. 1972. *Epilobium smyrnaeum* Boissier & Balansa in Boissier, Diagn. ser. II, 2: 52. 1856. *Epilobium roseum* β. *subsessile* Boissier, Fl. Orient. 2: 749. 1872.

Epilobium nervosum Boissier & Buhse, Nouv. Mém. Soc. Imp. Naturalistes Moscou 12: 88. 1860; Haussknecht, Monogr. Epil. 197. 1884; Steinberg in Shishkin & Bobrov, Fl. URSS 15: 589. 1949.

Epilobium almaatense Steinberg in Shishkin & Bobrov, Fl. URSS 15: 590. 1949, based on *E. laetum* M. Popov, Bull. Soc. Natl. Mosc. n. s. 44: 130. 1935, non *E. laetum* Wallich ex Haussknecht, 1884.

10. **Epilobium tianschanicum** Pavlov, Ucen. Zap. Moskovsk. Gosud. Univ. 2: 327. 1934; Steinberg in Shishkin & Bobrov, Fl. URSS. 15: 590. 1949.

11. **Epilobium cylindricum** D. Don, Prodr. Fl. Nepal. 222. 1825; Wallich, Numer. List 216, n. 6328.
N 1832; Haussknecht, Monogr. Epil. 200. 1884; H. Léveillé, Icon. Gen. Epil. t. 69. 1910; Raven, Bull. Brit. Mus. (Nat. Hist.) Bot. 2: 355. 1962; et in K. H. Rechinger, Fl. Iran 7: 13. 1964; Hara, Fl. E. Himal. 224. 1966; et in Hara & Williams, Enum. Fl. Pl. Nepal 2: 174. 1979; R. Stewart in Nasir & Ali, Ann. Cat. Vasc. Pl. W. Pakistan & Kashmir 505. 1972; Edit. Group Fl. Xizang, Enum. Vasc. Pl. Xizang (Tibet) 222. 1980; Hoch in Nasir et Ali, Fl. Pakistan 139: 19. 1981; C. Y. Wu, Index Fl. Yunnan. 1: 312. 1984; H. Li in C. Y. Wu, Fl. Xizang. 3: 359. 1986. *Epilobium roseum* var. *cylindricum* (D. Don) C. B. Clarke in J. D. Hooker, Fl. Brit. India 2: 585. 1879.

Epilobium beauverdianum H. Léveillé, Repert. Spec. Nov. Regni Veg. 8: 138. 1910; et Icon. Gen. Epil. t. 82. 1910.

Epilobium christii H. Léveillé, Repert. Spec. Nov. Regni Veg. 9: 19. 1910; et Icon. Gen. Epil. t. 72. 1910.

Epilobium sinense sensu Bartholomew et al., J. Arnold. Arb. 64: 74. 1983, pro parte; non H. Léveillé.

12. Epilobium tibetanum Haussknecht, Oesterr. Bot. Zeitschr. 29: 53. 1879; et Monogr. Epil. 201. 1884; H. Léveillé, Icon. Gen. Epil. t. 75. 1910; Raven, Bull Brit. Mus. (Nat. Hist.) Bot. 2: 357. 1962; et in K. H. Rechinger, Fl. Iran. 7: 11. 1964; R. Stewart, Annot. Cat. Vasc. Pl. W. Pakistan & Kashmir 507. 1972; Hoch in Nasir & Ali, Fl. Pakistan 139: 21. 1981.

Epilobium pseudobscurum Haussknecht, Oesterr. Bot. Zeitschr. 29: 53. 1879 ("pseudo-obscurum"); et Monogr. Epil. 221. 1884; H. Léveillé, Icon. Gen. Epil. t. 79. 1910; Raven, Bull. Brit. Mus. (Nat. Hist.) Bot. 2: 380. 1962; R. Stewart, Annot. Cat. Vasc. Pl. W. Pakistan & Kashmir 507. 1972; Hoch in Nasir & Ali, Fl. Pakistan 139: 21. 1981.

Epilobium leiospermum Haussknecht, Monogr. Epil. 206. 1884; H. Léveillé, Icon. Gen. Epil. t. 65. 1910; Raven, Bull. Brit. Mus. (Nat. Hist.) Bot. 2: 357. 1962; et in K. H. Rechinger, Fl. Iran. 7: 11. 1964; R. Stewart, Annot. Cat. Vasc. Pl. W. Pakistan & Kashmir 506. 1972.

Epilobium roseum var. *anagallidifolium* C. B. Clarke in J. D. Hooker, Fl. Brit. India 2: 585. 1879, pro parte, quoad pl. cit.

Epilobium nuristanicum K. H. Rechinger, Biol. Skr. 10(3): 61. 1958.

13. Epilobium sinense H. Léveillé, Bull. Herb. Boiss., sér. 2, 7: 590. 1907; et Icon. Gen. Epil. t. 74. 1910; et Fl. Kouy-Tcheou 301. 1915; Handel-Mazzetti, Symb. Sin. 7: 601. 1933; Bartholomew et al., J. Arnold Arb. 64: 74. 1983, pro parte.

14. Epilobium platystigmatosum C. B. Robinson, Philipp. J. Sci. 3: 210. 1908; Raven, Blumea 15: 272. 1967; et in H. L. Li et al., Fl. Taiwan 3: 888. 1977; C. Y. Wu, Index Fl. Yunnan. 1: 312. 1984.

Epilobium sp. nov.? Forbes & Hemsley, J. Linn. Soc. Bot. 23: 308. 1887, recorded from China, Hubei, Badong, *Henry 1043* (K).

Epilobium formosanum Masamune, Trans. Nat. Hist. Soc. Taiwan 29: 62. 1939; Hara, J. Jap. Bot. 18: 185. 1942; et Enum. Sperm. Jap. 266. 1954; et in Ohwi, Fl. Japan 821. 1953, (Engl. ed.) 657. 1965; S. S. Ying, Quart. J. Chin. Forest. 8: 121. 1975.

Epilobium cephalostigma var. *linearifolium* Hisauti, J. Jap. Bot. 14: 143. 1938.

Epilobium sohayakiense Koidzumi, Acta Phytotax. Geobot. 8: 61. 1939.

Epilobium sinense sensu Bartholomew et al., J. Arnold Arb. 64: 74. 1983, pro parte.

15. Epilobium hohuanense Ying, Quart. J. Chin. Forest. 8: 121. 1975.

16. Epilobium taiwanianum C. J. Chen, Hoch & Raven, sp. nov.

17. Epilobium royleanum Haussknecht, Oesterr. Bot. Zeitschr. 29: 55. 1879; et Monogr. Epil. 205. 1884; H. Léveillé, Icon. Gen. Epil. t. 66. 1910; Handel-Mazzetti, Symb. Sin. 7: 601. 1933; Raven, Bull. Brit. Mus. (Nat. Hist.) Bot. 2: 358. 1962; et in K. H. Rechinger, Fl. Iran. 7: 11. 1964; R. Stewart, Annot. Cat. Vasc. Pl. W. Pakistan & Kashmir 507. 1972; Hara in Hara & Williams, Enum. Fl. Pl. Nepal 2: 175. 1979; Edit. Group Fl. Xizang, Enum. Vasc. Pl. Xizang (Tibet) 222. 1980; Hoch in Nasir & Ali, Fl. Pakistan 139: 23. 1981; C. Y. Wu, Index Fl. Yunnan. 1: 312. 1984; H. Li in C. Y. Wu, Fl. Xizang. 3: 361. 1986.

Epilobium roseum var. *indicum* C. B. Clarke in J. D. Hooker, Fl. Brit. India 2: 584. 1879.

Epilobium roseum var. *dalhousieanum* C. B. Clarke in J. D. Hooker, Fl. Brit. India 2: 584. 1879.

Epilobium lividum Haussknecht, Monogr. Epil. 201. 1884; H. Léveillé, Icon. Gen. Epil. t. 73. 1910.

Epilobium himalayense Haussknecht, Monogr. Epil. 213. 1884.

Epilobium royleanum forma *glandulosum* Raven, Bull. Brit. Mus. (Nat. Hist.) Bot. 2: 360. 1962; R. Stewart, Annot. Cat. Vasc. Pl. W. Pakistan & Kashmir 507. 1972; Hoch in Nasir & Ali, Fl. Pakistan 139: 25. 1981.

Epilobium royleanum forma *glabrum* Raven, Bull. Brit. Mus. (Nat. Hist.) Bot. 2: 361. 1962; R. Stewart, Annot. Cat. Vasc. Pl. W. Pakistan & Kashmir 507. 1972; Hoch in Nasir & Ali, Fl. Pakistan 139: 25. 1981.

18. Epilobium kermodei Raven, Bull. Brit. Mus. (Nat. Hist.) Bot. 2: 364, pl. 33B. 1962.

Epilobium sp. nov.? Handel-Mazzetti, Symb. Sin. 7: 601. 1933, recorded from China, Hunan, Wugang, Yung Shan, Apr 1919, *T. H. Wang 56* (C, E, W).

19. Epilobium pannosum Haussknecht, Oesterr. Bot. Zeitschr. 29: 54. 1879 (Feb.); et Monogr. Epil. 209. 1884; H. Léveillé, Icon. Gen. Epil. t. 81. 1910; Handel-Mazzetti, Symb. Sin. 7: 601. 1933. *Epilobium khasianum* C. B. Clarke in J. D. Hooker, Fl. Brit. India 2: 585. 1879 (May); Barbey, Icon. Gen. Epil. t. 16. 1910; Kanjil & Das, Fl. Assam 2: 318. 1938. *Epilobium brevifolium* subsp. *pannosum* (Haussknecht) Raven, Bull. Brit. Mus. (Nat. Hist.) Bot. 2: 363. 1962; C. Y. Wu, Index Fl. Yunnan. 1: 311. 1984.

20. Epilobium brevifolium D. Don, Prodr. Fl. Nepal. 222. 1825; Haussknecht, Monogr. Epil. 207. 1884; H. Léveillé, Icon. Gen. Epil. t. 90. 1910; Raven, Bull Brit. Mus. (Nat. Hist.) Bot. 2: 361. 1962; Hara, Fl. East. Himal. 223. 1966; Hara in Hara & Williams, Enum. Fl. Pl. Nepal 2: 174. 1979; Edit. Group Fl. Xizang, Emun. Vasc. Pl. Xizang 222. 1980, exc. pl. Yadong; C. Y. Wu, Index Fl. Yunnan. 1: 311. 1984.

20a. Epilobium brevifolium subsp. **brevifolium.**
Epilobium trichoneurum var. *brachyphyllum* Haussknecht, Oesterr. Bot. Zeitschr. 29: 54. 1879.

20b. Epilobium brevifolium subsp. **trichoneurum** (Haussknecht) Raven, Bull. Brit. Mus. (Nat. Hist.) Bot. 2: 362. 1962; in H. L. Li, Fl. Taiwan 3: 887. 1977; Hara in Hara & Williams, Enum. Fl. Pl. Nepal 2: 174. 1979; Edit. Group Fl. Xizang, Enum. Vasc. Pl. Xizang (Tibet) 222. 1980. *Epilobium trichoneurum* Haussknecht, Oesterr. Bot. Zeitschr. 29: 54. (Feb.) 1879; et Monogr. Epil. 208. 1884; H. Léveillé, Icon. Gen. Epil. t. 84. 1910; Gagnepain in Lecomte, Fl. Indo-Chine 2: 991. 1921; Edit. Group Fl. Xizang, Enum. Vasc. Pl. Xizang 222. 1980, exc. pl. Qinghai et Xinjiang; C. Y. Wu, Index Fl. Yunnan. 1: 311. 1984; H. Li in C. Y. Wu, Fl. Xizang. 3: 360. 1986, excl. pl. Qinghai et Xinjiang. *Epilobium hookeri* C. B. Clarke in J. D. Hooker, Fl. Brit. India 2: 585. (May) 1879, nom. superfl.; Barbey, Icon. Gen. Epil. t. 15. 1885; Kanjil & Das, Fl. Assam 2: 317. 1938.

Epilobium esquirolii H. Léveillé, Bull. Herb. Boiss., sér. 2. 7: 590. 1907; et Icon. Gen. Epil. t. 146. 1910.

Epilobium cavalieri H. Léveillé, Bull. Herb. Boiss., sér. 2. 7: 590. 1907; et Icon. Gen. Epil. t. 83. 1910.

Epilobium cordouei H. Léveillé, Repert. Spec. Nov. Regni Veg. 6: 110. 1908; et Icon. Gen. Epil. t. 87. 1910; et Fl. Kouy-Tcheou 301. 1915; et Cat. Pl. Yunnan 185, f. 44. 1916.

Epilobium philippinense C. B. Robinson, Philipp. J. Sci., Sect. C, 3: 209. 1908.

Epilobium japonicum sensu Handel-Mazzetti, Symb. Sin. 7: 601. 1933, quoad pl. Yunnan et Hubei; non (Miquel) Haussknecht.

Epilobium pyrricholophum sensu Inst. Bot. Bor.-Occ. Fl. Tsinling 1: 355. 1981, pro parte, non Haussknecht.

21. Epilobium pyrricholophum Franchet & P. A. L. Savatier, Emun. Pl. Jap. 1: 168. 1875, nom. nud.; 2: 370. 1879; Haussknecht, Monogr. Epil. 210. 1884; H. Léveillé, Monde Pl. 4: 27. 1896; et Icon. Gen. Epil. t. 137. 1910; Nakai, Bot. Mag. (Tokyo) 22: 74, 81. 1908; et loc. cit. 25: 149. 1911; H. Hara, Bot. Mag. (Tokyo) 50: 307. 1936; et J. Jap. Bot. 18: 235. 1942; et in Ohwi, Fl. Japan 822. 1953, (Engl. ed.) 658. 1965; C. Pei et al., Handb. Seed Pl. S. Jiangsu 528. 1959; Inst. Bot. Acad. Sin., Icon. Corm. Sin. 2: 1018. 1972, pro parte; Inst. Bot. Bor.-Occid. Acad. Sin., Fl. Tsinling 1: 355. 1981, pro parte; Jiangsu Inst. Bot., Fl. Jiangsu 2: 548. 1982, pro parte.

Epilobium tetragonum var. *japonica* Miquel, Ann. Mus. Bot. Lugduno-Batavum 3: 94. 1867; et Prol. Fl. Jap. 258. 1867. *Epilobium japonicum* (Miquel) Haussknecht, Oesterr. Bot. Zeitschr. 29: 56. 1879; et Monogr. Epil. 209. 1884; Forbes & Hemsley, J. Linn. Soc. Bot. 23: 307. 1887; H. Léveillé, Monde Pl. 5: 124. 1896; et Icon. Gen. Epil. t. 147. 1910; Handel-Mazzetti, Symb. Sin. 7: 601. 1933, pro parte; Steinberg in Shishkin & Bobrov, Fl. URSS 15: 601. 1949. *Epilobium pyrricholophum* var. *japonicum* (Miquel) H. Hara, J. Jap. Bot. 18: 236. 1942.

Epilobium japonicum β *glanduloso-pubescens* Haussknecht, Oesterr. Bot. Zeitschr. 29: 56. 1879; et Monogr. Epil. 210. 1884.

Epilobium oligodontum Haussknecht, Oesterr. Bot. Zeitschr. 29: 58. 1879; et Monogr. Epil. 210. 1884.

Epilobium rouyeanum H. Léveillé, Bull. Acad. Int. Geogr. Bot. 9: 210. 1900; et Icon. Gen. Epil. t. 142. 1910. *Epilobium pyrricholophum* var. *curvatopilosum* H. Hara, J. Jap. Bot. 18: 237. 1942.

Epilobium hakkodense H. Léveillé, Bull. Acad. Int. Geogr. Bot. 10: 34. 1901; et Icon. Gen. Epil. t. 101. 1910.

Epilobium quadrangulum H. Léveillé, Bull. Soc. Agric. Sarthe 40: 72. 1905; et Repert. Spec. Nov. Regni Veg. 2: 173. 1906; et Icon. Gen. Epil. t. 104. 1910.

Epilobium makinoense H. Léveillé, Bull. Soc. Agric. Sarthe 40: 73. 1905; et Repert. Spec. Nov. Regni Veg. 2: 173. 1906; et Icon. Gen. Epil. t. 115. 1910.

Epilobium chrysocoma H. Léveillé, Bull. Herb. Boiss., sér. 2, 7: 589. 1907; et Icon. Gen. Epil. t. 98. 1910.

Epilobium arcuatum H. Léveillé, Bull. Herb. Boiss., sér. 2, 7: 589. 1907; et Icon. Gen. Epil. t. 141. 1910.

Epilobium prostratum H. Léveillé, Bull. Soc. Bot. France 54: 520. 1907, non Warburg 1893.

Epilobium kiusianum Nakai, Bot. Mag. (Tokyo) 22: 84. 1908.

Epilobium pyrricholophum var. *anuoleucholophum* H. Léveillé, Repert. Spec. Nov. Regni Veg. 6: 330. 1909.

Epilobium nakaianum H. Léveillé, Icon. Gen. Epil. 166; t. 140. 1910; ibid. 316. 1911.

Epilobium axillare Franchet ex Koidzumi, Fl. Symb. Orient.-Asiat. 85. 1930.

Epilobium myokoense Koidzumi, Fl. Symb. Orient.-Asiat. 86. 1930.

22. Epilobium amurense Haussknecht, Oesterr. Bot. Zeitschr. 29: 55. 1879; et Monogr. Epil. 203. 1884; Nakai, Fl. Kor. 1: 243. 1909; H. Léveillé, Icon. Gen. Epil. 112. 1910; Steinberg in Shishkin & Bobrov, Fl. URSS 15: 594. 1949; Hara in Ohwi, Fl. Japan 822. 1953, (Engl. ed.) 657. 1965; T. N. Liou et al., Clav. Pl. Chin. Bor.-Orient. 236. 1959, Fl. Pl. Herb. Chin. Bor.-Orient. 6: 151. 1977; Raven, Bull. Brit. Mus. (Nat. Hist.) Bot. 2: 367. 1962; et in H. L. Li et al., Fl. Taiwan 3: 885. 1977; Inst. Bot. Acad. Sin., Icon. Corm. Sin. 2: 1020. 1972; C. J. Wu in H. C. Fu et al., Fl. Intermongol. 4: 133. 1979; C. Y. Wu, Index Fl. Yunnan. 1: 311. 1984. *Epilobium origanifolium* var. *pubescens* Maximowicz, Prim. Fl. Amur. 105. 1859.

22a. Epilobium amurense subsp. **amurense.**

Epilobium laetum Wallich ex Haussknecht, Monogr. Epil. 218. 1884. *Epilobium laetum* Wallich, Numer. List 216, n. 6329. 1832, nom. nud.; H. Léveillé, Icon. Gen. Epil. t. 97. 1910. *Epilobium amurense* subsp. *laetum* (Wallich ex Haussknecht) Raven, Bull. Brit. Mus. (Nat. Hist.) Bot. 2: 367. 1962; Hara, Fl. East. Himal. 2: 89. 1971; et in Hara & Williams, Enum. Fl. Pl. Nepal 2: 174. 1979; Edit. Group Fl. Xizang, Enum. Vasc. Pl. Xizang 222. 1980; H. Li in C. Y. Wu, Fl. Xizang. 3: 361. 1986.

Epilobium nepalense Haussknecht, Oesterr. Bot. Zeitschr. 29: 53. 1879; et Monogr. Epil. 218. 1884.

Epilobium tetragonum sensu C. B. Clarke in J. D. Hooker, Fl. Brit. India 2: 586. 1879, pro parte, non Linnaeus.

Epilobium tenue Komarov, Trudy Imp. S.-Peterburgsk. Bot. Sada 25: 95. 1905; Steinberg in Shishkin & Bobrov, Fl. URSS 15: 598. 1949; T. N. Liou et al., Clav. Pl. Chin. Bot.-Orient. 236. 1959; et Fl. Pl. Herb. Chin. Bor.-Orient. 6: 147. 1977.

Epilobium yabei H. Léveillé, Bull. Soc. Agric. Sarthe 40: 72. 1905; et Repert. Spec. Nov. Regni Veg. 2: 173. 1906.

Epilobium gansuense H. Léveillé, Bull. Herb. Boiss., sér. 2. 7: 590. 1907.

Epilobium miyabe H. Léveillé, Repert. Spec. Nov. Regni Veg. 5: 8. 1908; et Icon. Gen. Epil. t. 132. 1910.

Epilobium ovale Takeda, J. Linn. Soc., Bot. 42: 466. 1914.

22b. Epilobium amurense subsp. **cephalostigma** (Haussknecht) C. J. Chen, Hoch & Raven, comb. et stat. nov. *Epilobium cephalostigma* Haussknecht, Oesterr. Bot. Zeitschr. 29: 57. 1879; et Monogr. Epil. 195. 1884; Nakai, Fl. Kor. 1: 243. 1909; H. Léveillé, Icon. Epil. t. 63. 1910; Kitagawa, Lineam. Fl. Mansh. 328. 1939, excl. syn.; Hara, J. Jap. Bot. 18: 176, 232. 1942; et Enum. Sperm. Jap. 3: 246. 1954; et in Ohwi, Fl. Japan 822. 1953, (Engl. transl.) 657. 1965; Steinberg in Shishkin & Bobrov, Fl. URSS 15: 596. 1949, (Engl. transl.) 15: 446. 1974; T. N. Liou et al., Clav. Pl. Chin. Bor.-Orient. 236. 1959, excl. fig.; et Fl. Herb. Chin. Bor.-Orient. 6: 150. 1977; C. J. Wu in H. C. Fu et al., Fl. Intramongol. 4: 135. 1977, pro parte; Inst. Bot.

Bor.-Occid. Acad. Sin., Fl. Tsinling 1(3): 354. 1981; Q. H. Xing in S. Y. He et al., Fl. Beijing 1: 625. 1984.

Epilobium affine sensu Maximowicz, India Hort. Petrop. 16. 1869, pro parte; non Bongard, 1832.

Epilobium calycinum Haussknecht, Monogr. Epil. 196. 1884; Nakai, Fl. Kor. 1: 245. 1909; Steinberg in Shishkin & Bobrov, Fl. URSS 15: 596. 1949.

Epilobium nudicarpum Komarov, Trudy Imp. S.-Peterburgsk. Bot. Sada 18: 432. 1901; et Fl. Mansh. III, 1: 94. 1905; Steinberg in Shishkin & Bobrov, Fl. URSS. 15: 597. 1949; T. N. Liou et al., Fl. Pl. Herb. Chin. Bor.-Orient. 6: 145. 1977. *Epilobium cephalostigma* var. *nudicarpum* (Komarov) H. Hara, J. Jap. Bot. 18: 234. 1942; et in Ohwi, Fl. Japan 658. 1953, (Engl. ed.) 658. 1965.

Epilobium angulatum Komarov, Trudy Imp. S.-Peterburgsk. Bot. Sada 18: 432. 1901; et Fl. Mansh. III, 1: 94. 1905; Steinberg in Shishkin & Bobrov, Fl. URSS 15: 597. 1949; T. N. Liou et al., Fl. Pl. Herb. Chin. Bor.-Orient. 6: 145. 1977.

Epilobium cylindrostigma Komarov, Trudy Imp. S.-Peterburgsk. Bot. Sada 25: 95. 1905; Steinberg in Shishkin & Bobrov, Fl. URSS 15: 596. 1949.

Epilobium coreanum H. Léveillé, Bull. Herb. Boiss., sér. 2, 7: 590. 1907; et Icon. Gen. Epil. t. 59, 60. 1910; Nakai, Fl. Kor. 1: 245. 1909.

Epilobium consimile var. *japonicum* Nakai, Bot. Mag. (Tokyo) 25: 148. 1911.

Epilobium sugaharai Koidzumi, Acta Phytotax. Geobot. 5: 121. 1936.

23. Epilobium kingdonii Raven, Bull. Brit. Mus. (Nat. Hist.) Bot. 2: 377. 1962; Edit. Group Fl. Xizang, Enum. Vasc. Pl. Xizang (Tibet) 222. 1980, pro parte, excl. pl. Zayu.

24. Epilobium wallichianum Haussknecht, Oesterr. Bot. Zeitschr. 29: 54. 1879; et Monogr. Epil. 218. 1884; H. Léveillé, Icon. Gen. Epil. t. 128. 1910; Raven, Bull. Brit. Mus. (Nat. Hist.) Bot. 2: 365. 1962; Hara, Fl. East. Himal. 1: 224. 1966; et 2: 89. 1971; et in Hara & Williams, Enum. Fl. Pl. Nepal 2: 175. 1979; Edit. Group Fl. Xizang, Enum. Vasc. Pl. Xizang (Tibet) 223. 1980; C. Y. Wu, Index Fl. Yunnan. 1: 313. 1984; H. Li in C. Y. Wu, Fl. Xizang. 3: 360. 1986.

Epilobium nepalense Haussknecht, Oesterr. Bot. Zeitschr. 29: 53. 1879, pro parte (as to syntype collection from Assam, Mairang, Khasi Hills, Jul 1850, *Hooker & Thomson 1945,* K). As noted by Raven (1962a), Haussknecht cited collections of two species among the syntypes he designated for this name; Raven chose a lectotype that brings this name into synonymy with *Epilobium laetum* Wallich ex Haussknecht (=*Epilobium amurense* Haussknecht subsp. *amurense*).

Epilobium tanguticum Haussknecht, Oesterr. Bot. Zeitschr. 29: 56. 1879; et Monogr. Epil. 196. 1884; Handel-Mazzetti, Symb. Sin. 7: 601. 1933.

Epilobium tetragonum sensu C. B. Clarke in J. D. Hooker, Fl. Brit. India 2: 586. 1879, pro parte, non Linnaeus.

Epilobium souliei H. Léveillé, Bull. Herb. Boiss., sér. 2, 7: 588. 1907; et Icon. Gen. Epil. t. 121. 1910. *Epilobium wallichianum* subsp. *souliei* (H. Léveillé) Raven, Bull. Brit. Mus. (Nat. Hist.) Bot. 2: 366. 1962; Hara in Hara & Williams, Enum. Fl. Pl. Nepal 2: 175. 1979; Edit. Group Fl. Xizang, Enum. Vasc. Pl. Xizang (Tibet) 223. 1980; Chowdhery & Wadhwa, India J. Forest. 4: 329. 1981.

Epilobium duclouxii H. Léveillé, Repert. Spec. Nov. Regni Veg. 6: 110. 1908; et Icon. Gen. Epil. t. 144. 1910; et Cat. Pl. Yunnan 193, f. 45. 1916.

Epilobium mairei H. Léveillé, Repert. Spec. Nov. Regni Veg. 12: 283. 1913; et Cat. Pl. Yunnan 193, f. 49. 1916.

Epilobium sykesii Raven, Bull. Brit. Mus. (Nat. Hist.) Bot. 2: 366. 1962.

Epilobium himalayense sensu Handel-Mazzetti, Symb. Sin. 7: 601. 1933; non Haussknecht.

25. Epilobium laxum Royle, Illus. Bot. Himal. Mount. 211, t. 43, fig. 2. 1835; Raven, Bull. Brit. Mus. (Nat. Hist.) Bot. 2: 369. 1962; et in K. H. Rechinger, Fl. Iran. 7: 17. 1964; R. Stewart, Ann. Cat. Vasc. Pl. W. Pakistan & Kashmir 506. 1972; Hoch in Nasir & Ali, Fl. Pakistan 139: 28. 1981.

Epilobium tetragonum Linnaeus var. ?*amplectens* Bentham ex C. B. Clarke in J. D. Hooker, Fl. Brit. India 2: 587. 1879. *Epilobium amplectens* Bentham in Wallich, Numer. List 216, n. 6330. 1832, nom. nud. *Epilobium amplectens* (Bentham ex C. B. Clarke) Haussknecht, Monogr. Epil. 208. 1884; H. Léveillé, Icon. Gen. Epil. t. 102. 1910.

Epilobium duthiei Haussknecht, Monogr. Epil. 205, t. 9, fig. 54. 1884; H. Léveillé, Icon. Gen. Epil. t. 106. 1910.

Epilobium sadae H. Léveillé, Bull. Herb. Boiss., sér. 2, 7: 588. 1907; et Icon. Gen. Epil. t. 85. 1910.

Epilobium subnivale Popov ex Pavlov, Wiss. Ber. Moskau. Staatsuniv. 2: 329. 1934; Steinberg in Shishkin & Bobrov, Fl. URSS 15: 592. 1949.

26. Epilobium sikkimense Haussknecht, Oesterr. Bot. Zeitschr. 29: 52. 1879; Monogr. Epil. 204. 1884; H. Léveillé, Icon. Gen. Epil. t. 88. 1910; Handel-Mazzetti, Symb. Sin. 7: 601. 1933, pro parte; Hara in Hara & Williams, Enum. Fl. Pl. Nepal 2: 175. 1979; Edit. Group Fl. Xizang, Enum. Vasc. Pl. Xizang (Tibet) 223. 1980; C. Y. Wu, Index Fl. Yunnan. 1: 311. 1984; H. Li in C. Y. Wu, Fl. Xizang. 3: 362. 1986.

Epilobium alsinifolium sensu C. B. Clarke in J. D. Hooker, Fl. Brit. India 2: 586. 1879, pro parte; non Villars.

Epilobium sikkimense subsp. *ludlowianum* Raven, Bull. Brit. Mus. (Nat. Hist.) Bot. 2: 373. 1962; Hara, Fl. East. Himal. 224. 1966; et in Hara & Williams, Enum. Fl. Pl. Nepal 2: 175. 1979.

Epilobium trilectorum Raven, Bull. Brit. Mus. (Nat. Hist.) Bot. 2: 374. 1962.

Epilobium soboliferum Raven, Bull. Brit. Mus. (Nat. Hist.) Bot. 2: 375. 1962.

Epilobium squamosum Raven, Bull. Brit. Mus. (Nat. Hist.) Bot. 2: 380. 1962.

27. Epilobium williamsii Raven, Bull. Brit. Mus. (Nat. Hist.) Bot. 2: 378. 1962; Hara in Hara & Williams, Enum. Fl. Pl. Nepal 2: 175. 1979; Edit. Group Fl. Xizang, Enum. Vasc. Pl. Xizang (Tibet) 223. 1980; H. Li in C. Y. Wu, Fl. Xizang. 3: 362. 1986.

Epilobium sikkimense sensu Handel-Mazzetti, Symb. Sin. 7: 601. 1933, pro parte, quoad pl. Sichuan, non Haussknecht.

28. Epilobium subcoriaceum Haussknecht, Oesterr. Bot. Zeitschr. 29: 56. 1879; et Monogr. Epil. 196. 1884.

29. Epilobium gouldii Raven, Bull. Brit. Mus. (Nat. Hist.) Bot. 2: 371, pl. 35b. 1962, exclud. spec. *King's collector 146;* Edit. Group Fl. Xizang, Enum. Vasc. Pl. Xizang (Tibet) 222. 1980; H. Li in C. Y. Wu, Fl. Xizang. 3: 362. 1986.

30. Epilobium fangii C. J. Chen, Hoch & Raven, sp. nov.

31. Epilobium palustre Linnaeus, Sp. Pl. 1: 348. 1753; Maximowicz, Prim. Fl. Amur. 105. 1859; Boissier, Fl. Orient. 2: 748. 1872; C. B. Clarke in J. D. Hooker, Fl. Brit. India 2: 585. 1879; Haussknecht, Monogr. Epil. 128. 1884; Forbes & Hemsley, J. Linn. Soc. Bot. 23: 308. 1887; Komarov, Fl. Mansh. 3: 90. 1907; et Fl. Penin. Kamch. 2: 327. 1929; Nakai, Fl. Kor. 1: 243. 1909; H. Léveillé, Icon. Gen. Epil. tt. 260–261. 1911; Hegi, Ill. Fl. Mitt.-Europ. 5: 839. 1926; Kitagawa, Lineam. Fl. Mansh. 329. 1939; Hara, J. Jap. Bot. 18: 840. 1942; et Enum. Sperm. Jap. 268. 1954; et in Hara & Williams, Enum. Fl. Pl. Nepal 175. 1979; Steinberg in Shishkin & Bobrov, Fl. URSS 15: 613. 1949; T. N. Liou et al., Clav. Pl. Chin. Bor.-Orient. 263. 1959; et Fl. Pl. Herb. Chin. Bor.-Orient. 6: 147. 1977; Raven, Bull. Brit. Mus. (Nat. Hist.) Bot. 2: 381. 1962; et Notes Royal Bot. Gard. Edinburgh 24: 195. 1964; et in K. H. Rechinger, Fl. Iran. 7: 12. 1964; et in Tutin et al., Fl. Europ. 2: 310. 1968; Dept. Biol. Beijing Norm. Univ., Fl. Beijing 2: 667. 1964; Q. H. Xing in S. Y. He et al., ibid. rev. ed. 1: 625. 1984; Chamberlain & Raven in Davis, Fl. Turkey 4: 192. 1972; R. Stewart, Annot. Cat. Vasc. Pl. W. Pakistan & Kashmir 507. 1972; Inst. Bot. Acad. Sin., Icon. Corm. Sin. 2: 1019. 1972; C. J. Wu in H. C. Fu et al., Fl. Intramongol. 4: 131. 1979, pro parte; Edit. Group Fl. Xizang, Enum. Vasc. Pl. Xizang (Tibet) 222. 1980; Inst. Bot. Bor.-Occid. Acad. Sin., Fl. Tsinling. 1: 354. 1981, pro parte; Hoch in Nasir & Ali, Fl. Pakistan 139: 33. 1981; B. Grubov, Man. Vasc. Pl. Mongol. 185. 1982. *Epilobium palustre* var. *typicum* C. B. Clarke in J. D. Hooker, Fl. Brit. India 2: 585. 1879.

Epilobium rhynchocarpum Boissier, Diagn. ser. 2, 2: 53. 1856. *Epilobium fischerianum* Pavlov, Bjull. Moskovsk. Obsc. Isp. Prir., Otd. Biol., n.s., 38: 105. 1929, nom. superfl., based on *E. palustre* "forma *fischerianum*" Haussknecht, Monogr. Epil. 133. 1884.

Epilobium palustre var. *majus* C. B. Clarke in J. D. Hooker, Fl. Brit. India 2: 585. 1879.
Epilobium palustre var. *minimum* C. B. Clarke in J. D. Hooker, Fl. Brit. India 2: 586. 1879.

32. Epilobium fastigiatoramosum Nakai, Bot. Mag. (Tokyo) 33: 9. 1919; Kitagawa, Rep. Inst. Sci. Res. Manchuria 1: 276. 1937; et Lineam. Fl. Manshuria 328. 1939; Hara, J. Jap. Bot. 18: 183. 1942; et in Ohwi, Fl. Japan (Engl. ed.) 657. 1965; T. N. Liou et al., Clav. Pl. Chin. Bor-Orient. 236. 1959; Fl. Pl. Herb. Chin. Bor.-Orient. 6: 149. 1977.
Epilobium baicalense Popov, Bot. Mater. Gerb. Bot. Inst. Komarova Akad. Nauk SSSR 18: 6. 1957.
Epilobium palustre sensu C. J. Wu in H. C. Fu et al., Fl. Intramongol. 4: 131. 1979, pro parte; Inst. Bot. Bor.-Occ. Acad. Sin., Fl. Tsinling 1: 354. 1981, pro parte, non Linnaeus.

33. Epilobium minutiflorum Haussknecht, Oesterr. Bot. Zeitschr. 29: 55. 1879; et Monogr. Epil. 212. 1884; Aitchison, J. Linn. Soc. Bot. 29: 163. 1892; Nakai, Fl. Kor. 1: 245. 1909; H. Léveillé, Icon. Gen. Epil. t. 135. 1910; Steinberg in Shishkin & Bobrov, Fl. URSS 15: 620. 1949; Raven, Bull. Brit. Mus. (Nat. Hist.) Bot. 2: 382. 1962; et Notes Royal Bot. Gard. Edinburgh 24: 195. 1962; et in K. H. Rechinger, Fl. Iran. 7: 13. 1964; Chamberlain & Raven in Davis, Fl. Turkey 4: 180. 1972; R. Steward, Annot. Cat. Vasc. Pl. W. Pakistan & Kashmir 506. 1972; Edit. Group Fl. Xizang, Enum. Vasc. Pl. Xizang (Tibet) 222. 1980; Hoch in Nasir & Ali, Fl. Pakistan 139: 34. 1981; Grubov, Man. Vasc. Pl. Mongol. 185. 1982; H. Li in C. Y. Wu, Fl. Xizang. 3: 359. 1986. *Epilobium tetragonum* β. *minutiflorum* (Haussknecht) Boissier, Fl. Orient. Suppl. 240. 1888.
Epilobium modestum Haussknecht, Oesterr. Bot. Zeitschr. 29: 55. 1879; et Monogr. Epil. 211. 1884; H. Léveillé, Icon. Gen. Epil. t. 134, 1910; Steinberg in Shishkin & Bobrov, Fl. URSS 15: 620. 1949.
Epilobium palustre sensu C. B. Clarke in J. D. Hooker, Fl. Brit. India 2: 585. 1879, pro parte; H. R. Wu in H. C. Fu et al., Fl. Intramongol. 4: 131. 1979, pro parte; Inst. Bot. Bor.-Occ. Acad. Sin., Fl. Tsinling 1: 353. 1981, pro parte; non Linnaeus.
Epilobium tetragonum sensu Aitchison, J. Linn. Soc. Bot. 18: 60. 1880, pro parte; Franchet, Pl. David. 1: 134. 1884; Komarov, Fl. Mansh. 3: 93. 1905; non Linnaeus.
Epilobium propinquum Haussknecht, Monogr. Epil. 213. 1884, nom. nov. H. Léveillé, Icon. Gen. Epil. t. 76. 1910; Kitagawa, Lineam. Fl. Mansh. 329. 1939; T. N. Liou et al., Clav. Pl. Chin. Bor.-Orient. 236. 1959; et Fl. Pl. Herb. Chin. Bor.-Orient. 6: 153. 1977. *Epilobium decipiens* Haussknecht, Oesterr. Bot. Zeitschr. 29: 57. 1879; non F. Schultz, 1861.
Epilobium cephalostigma sensu C. J. Wu in H. C. Fu et al., Fl. Intramongol. 4: 62. 1979, pro parte; non Haussknecht.

34. Epilobium ciliatum Rafinesque, Med. Repos. II. 5: 361. 1808. *Epilobium ciliatum* subsp. *ciliatum.*
Epilobium maximowiczii Haussknecht, Oesterr. Bot. Zeitschr. 29: 57. 1879.
Epilobium punctatum H. Léveillé, Bull. Acad. Geogr. Bot. 11: 316. 1902; et Icon. Gen. Epil. t. 145. 1910.
Epilobium kurilense Nakai, Bot. Mag. (Tokyo) 22: 83. 1908. *Epilobium glandulosum* var. *kurilense* (Nakai) H. Hara, J. Jap. Bot. 18: 238. 1942.
Epilobium glandulosum var. *asiaticum* H. Hara, J. Jap. Bot. 18: 241. 1942; et in Ohwi, Fl. Japan 853. 1953, et (English ed.) 658. 1965; et Enum. Sperm. Jap. 266. 1954.
Epilobium glandulosum sensu Y. L. Chang in T. N. Liou et al., Clav. Pl. Chin. Bor.-Orient. 1959; et Fl. Pl. Herb. Chin. Bor.-Orient. 6: 153. 1977; non Lehmann.
Epilobium tenue sensu Y. L. Chang in T. N. Liou et al., Clav. Pl. Chin. Bor.-Orient. 236. 1959; et Fl. Pl. Herb. Chin. Bor.-Orient. 6: 153. 1977; non Komarov.
Epilobium cylindrostigma sensu Y. L. Chang in T. N. Liou et al., Clav. Pl. Chin. Bor.-Orient. 236. 1959; et Fl. Pl. Herb. Chin. Bor.-Orient. 6: 151. 1977; non Komarov.

35. Epilobium clarkeanum Haussknecht, Monogr. Epil. 220. 1884; H. Léveillé, Icon. Gen. Epil. t. 113. 1910; Raven, Bull. Brit. Mus. (Nat. Hist.) Bot. 2: 381. 1962; Chowdhery & Wadhwa, India J. For. 4: 329. 1981.
Epilobium alpinum sensu C. B. Clarke in J. D. Hooker, Fl. Brit. India 2: 586. 1879, non Linnaeus.

36. Epilobium pengii C. J. Chen, Hoch & Raven, sp. nov.

37. Epilobium anagallidifolium Lamarck, Encycl. 2: 376. 1786; Haussknecht, Monogr. Epil. 152. 1884; Komarov, Fl. Penin. Kamch. 2: 322. 1929.

Epilobium alpinum sensu Steinberg in Shishkin & Bobrov, Fl. URSS 15: 607. 1949, pro parte; non Linnaeus.

Epilobium dielsii H. Léveillé, Repert. Spec. Nov. Regni Veg. 3: 20. 1906; et Icon. Gen. Epil. t. 116. 1910; Nakai, Bot. Mag. (Tokyo) 22: 76. 1908; Hara, J. Jap. Bot. 18: 248. 1942; et in Ohwi, Fl. Japan (Engl. ed.) 658. 1965; Shimizu, New Alpine Flora Japan 178. 1983.

NUMERICAL LIST OF SPECIES

1. *Epilobium latifolium*
2. *Epilobium speciosum*
3. *Epilobium conspersum*
4a. *Epilobium angustifolium* subsp. *angustifolium*
4b. *Epilobium angustifolium* subsp. *circumvagum*
5. *Epilobium hirsutum*
6. *Epilobium parviflorum*
7. *Epilobium blinii*
8. *Epilobium nankotaizanense*
9a. *Epilobium roseum* subsp. *roseum*
9b. *Epilobium roseum* subsp. *subsessile*
10. *Epilobium tianschanicum*
11. *Epilobium cylindricum*
12. *Epilobium tibetanum*
13. *Epilobium sinense*
14. *Epilobium platystigmatosum*
15. *Epilobium hohuanense*
16. *Epilobium taiwanianum*
17. *Epilobium royleanum*
18. *Epilobium kermodei*
19. *Epilobium pannosum*
20a. *Epilobium brevifolium* subsp. *brevifolium*
20b. *Epilobium brevifolium* subsp. *trichoneurum*
21. *Epilobium pyrricholophum*
22a. *Epilobium amurense* subsp. *amurense*
22b. *Epilobium amurense* subsp. *cephalostigma*
23. *Epilobium kingdonii*
24. *Epilobium wallichianum*
25. *Epilobium laxum*
26. *Epilobium sikkimense*
27. *Epilobium williamsii*
28. *Epilobium subcoriaceum*
29. *Epilobium gouldii*
30. *Epilobium fangii*
31. *Epilobium palustre*
32. *Epilobium fastigiatoramosum*
33. *Epilobium minutiflorum*
34. *Epilobium ciliatum*
35. *Epilobium clarkeanum*
36. *Epilobium pengii*
37. *Epilobium anagallidifolium*

INDEX TO NUMBERED COLLECTIONS CITED

The numbers in parentheses refer to the corresponding species in the text and in the Numerical List of Species presented above.

Boissier & Buhse 673 (9b).
Bonati 391 (5); 478 (5).
Bretschneider 1868 (5).
British Mus. Exped. 12399a (24).
Butuo Pl. Exped. 13739 (22a).
Cao, Z. Y. 126 (33); 555 (5); 757 (5).
Carles 28 (20b).
Cavalerie 89 (5); 208 (5); 352 (20b); 992 (24); 1784 (13); 2438 (13); 2924 (24); 3036 (20b); 3075 (5); 3151 (20b); 3152 (20b); 3982 (6); 4123 (19); 4390 (13); 7622 (20b).
Chaeden 15 (4a).
Chanet 917 (4b); 1039 (4b).
Chanet & Serre 177 (6); 674 (6); 679 (6); 917 (4b); 2600 (4b).
Chang & Cheng 924 (5).
Chang, C. W. 3347 (10); 4042 (5).
Chang, C. W., et al. 3474 (10); 3482 (10); 4363 (33); 4518 (33).
Chang, G. L., & X. Pu 79104 (6).
Chang, G. L., et al. 703 (5).
Chang, H. J. 2008 (14).
Chang, H. T. 4716 (20b).
Chang, J. 2009 (5).
Chang, P. F. 621 (5).
Chang, S. S. 5713 (18); 5970 (4a).
Chang, S. S., & Ren, Y. X. 4832 (22a); 5464 (22a); 5709 (4b); 5823 (26); 5873 (4b); 5995 (4b); 6050 (4a); 6170 (4a); 6172 (4a); 6234 (26); 6273 (22a); 6402 (4b); 6496 (4b); 7165 (22a); 7186 (22a); 7251 (4b); 7519 (22a); 7556 (22a).
Chang, S. S., et al. 5114 (18); 5464 (30).
Chang, T. P. 303 (5).
Chang, W. D. 110 (5).
Chang, W. H. 381 (5).
Chang, X. M. 4 (5); 497 (6); 847 (5).
Chang, Y. L. 2113 (34).
Chang, Y. P. 577 (27); 739 (5).
Chang, Y. T., & Lang, K. L. 478 (24); 951 (24); 1067 (24); 1131 (24); 1593 (24).
Chang, Y. T., & Lang, K. Y. 7 (30); 31 (4a); 405 (27); 478 (22a); 651 (4b); 705 (26); 854 (11); 951 (17); 1131 (26); 1135 (4b); 1593 (27).
Chang, Z. Y. 57 (5); 6057 (5); 6138 (6); 6156 (5); 6229 (33); 7037 (5);7433 (6); 7961 (5); 7491 (11); 8056 (5); 8201 (5); 13454 (6); 15161 (33); 15390 (5); 25756 (6).
Chang, Z. Y., & Zhou, H. F. 22770 (17).
Changbai Shan Exped. 150 (4a); 667 (4a); 954 (22a).
Chao, J. M., & Kao, M. T. 6181 (22a).
Chapman 88 (27).
Chen, C. J. 83001 (5); 83007 (5); 83010 (5); 89522 (21).
Chen, C. J., & Gu, H. Y. 88050 (21); 88055 (21).
Chen, C. J., & Hoch 86024 (5); 86026 (31); 86027 (33); 86052 (6); 86053.(31); 86054 (33); 86065 (4a); 86070 (9b); 86071 (9a); 86080 (37); 86081 (31); 86083 (9b); 86084 (9a); 86099 (4a); 86132 (31); 86139 (4a); 86143 (9b); 86144 (9a); 86157 (4a); 86170 (33); 86208 (4a); 86235 (25); 86262 (33); 86263 (31); 86275 (1); 86276 (4a); 86311 (10); 86312 (5); 86322 (10); 86329 (4a); 86331 (9b).
Chen, C. J., & Liu, Z. Y. 83062 (24); 83068 (24); 83069 (22b); 83070 (22b); 83074 (20b); 83076 (22b); 83081 (22b); 83087 (20b); 83091 (18); 83096 (22b).
Chen, C. J., & Zhu, X. Y. 89077 (4b).
Chen, F. H. 144 (4a, 4b); 355 (31); 430 (4a, 4b).
Chen, H. 22133 (20b).
Chen, J. 910 (13).
Chen, J. L., & Zhou, B. K. 9 (4a).
Chen, J. Z. 51413 (21).
Chen, K. L. 184 (22a).
Chen, L. G. 42977 (5).
Chen, L. Q., et al. 1829 (5).
Chen, M. 2104 (20b).
Chen, Q. L., et al. 1820 (20b).
Chen, Q. N. 2051 (14).
Chen, S. Y., et al. 4158 (4b); 5316 (4b).
Chen, T. H. 430 (4b).
Chen, W. C. 4190 (22b).
Chen, W. L. 10686 (22a).
Chen, W. L., & Chen, C. J. 5587 (5); 5607 (22a, 26); 5662 (11); 5692 (17); 5698 (22a); 5776 (28); 5956 (26); 5979 (26); 6017 (11); 6018 (11); 6023 (12); 6035 (12); 6123 (26); 6127 (12); 6128 (22a); 6138 (5); 6184 (11); 6189 (22a); 6190 (26); 6229 (30); 6300 (27); 6305 (4a); 6343 (4a); 6358 (27); 6392 (12); 6402 (26); 6418 (11); 6420 (12); 6504 (24); 6551 (4a); 6589 (27); 6642 (26); 6672 (24); 6789 (4a); 6856 (24); 6878 (26); 6933 (26, 27); 6934 (31); 7066 (11); 7096 (26); 7108 (26); 7358 (31); 7368 (26); 7492 (31); 7503 (4a); 7507 (17); 7508 (32); 7509 (17); 7569 (17); 7877 (24); 7878 (26); 7880 (26); 7881 (26); 7909 (31); 7910 (22a); 7935 (4a); 7977 (4a); 7981 (4b); 8208 (28); 8212 (31); 8217 (32); 8243 (31); 8313 (4a); 8314 (28); 8315 (22a); 8405 (26); 8408 (26); 8415 (18); 8450 (4b); 8477 (4a); 8478 (26); 8480 (22a); 8547 (22a); 8548 (22a); 8549 (22a); 8550 (4b); 8582 (6); 8584 (22a); 8585 (22a); 8635 (4a); 8695 (22a); 8715 (5); 8742 (5); 8743 (22a); 8772 (5); 8801 (17); 8802 (6); 8803 (14); 8807 (4b); 8813 (11); 8818 (22a); 8829 (22a); 8834 (20b); 8840 (22b); 8846 (18); 8847 (31); 8849 (4a); 8852 (4b); 8853 (18); 8854 (31); 8855 (4b); 8856 (28).

Chen, Y. 593 (20b); 692 (20b); 990 (18); 2011 (5); 2017 (22b); 2025 18, 22a); 2919 (13); 7133 (4b); 7147 (4b).
Chen, Y. L. 1448 (4a); 1772 (4b).
Chen, Z. Z. 51173 (22b); 52432 (20b).
Cheng, S. Z., & Li, C. S. 126 (26); 284 (26).
Cheng, W. C., & Hwa, C. T. 712 (13).
Cheng, X. Z., & Li, P. S. 74 (3).
Cheng, Z. Z. 3 (33).
Chengde Exped. 1622 (32).
Chengdu Inst. Biol. 2523 (5); 8639 (5); 9302 (18); 11047 (22a); 12304 (20b); 13030 (5); 14013 (5); 14174 (5); 14330 (19).
Cheo & Hsu 328 (4b).
Cheo, T. Y., & Yen, L. 181 (5); 298 (5).
Chiao, C. Y. 1647 (4b); 2851 (5).
Chiao, C. Y., & Fan, C. S. 330 (20b); 818 (4b).
Chien, C. C. 462 (22b).
Chien, J. J. 450 (4b); 1031 (31).
Chien, M. Z. 1028 (5); 1052 (22b); 1487 (20b).
Chien, S. S. 1284 (22b).
Chin, P. Y. 45 (7).
Ching, R. C. 147 (33); 401 (4a); 438 (4b); 443 (5); 575 (22a); 871 (4a); 877 (4a); 891 (1); 909 (22a, 24, 28); 1023 (1); 1064 (4a); 1099 (1); 1275 (4a); 1310 (4a); 1456 (4a); 1666 (33); 1853 (4a); 2464 (5); 2490 (20b, 21); 3029 (4a); 3279a (4a); 3325 (4a); 3457 (9a); 4183 (22b); 4506 (20b); 10173 (21); 21093 (22a); 21165 (5); 22070 (22a); 22610 (18); 22707 (7); 23202 (5); 23258 (22a); 23813 (24); 23993 (4b); 24463 (22a); 33202 (6); 99610 (18).
Ching, Z. L. 1302 (31).
Chou, K. S., & Dong, H. M. 76003 (13).
Chou, L. H., & Sun, L. N. 1941 (17).
Chou, T. Y. 400 (5); 5328 (5); 7358 (32); 7533 (32); 651198 (10); 652047 (5); 652051 (33); 652123 (31).
Chou, T. Y., et al. 650903 (9b); 650904 (9b, 10); 650907 (4a); 651037 (4a); 651105 (1); 651180 (4a); 651185 (4a); 651201 (31); 652073 (4a).
Chow, H. C. 7672 (18); 9875 (30); 9931 (18); 12190 (18); 12265 (22a).
Chow, H. F. 40527 (5); 40579 (4b); 41564 (5).
Chow, S. 1282 (7).
Chow, T. L., & Ho, C. 13244 (22a).
Chu, C. H. 1112 (22a); 1336 (18).
Chu, G. L. 5448 (5).
Chu, G. L., & Xu, W. Y. 5431 (33).
Chu, K. L. 1115 (18); 1781 (5); 2070 (4b); 2102 (5); 2587 (30); 2923 (26); 2929 (18); 3182 (18, 22a); 3491 (22a); 3597 (4b); 3834 (5); 3947 (24); 3994 (20b); 6358 (20b); 7488 (24).
Chu, S. C., & Zhu, Y. C. 422 (22b).
Chu, T. S. 326 (21).
Chuang, C. C. 2952 (22a).
Chuang, C. C., & Chuma 8862 (22a).
Chuang, C. C., & Kao, M. T. 4219 (22a); 4257 (15); 4450 (20b); 4454 (14).
Chun, S. H. 2478 (21); 16631 (21).
Chun, W. Y. 5338 (13); 8867 (5).
Chun, W. Y., & Chien, S. S. 8155 (5).
Chung, Z. S. 83386 (22b); 83512 (22a, b); 83513 (14).
Clarke 22190 (17); 29816 (31).
Clemens 1448 (22a); 6277 (4b); 6277a (4b).
Couduback 5589 (4b).
Cui, S. C. 212 (5).
Cunningham 167 (4a); 224 (3); 482 (4b).
Cuo, X. L. 92 (32).
Dai, L. Y. 1250 (13).
Dai, L. Y., & Qian, C. H. 621 (13); 821 (13); 1219 (14); 1135 (20b).
Dai, T. L. 101050 (4b); 101179 (20b); 101291 (22b); 101656 (6); 101750 (22a); 101810 (4b); 101813 (4b); 102161 (22b); 102259 (5); 102585 (17, 22b); 102673 (20b, 22b); 103116 (13); 103354 (5); 103409 (5); 103669 (13); 103867 (14, 22a); 104277 (13); 104681 (17); 104597 (13); 104773 (11, 13); 104800 (11); 104860 (11); 105622 (4b); 105664 (4b); 105811 (22a, 22b); 105906 (11, 17, 22b); 106539 (14); 106863 (17); 107074 (17); 107393 (14); 107406 (20b); 109622 (4b).
Dali Exped. 205 (26).
Dang, P. 574 (20b).
Daniels 268 (4b).
Dao, D. D. 225 (20b); 1126 (20b); 11281 (4a).
Dao, G. D. 13524 (6).
Dao, Y. H. 50405 (30); 50442 (4a); 51641 (4b).
Daocheng Pl. Exped. 2041 (24); 2424 (11); 2524 (12).
David 427 (5); 553 (4a); 1975 (33); 1990 (32); 2242 (4b).
Dawu Exped. 106 (4a); 192 (4a); 5859 (24).
Delavay 2 (19); 3 (5); 4 (24, 26); 5 (5); 931 (13, 22a); 932 (7, 20b); 85 (26); 1223 (5); 1224 (20b); 1503 (7); 2003 (19); 2004 (19); 2005 (20b); 2006 (7); 2007 (24); 2146 (20b); 2864 (5); 2864 (5); 2937 (19); 2942 (5); 3847 (4b); 5145 (24, 35).
Delessent 2284 (22a).
Deng, D. Z. 2003 (26).
Deng, L. 2382 (20b); 5710 (22b); 7686 (21).
Deng, L. K. 244 (6).
Deng, X. F. 181 (21); 791221 (20b); 791311 (3); 791316 (26); 791330 (24); 791338 (24); 791411 (24); 791420 (26); 791435 (22a); 791456 (22a); 791510 (26); 791533 (27); 791559 (26);
Deqen Pl. Exped. 23962 (13).

Diau, Z. S. 14851 (11).
Ding, J. L. 417 (5).
Ding, Q. X. 65 (5); 203 (14).
Dongchuan Exped. 63074 (17); 63106 (4b); 63174 (28).
Dongling Exped. 109 (4b); 1092 (4b).
Dorsett & Dorsett 3470 (4a).
Ducloux 485 (24); 500 (5); 522 (22a); 603 (13); 2283 (7); 2684 (5); 2685(13, 24); 2686 (24); 2687 (24); 2853 (24); 3163 (24); 3820 (13); 4157 (7); 4158 (22a); 4378 (24); 4567 (5); 5336 (4a); 5639 (18); 7079 (7); 7089 (22a); 7263 (20b); 7269 (20b); 7310 (24); 7536 (5); 7573 (7); 7624 (5).
Duolun Pl. Exped. 4 (31).
Duthie 1046 (25).
East China Bot. Station 6052 (21); 6404 (22b); 6881 (20b); 6886 (20b).
East Sichuan Exped. 3952 (20b).
Ecol. Group Xizang Exped. 11535 (24); 14717 (3); 15548 (4a); 15566 (22a).
Emei Bot. Exped. 104 (4b).
Esquirol 84 (20b); 1097 (5); 2501 (20b); 3007 (20b).
Faber 173 (20b); 285 (20b).
Fan, C. S., & Li, Y. Y. 627 (21).
Fang, M. Y. 24667 (5).
Fang, W. P. 1075 (18); 1309 (18, 22a); 1460 (4b); 1469 (4b); 2647 (22a); 2715 (22a); 2717 (22a); 2822 (4b); 2823 (22a); 2884 (18, 22a); 2885 (22a); 2941 (4b); 3489 (5); 3662 (4a, 4b); 3667 (4a, 4b); 3719 (5); 4106 (26); 4190 (11, 22a); 4309 (4a, 4b); 4542 (30); 4971 (5); 5100 (5); 5377 (5); 6058 (22a, 28); 6635 (22a); 7304 (22a); 7696 (30); 7697 (30); 8466 (22a); 8589 (13); 8911 (22a); 9280 (5); 11104 (22a); 11478 (5); 12794 (22a); 12811 (18); 12819 (22a); 12975 (24); 12985 (18); 12993 (4b); 12998 (18); 15077 (14); 17129 (22a); 17580 (18); 17694 (22a); 20383 (22a); 20434 (4a); 36710 (4b); 36915 (4b).
Fang, W. P., & Hu, W. K. 34756 (24); 51479 (4a); 53977 (22a).
Fang, W. P., et al. 20452 (18); 20453 (30).
Fang, W. Z. 495 (31).
Fang, Z. F. 882 (4b); 959 (22a).
Fangjing Shan Exped. 362 (18, 21); 907 (18).
Farges 314 (4b); 726 (22a); 105811 (22a); 106863 (22a); 107074 (22a).
Farrer & Purdom 717 (4a).
Faurie 286 (22b); 559 (22a); 560 (14); 561 (22a); 1098 (21); 1307 (22a); 1317 (21); 3731 (21); 5101 (21); 5396 (22a); 6314 (21); 6663 (21); 6672 (37).
Feng, J. H. 6434 (24).
Feng, K. M. 811 (5, 19); 1694 (4b); 1958 (22a, 26); 2476 (4a); 2565 (5); 44410 (18); 4777 (7); 5386 (26); 5416 (3); 5563 (4b); 5908 (22a); 6161 (26); 6250 (4b); 6805 (4b); 6598 (26); 6861 (26); 7130 (11); 7635 (3); 7839 (27); 7857 (27); 8402 (22a, 24); 8563 (11); 8806 (22a); 9007 (4b); 10230 (5); 10232 (5); 10681 (20b); 11180 (20b); 12669 (20b); 22332 (18); 22692 (22a); 22707 (7); 23304 (4b); 23681 (20b); 23959 (14); 23962 (11); 23990 (22a); 23991 (11); 24605 (20b).
Feng, K. M., et al. 80–27 (26); 80–64 (20b).
Feng, N. N. 265 (5).
Feng, R. Q. 21165 (5).
Fenzel 183 (4a); 277 (4a); 2525 (4b); 2555 (31); 2635 (31).
First Aug. Coll. Agr. 50 (33).
Foping Shan Exped. A124 (22b); A416 (6); 1301 (22b).
Forrest 2490 (20b); 2491 (26); 2589 (22a); 2759 (4b); 2820 (22a); 2894 (17); 2975 (19); 4457 (4b); 4458 (24); 4459 (26); 4460 (7); 4461 (5); 4463 (24); 4464 (22a); 5809 (26); 6909 (24); 6971 (26); 7146 (7); 7508 (7); 8298 (22a); 9275 (19); 10283 (4b); 10328 (11); 11941 (24); 14320 (4a); 14443 (1); 28054 (4b).
Fougnia 446 (24).
Fu, D. Z. 462 (31).
Fu, D. F., & Cao, Y. L. 1114 (13).
Fu, D. Z., & Xiang, Q. Y. 83055 (5).
Fu, D. Z., & Zhai, Z. L. 87–2202 (26).
Fu, F. D., & Cao, Y. L. 0125 (18); 0149 (18); 0282 (18); 649 (22b); 1008 (22b); 1133 (14).
Fu, F. D., et al. 9115 (4a).
Fu, G. X., & Zhang, Z. S. 907 (22b); 1030 (4b); 1314 (4b); 1589 (20b, 22a); 1661 (21); 3972 (4a); 4025 (1).
Fu, J. Q. 108 (22a); 912 (5); 2108 (17); 2889 (20b).
Fu, K. C. 5169 (22b).
Fu, K. H. 901 (24); 1182 (22a).
Fu, K. H., et al. 1100 (24); 1589 (24).
Fu, K. T. 1013 (4b); 1160 (28); 1318 (4a); 1911 (20b); 1970 (4a); 3442 (5); 4888 (22b); 5148 (6); 5169 (22a); 5354 (5); 5519 (6); 5778 (20b); 5779 (14); 5819 (22a); 6228 (13); 6789 (22b); 6825 (6); 6928 (5); 6995 (6); 7113 (5); 9564 (3); 9867 (20b); 10943 (21); 10953 (14); 11483 (20b); 11658 (5); 12572 (4b); 12956 (5); 13424 (4b); 13627 (6); 13809 (6); 14392 (22a); 14556 (22a); 16055 (4b); 17945 (4b).
Fu, P. Y. 333 (34); 770 (4b); 952 (33); 2321 (31); 3272 (4b); 4415 (4a).
Fu, P. Y., & Zhang, Z. H. 18 (4a).
Fu, P. Y., et al. 462 (22b); 952 (34); 1012 (34); 1588 (31); 2475 (32); 3273 (34); 3426 (22b); 3796 (4b).

Fu, W. G. 8257 (22a); 34751 (18).
Fu, Z. X., & Kuo, Y. H. 783 (13).
Fujian Exped. 377 (21).
Fukuyama & Suzuki-Tokio 15091 (36).
Gang, S. S. 169 (22b).
Gansu Exped. 2068 (33).
Gao, B. S., & Wu, Y. T. 111539 (4a).
Gao, T. S. 1511 (21).
Gao, W. L. 102 (4b).
Gao, X. P. 53407 (5).
Ge, B. Z., & Wang, W. Y. 22227 (31).
Geng, G. C. 597 (33).
Go, B. Z. 400 (4b); 468 (4a); 890 (4b); 898 (22b); 1402 (4b); 6643 (4b); 9547 (4b); 23300 (28).
Go, B. Z., & Wang, W. Y. 12261 (4a); 21647 (4a); 22878 (27); 22925 (27); 22995 (26).
Go, B. Z., et al. 20177 (18).
Go, Q. H. 706 (6).
Golog Exped. 553 (4a); 574 (31).
Gong, B. Z., & Wang, W. I. 8429 (31).
Gong, Z. Y. 1063 (33).
Gou, B. Z. 6643 (4a).
Gould 409 (29); 1452 (29); 2357 (26).
Gressitt 108 (22a); 140 (22a); 382 (22a).
Guan, Z. T. 6623 (22a); 6981 (22a); 7477 (22a); 8277 (18); 8714 (30); 8991 (18); 9086 (22a); 9665 (18).
Guang, Z. T. 6624 (4b).
Guangdong 73rd Exped. D1161 (22b).
Guangfu Exped. 899 (22b).
Guichi Exped. 7184 (5).
Guizhou Exped. 335 (20b); 510 (5); 3625 (21); 5760 (5); 6388 (5); 7185 (13); 7537 (5); 8428 (5).
Guizhou Inst. Chin. Med. 5 (11); 149 (11).
Guo, B. Z. 632 (22a); 827 (22a); 1728 (22a); 6219 (31); 6591 (31); 9976 (31).
Guo, B. Z., & Wang, W. I. 12043 (31); 12206 (31).
Hamet-Ahti 3231 (20b).
Hancock 42 (20b); 69 (22a); 316 (5); 1876 (4a).
Handel-Mazzetti 643 (24); 782 (24); 2568 (22b); 3156 (4b); 3515 (26); 637 (26); 4904 (24); 5164 (5); 6211 (7); 7308 (27); 7646 (20b); 9863 (3); 11233 (22b); 11263 (6).
Hao, K. S. 75 (4b); 631 (4b); 1069 (4a); 3939 (5); 4004 (4b); 4008 (22b); 4157 (13); 4159 (5); 4210 (6); 4325 (17).
Hao, S. Y. 4171 (22a).
Haussknecht 812 (33).
Hay 124 (2).
He Exped. 3154 (22a); 3322 (22a); 5981 (22a).
He, D. P. 4406 (30); 44413 (4b); 44646 (22a); 45122 (4b); 45262 (4b); 45381 (4b); 45581 (22a); 46043 (20b); 46527 (4b).
He, G. Z. 4861 (20b); 4862 (20b).
He, S. X. 403 (20b).
He, W. Y. 2957 (4a).
He, X. Y. 6055 (24); 6091 (4b); 23939 (22b); 25426 (22b); 31174 (22b).
He, Y. 194 (31); 264 (4a).
He, Y. Q. 883 (5); 1179 (5); 1691 (6); 3653 (4a); 4151 (31); 4507 (4a); 536 (32); 4563 (32); 5090 (22b); 5154 (4b); 5209 (22b); 5356 (22b); 5886 (4b); 5978 (4b); 6012 (22b); 7278 (33).
He, Y. X. 27953 (21).
He, Z. 11601 (4b); 13246 (4b); 14241 (4b).
He, Z., & Zhou, Z. L. 13246 (4b); 13386 (6); 13554 (20b); 13662 (24); 13890 (5); 14071 (22a); 14241 (4a, 4b).
He, Z. G. 559 (20b); 586 (5); 590 (5); 598 (5).
Hebei Exped. 19 (6); 78 (6); 915 (4b); 1723 (4b); 3518 (4b); 3731 (4b); 4153 (4b); 4282 (4b); 4286 (4b); 6548 (4b); 7670 (4b); 8574 (31, 32); 64575 (32).
Hebei Med. Pl. Exped. 121 (22b); 147 (22b); 760 (5).
Hebei Pl. Exped. 79 (14); V538 (14); 796 (32); 1674 (4a); 2095 (4a); 2353 (32); 3735 (32).
Hebei Univ. Agr. 556 (6); 2114 (4b); 3054 (4b); 3903 (33); 42252 (22b); 42738 (31); 42439 (32); 42740 (31); 43613 (22a, b); 43725 (5).
Height 5839 (6).
Heilongjiang Econ. Pl. Exped. 626 (32).
Helan Shan Exped. 262 (4a).
Hemsley 897 (21); 898 (21).
Henan Exped. 106 (5); 935 (6); 991 (5, 6); 1209 (5); 1305 (6); 1327 (6); 1740 (6); 2009 (5); 2197 (14); 2365 (22b); 4204 (17); 4218 (6); 4250 (6); 6084 (5); 14301 (5); 14766 (5); 15002 (5, 6); 20031 (5); 20135 (22b); 20164 (22a); 20726 (13); 21087 (20b, 21); 22054 (5); 22095 (5); 22609 (22b); 23549 (5, 6); 23974 (5); 34337 (6); 34596 (4b); 34798 (22b); 34878 (5); 35038 (5); 35186 (5); 35285 (5); 56069 (5).
Henan Forest. Dept. 453 (5); 640 (5); 951 (20b); 1163 (22b).
Hengduan Shan Exped. 59 (18); 365 (18); 657 (5); 996 (4a); 1263 (18); 482 (4a); 1571 (22a); 1582 (11); 1612 (22a); 1817 (17, 22a); 2071 (5); 2138 (26); 2248 (26); 2282 (4a); 2305 (22a); 2307 (22a); 2344 (22a); 2389 (26); 2508 (26); 2562 (22a); 2579 (22a); 2745 (5); 2793 (26); 2800 (26); 2976 (5); 2977 (11); 2978 (22a); 3068 (5); 3226 (26); 3621 (4a); 3681 (4a); 3710 (32); 3788 (26); 3839 (26).
Henry 59 (18); 173 (20b); 285 (20b); 287 (13); 1037 (14); 1043 (14); 1747 18); 1803 (22b); 1868 (22b); 2257 (5); 2445 (22b); 2753 (5); 2950 (6); 4843 (6); 5082 (22b); 6031 (4b);

6031a (4b); 6141 (18, 22a); 6409 (22a); 6429 (11); 6453 (13); 6808 (14); 6941 (22a, 24); 7320 (22a); 7521 (5); 8945 (4b, 18); 8948 (18); 9481 (22a); 9642 (20b); 10638 (18, 22b); 12627 (19); 13457 (19); 13723 (5).
Herb. Beijing Inst. Bot. 397 (4a); 787 (4a); 915 (4a); 1457 (4a); 1474 (4a); 1555 (4a); 1660 (6); 1806 (4a); 2152 (4b); 2336 (4a); 5236 (4a); 7298 (4b); 10178 (4a); V1057 (4b).
Herb. Hunan Teacher's Coll. 9248 (21); 9249 (5).
Hezhang Exped. 453 (20b).
Hidetaro 1337 (8).
Ho, G. 2859 (21).
Ho, H. Y. 2959 (5).
Ho, X. Y. 1088 (21).
Hoch 3269 (36); 3288 (15); 3303 (15); 3306 (15); 3310 (15); 3311 (16); 3382 (15).
Hoch & Peng, C.-I 3381 (36); 3406 (16).
Hong, D. Y., & Ying, J. S. 0729 (22a); 1225 (22a).
Hong, Q., & Huang, Z. S. 46 (4b).
Hoo, S. Y. 5917 (22a); 6056 (22a); 6152 (22a).
Hooker & Thomson 2322 (19).
Hop 16128 (4b).
Hopkinson 491 (4b); 815 (4b); 1256 (31).
Hou, Y. X. 82 (31).
Howell 83 (19); 221 (20b); 222 (18); 223 (22a).
Hsia, W. Y. 294 (5); 1275 (5); 1316 (33); 1393 (4b); 1575 (22a); 2001 (5); 2029 (4a); 2210 (32); 2415 (4b); 2957 (4a); 3109 (31); 3622 (33); 3687 (5); 5517 (4b); 5792 (5); 5879 (22b); 6309 (5); 6329 (14); 6390 (5).
Hsiang, C. H., et al. 93309 (18).
Hsiang, K. S., & Lang, K. Y. 41 (18).
Hsieh, C. C. 41458 (22a); 41462 (18); 41776 (5); 42516 (22a); 42633 22a); 42677 (22a); 42808 (5).
Hsieh, C. F. 1257 (14); 1332 (22a).
Hsing, C. H., & Zhou, Z. L. 93862 (14).
Hsing, C. H., et al. 32532 (14).
Hsiung, C. H. 91865 (13).
Hsiung, C. H., et al. 30490 (30).
Hsiung, Y. K. 756 (22b).
Hsu, C. C. 3289 (14); 5368 (22a); 5801 (14); 5916 (20b, 22a); 5983 (8); 5998 (15); 6110 (14); 6111 (22a); 6238 (20b); 9202 (22a).
Hsu, C. C., & Hsu, R. 3767 (14); 3950 (20b, 22a); 4009 (20b); 9291 (22a).
Hu, C. M. 98 (5); 4848 (20b).
Hu, L. C. 50148 (18).
Hu, S. Y. 1164 (11); 1309 (4b).
Hu, S. Z. T255(1).
Hu, W. G., & He, Z. 10012 (26); 10181 (18); 10200 (18); 10460 (4a); 10611 (4a); 10680 (4b); 10818 (4a); 10981 (4a); 10984 (4a).
Hu, W. K. 8257 (28); 34657 (22a); 34798 (4b); 35383 (4b); 35421 (4b); 37718 (4b); 37799 (22a).
Hu, W. K., & He, Z. 10268 (4b); 10474 (24).
Huan, M. X. 113064 (5).
Huan, Z. P., et al. 780 (4a); 1619 (4a); 1661 (4a).
Huang, Z. P., et al. 2336 (17).
Huang He Exped. 315 (5); 376 (5); 850 (31); 968 (5); 1699 (5); 1716 (4b); 1915 (31); 1940 (5); 2235 (4b); 2391 (5); 2425 (4b); 2492 (4b); 2544 (4a); 2566 (32); 2606 (4b); 2718 (4b); 2903 (22a); 3135 (4b); 3154 (22a); 3222 (24); 3297 (4b); 3322 (22a); 3442 (5); 3471 (4b); 5981 (22a); 7993 (33); 8718 (32).
Huang, C. H. 83–001 (20b).
Huang, K., et al. 633 (22a).
Huang, M. X. 11986 (20b); 112191 (20b).
Huang, R. H. 142 (22a); 321 (4b).
Huang, T. C. 1343 (22a); 4206 (22a); 4226 (14); 5774 (22a); 6521 (22a); 7078 (22a); 7080 (20b); 7123 (20b); 7126 (20b); 7130 (20b); 8513 (22a).
Huang, T. C., & Hsieh, C. F. 7174 (20b).
Huang, T. C., & Kao, M. T. 5731 (22a).
Huang, T. C., & Yang, K. C. 1845 (15).
Huang, T. C., et al. 5730 (22a); 8906 (22a).
Huang, Y. X. 2279 (21).
Huang, Z., et al. 1341 (18).
Huang, Z. P. 637 (11).
Huang, Z. P., et al. 538 (22a); 633 (24); 2326 (22a).
Hugel 1153 (4a); 1192 (5).
Hugh 32 (5, 6); 51 (14); 81 (5).
Hulun Buir Meng Exped. 325 (4a).
Hunan Exped. 596 (13); 12235 (20b).
Hunan Med. Exped. 63476 (20b).
Hunan Veg. Group 515 (5).
Hunan-Guizhou Exped. 2611 (6); 2698 (22b); 3403 (22b); 3439 (6); 3562 (6); 3581 (6); 3547 (6).
Huo, S. Y. 20483 (32).
Huo, X. Y. 5031 (18).
Huping Shan Exped. A25 (20b); A48 (20b); A109 (20b).
HWNP Xizang Exped. 1898 (26).
Ili Exped. 21 (5); 25 (4a).
Inst. Bot. 6748 (5); 6796 (5).
Inst. Wild Plants 7422 (21).
Jacquemont 1739 (2).
Jang, S., et al. 2631 (26).
Ji, Z. S. 329 (5).
Jian, S. 5301 (4a).
Jian, X. L. 36915 (4b).
Jian, X. L., & Xiong, J. H. 34264 (26); 34589 (26).
Jiang, N., & Zhao, C. F. 218 (17).
Jiang, S. 1580 (4b); 1849 (4b); 5328 (26); 6159

(11); 6220 (4b); 6221 (12); 7574 (5); 7727 (4b); 9629 (17).
Jiang, S., & Jing, C. L. 1801 (22a); 1817 (22a); 1869 (18); 2115 (4a).
Jiang, S., & Li, C. 2759 (17).
Jiang, S., & Zheng, L. 104 (3).
Jiang, X. C. 290 (32).
Jiang, X. L. 36317 (4a); 36404 (4a); 37969 (22a).
Jiang, X. N. 36261 (31).
Jiangan School of Agr. 1335 (22b).
Jiangxi Pl. Exped. 222 (21); 552 (21); 838 (21); 890 (21); 1021 (20b).
Jiao, F. X. 2433 (24).
Jilin Teacher's Univ. 657 (31).
Jin, D. F. 191 (5); 369 (22a).
Jing, D. F. 247 (4b); 364 (4b); 521 (4b).
Jing, S. Y. 7 (31).
Jingfo Shan Exped. 1013 (18); 1887 (5); 1897 (13); 2025 (24); 2031 (22b); 2130 (6); 2166 (24); 2310 (5); 2311 (14); 2439 (24).
Jinyang Pl. Exped. 13468 (22a).
Jiulian Shan Tree Farm 781304 (21).
Ju Ud Meng Exped. 159 (4a).
Junatov 11905 (1).
Kanehira 419 (20b).
Kanehira & Sasaki 21782 (20b).
Kanehit 21160 (22a).
Kang Xian Exped. 387 (6).
Kao, M. T. 3160 (14); 4125 (22a); 4899 (22a); 5007 (14); 5241 (22a); 7581 (22a); 7767 (14); 7772 (20b); 7802 (15); 8506 (22a); 8601 (20b).
Kaproon 4 (9a); 12 (9a); 96 (9a).
Kato 418 (14).
Kawakami 81728 (22a).
Keng 127 (5).
Keng & Kao 2619 (14); 2634 (14).
Keng, Y. L. 637 (33).
Kermode 17233 (18).
Kersten 1382 (12).
Kimura et al. 57 (22a); 77 (22a); 78 (22a).
Kinashi 7 (34).
King's collector 146 (27).
King, T. F. 27 (4b); 135 (4b); 364 (4a); 521 (4b).
Kingdon-Ward 754 (26); 3296 (26); 4304 (27); 4380 (26); 10034 (17); 10118 (2); 11902 (26, 27); 12103 (11); 12372 (23).
Komarov 1124 (4a).
Kong, L. Q. 1526 (20b).
Konta, F., & Lang, K. Y. 86 (31); 163 (31); 324 (5); 524 (33).
Kotschy 616 (31).
Koyama 23875 (15).
Koyama et al. 1187 (19).
Kuan, K. C. 1176 (5); 1288 (1); 1316 (33); 1382 (9b); 1589 (33); 1622 (22a); 1646 (22a); 1767 (1); 1842 (22a); 1853 (4a); 1954 (31); 2085 (22a); 2095 (1); 2247 (4a); 2424 (33); 2446 (31); 2574 (4a); 2685 (31); 2820 (5); 2838 (31); 2974X (14); 3078 (33); 3082 (5); 3358 (4a); 3639 (5); 3971 (4a); 4264 (31); 4374 (10); 4570 (31); 7043 (5); 21374 (10); 27004 (22a); 74407 (22b); 75335 (22b); 75433 (22b); 77278 (31).
Kuan, K. C., & Chen, Y. L. 800 (22a); 837 (4b); 1081 (4b); 2049 (33); 2158 (32); 2417 (31, 32); 2467 (4b); 2542 (5); 2549 (5).
Kuan, K. C., & Dai, T. L. 671 (5).
Kuan, K. C., & Yang, B. M. 4 (22b); 80464 (22b).
Kuan, K. C., et al. 86 (22a, 24); 241 (4b); 251 (22b); 373 (26); 505 (30); 649 (17); 729 (4a); 1150 (18); 1561 (20b); 1587 (14); 1592 (5, 6); 1835 (4b); 1907 (14); 2195 (22b); 2220 (18); 2229 (14); 2306 (20b); 2510 (14); 2963 (5); 4008 (24, 26).
Kung, H. S. 6007 (22a, 24); 6008 (4a, 4b); 6009 (3); 6049 (17); 6050 (4a); 6050 (4b); 6052 (22a); 6058 (31, 32); 6093 (22a); 6203 (22a); 6218 (5); 6268 (17); 6269 (3, 17).
Kung, H. S., et al. 20 (4a); 40 (17); 44 (14).
Kung, H. W. 554 (4b); 930 (4a); 1085 (32, 33); 1216 (22a); 1773 (4b); 1830 (4b); 1964 (31); 2178 (31); 2978 (5); 3009 (4b).
Kung, S. H. 6218 (11).
Kunming Bot. Inst. 235 (28); 24070 (27); 21272 (20b); 22454 (20b); 23990 (22a); 64601 (18); 64602 (20b).
Kuo 8332 (22a); 8429 (22a).
Kuo, B. Z., & Yang, Y. C. 9728 (14).
Kuo, C. M. 3983 (20b).
Kuo, C. S. 5738 (22a); 80405 (22a); 80293 (22a).
Kuo, C. S., et al. 7054 (20b).
Kuo, P. C. 423 (17); 1623 (17); 9728 (17).
Kuo, P. C., & Wang, W. Y. 21938 (11).
Kuo, Y. S., et al. 538 (4b).
Kuoh, C. S. 1255 (20b); 3009 (8); 3011 (8); 3045 (22a); 3222 (14); 3262 (16).
Lai, S. K., et al. 343 (20b).
Lai, S. S. 1640 (21); 2911 (21); 3704 (21).
Lai, S. S., et al. 4540 (21); 4672 (5).
Landa 112 (22b).
Lang et al. 805 (12).
Lang, K. Y. 918 (18); 2827 (27).
Lang, K. Y., et al. 325 (30); 805 (17); 871 (27); 992 (24); 1218 (22a); 1254 (14); 1272 (4b); 1488 (17); 2262 (4a); 2426 (26); 2451 (4a); 2647 (4a); 2664 (24); 2705 (4a); 2967 (26).
Lang, S. B. 30 (22a).
Lanping Exped. 80006 (5).
Lanzhou Univ. Biol. Dept. 719 (33).
Lanzhou Univ. Exped. 109 (4b); 129 (4a); 177 (22a); 183 (4b); 211 (4b); 223 (22a); 236

Ling, Q. 791675 (22a).
Ling, Q., & Deng, X. F. 700588 (26); 790447 (26).
Ling, R. 3175 (21).
Ling, Y. 15199 (21).
Ling, Y. R. 74214 (9a); 74370 (10); 74854 (5); 74885 (10); 741149 (9a); 741311 (9a).
Ling, Y. R., et al. 80–158 (22a).
Liou, K. M. 696 (4b); 413 (4b); 497 (5); 505 (6); 524 (6); 1280 (4b); 2649 (5); 2686 (4b); 2848 (5, 6); 2923 (6); 2830 (5); 3054 (4a); 3205 (5); 3207 (6); 3286 (4a); 3290 (32); 3337 (5); 3500 (4b); 3508 (4b); 3782 (32); 4584 (22a); 4703 (17); 4752 (5); 4797 (6); 4801 (4b); 4921 (13); 5124 (22a); 5289 (4b); 5614 (5); 5956 (4b); 6139 (22a); 6384 (4b); 6576 (31); 6885 (4a); 6962 (5); 6970 (4b); 7830 (6); 8946 (5); 9298 (5, 6); 10179 (5); 13755 (5); 16505 (5); 16518 (5); 16860 (5); 19499 (6); 20453 (5); 20726 (5); 20867 (5); 22938 (5).
Liou, T. H., & Tsoong, P. C. 3229 (22a).
Liou, T. H., et al. 7776 (31); 7872 (31); 10184 (31).
Liou, T. N. 1516 (22a); 2855 (31, 33); 3032 (31); 3394 (10); 3506 (31); 3886 (34); 3891 (34); 4094 (34); 4103 (34); 4334 (32); 4697 (32); 8830 (31); 10301 (30); 14252 (20b); 14292 (19); 14379 (17); 14656 (19); 16981 (20b); 17441 (19); 17641 (17); 20277 (19); 20664 (19); 20967 (19); 21102 (20b); L.5538 (17).
Liou, T. N., & Liou, K. M. 1214 (21); 1343 (21).
Liou, T. N., & Tsoong, P. C. 766 (22a); 943 (22a); 1522 (22a); 1554 (17); 2623 (22b); 2736 (32); 2751 (26); 3205 (6); 3212 (13); 3449 (6).
Liou, T. N., & Wang, C. 1092 (22a); 1208 (30); 1237 (4b).
Liou, T. N., et al. 803 (31); 1148 (4b); 1516 (34); 1638 (22b, 31, 34); 1785 (31); 4188 (34); 5984 (34); 6035 (34); 6370 (5); 7204 (4a); 8087 (4a); 8827 (4a); 10184 (32); 10193 (34).
Liu, C. 288 (4a).
Liu, C. S. 467 (24); 849 (4b); 1000 (24); 1061 (22a); 1062 (22a, 27); 1185 (24); 1277 (18).
Liu, G. J. 66 (10).
Liu, G. L., et al. 172 (4a); 260 (4a); 311 (22a); 454 (4a); 593 (4a); 618 (4a); 680 (33); 685 (4a); 1091 (10); 6523 (31).
Liu, J. C. 147 (33); 911 (31); 1092 (5); 1113 (22a); 1131 (22a); 1166 (6); 1167 (5); 1333 (5); 1469 (4b); 2303 (4b).
Liu, J. M. 1602 (33); 6139a (31); 10030 (5); 10481 (22a); 10617 (4b).
Liu, J. S. 7627 (5, 11).
Liu, L. F., & Yang, S. F. 5514 (18).
Liu, L. H. 283 (20b); 757 (20b); 8372 (21); 9746 (21); 9789 (5); 11151 (20b).
Liu, L. H., & He, G. X. 15144 (20b); 16621 (20b, 21).
Liu, P. 2044 (6).
Liu, P. L., & Jiang, T. H. 67 (5).
Liu, Q. Y. 891 (5).
Liu, S. W. 423 (31, 32); 2003 (4a); 2084 (4b); 2445 (4a).
Liu, S. W., et al. 2945 (4b).
Liu, S. W., & Chen, C. J. 2726 (3).
Liu, S. W., & Huang, Y. F. 869 (31).
Liu, S. W., & Lio, D. S. 1395 (4b).
Liu, S. W., & Lo, D. S. 1195 (31); 1392 (27); 2–1392 (28).
Liu, T. S., et al. 290 (22a).
Liu, X. X. 24173 (21); 25082 (5); 29173 (21).
Liu, X. Y. 547 (6); 603 (33); 793 (4b); 1004 (14); 13103 (4b); 21045 (32); 21449 (4b); 21497 (22a).
Liu, Y. 512 (21); 10601 (4a); 10655 (33); 10773 (4a); 10961 (4a); 11520 (4b); 11682 (4b); 11813 (4b); 11832 (4b); 11860 (4b); 11920 (33); 12506 (4a); 13168 (33); 13170 (6); 13323 (14); 13581 (14); 13646 (5); 15511 (5).
Liu, Y. G. 619 (21).
Liu, Y. L., et al. 341 (4a).
Liu, Y. X. 28675 (20b); 28854 (20b).
Liu, Y. Z. 1 (4a); 2 (31); 8 (4a).
Liu, Z. A. 2304 (17); 22582 (3).
Liu, Z. G. 285 (22a); 1029 (24); 1438 (24); 1440 (4a).
Liu, Z. G., et al. 50 (24).
Liu, Z. K. 215 (4a).
Liu, Z. S. 189 (3); 810 (4a); 846 (4a); 849 (4a); 994 (26); 1000 (26); 1063 (26).
Liu, Z. Y. 267 (13).
Liuzhou Exped. 2594 (22b).
Lixian Exped. 697 (22b).
Lo, D. S., & Zuo, Z. C. 800445 (4a); 801405 (4a); 801447 (4a).
Lo, D. S., et al. 80–0942 (31); 80–1301 (31); 80–1382 (31).
Lo, S. Q. 1334 (4a).
Long, R. Z., et al. 40 (20b).
Longgan Exped. 6648 (21).
Lu Shan Exped. 188 (22b).
Lu, C. Z. 63–20096 (20b).
Lu, Q. H. 2485 (22b).
Lu, T. H. 274 (22a).
Lu, Z. A. 464 (5).
Luding Pl. Exped. 23113 (22a).
Ludlow 663 (1).
Ludlow & Sherriff 510 (4b); 1962 (23); 2066 (4a); 2182 (26); 2322 (4b); 8851 (11, 27);

Shanxi Exped. 95 (4b); 116 (5); 241 (4a); 618 (4a, 4b); 1347 (32); 1965 (22a).
Shanxi Teacher's Coll. 235 (4b).
Shao, W. S. 2 (6).
Shennongjia Pl. Exped. 10582 (18); 10766 (4b); 10816 (22a); 10899 (13); 11062 (22a); 11481 (21); 11736 (13); 11796 (22b); 21273 (6); 21433 (4b); 21434 (4b); 21569 (13); 21629 (11, 22b); 21670 (6); 21796 (13); 21957 (22b); 22467 (22a); 22500 (20b); 22723 (22b); 22746 (13); 22840 (20b); 22844 (4b); 22870 (4b); 30978 (6); 31061 (13); 31145 (22b); 31240 (11); 31478 (6); 31514 (11); 31536 (4b); 31551 (13); 31650 (22a); 31712 (4b); 31784 (22a); 32213 (22a); 32235 (14); 32280 (4b); 32266 (22a); 32274 (22a); 32825 (14).
Shimizu 20253 (22a); 20358 (22a); 20377 (15); 20379 (22a).
Shimizu & Chuang, C. C. 20190 (22a); 20204 (22a); 20454 (14).
Shing, K. S., & Lang, K. Y. 34 (4b); 35 (4b).
Shizhu Pl. Exped. 3384 (14).
Shuicheng Exped. 525 (13).
Sichuan Agr. College 667 (18); 6273 (17); 7246 (24).
Sichuan Econ. Pl. Exped. 55 (20b); 249 (31); 274 (11); 348 (4a); 460 (18); 662 (11); 738 (22a); 831 (30); 894 (3, 4a); 975 (30).1290 (18); 1551 (4b); 1564 (4b); 1718 (24); 1727 (22a); 1993 (4b); 4033 (22a); 4144 (4b); 4161 (22a); 4175 (11); 4176 (22a); 4561 (22a); 5913 (22a).
Sichuan Exped. 1435 (26); 2324 (17); 2424 (17); 2631 (22a); 3018 (24); 5001 (7); 7533 (31); 8220 (24); 10379 (4a); 13071 (20b); 13477 (7); 15474 (24); 22202 (24).
Sichuan Forest. Exped. 125 (22a); 195 (4b).
Sichuan Grassland Exped. 22458 (22a); 22482 (22a).
Sichuan Inst. Biology 8220 (26); 21038 (27).
Sichuan Med. Exped. 196 (22a); 12413 (18); 15411 (26); 15420 (4b); 20145 (4a); 20145a (4a); 22331 (4b); 22329 (4a); 22363 (26); 23026 (20b); 25370 (17); 26797 (19).
Sichuan Med. Pl. Exped. 12731 (22a); 15005 (11); 15205 (22a); 15414 (31); 15531 (11); 15627 (31); 15963 (5); 22143 (11); 22520 (31); 25025 (30); 27851 (5).
Sichuan Test. Ground Forest. 146 (4a).
Sichuan Univ. 23603 (22a).
Sichuan Veget. Exped. 678 (4a); 948 (4a, 4b); 1716 (4a); 2041 (22a); 2146 (4a); 2422 (4a); 2525 (22a); 2630 (4a); 2715 (4a); 3018 (22a); 3538 (17); 3629 (18); 5775 (4a); 5881 (4b); 7251 (4b); 7708 (4a); 9362 (4a); 10298 (31); 10402 (31); 10411 (4a); 12096 (22a); 13074 (18); 13218 (4b); 13739 (18).
Sichuan-Guizhou Exped. 1275 (20b).
Silvestri 1578 (6); 6169 (14); 6171 (5); 6174 (5).
Simeon Ten 1185 (13).
Sin, S. S. 471 (20b).
Sing, B. X. 11 (22a); 69 (22a); 97 (22a); 888 (17, 22a).
Sino-American Bot. Exped. 715 (22a); 745 (5); 745a (19); 860a (18); 1019 (7);1132 (26); 1353 (5); 1398 (22a); 1560 (5); 8606 (22a).
Sino-American Exped. 24 (13); 318 (17); 319 (14); 380 (14); 599 (13); 722 (14); 1013 (14); 1160 (6); 1161 (6); 1831 (11); 1832 (6); 1835 (6).
Sino-American 1980 Exped. 85 (22b); 119 (22b).
Sino-American Guizhou Bot. Exped. 419 (22a); 432 (22a); 481 (22a); 1190 (21); 1674 (21); 1831 (24); 2402 (6); 2404 (6).
Sino-British Exped. to Cangshan 0474 (7).
Sino-English Fanjing Shan Exped. F066 (22a).
Sino-German Exped. 8495 (31); 8827 (4a).
Sino-U.S.S.R. Bot. Exped. 980 (30); 1571 (18); 1947 (18); 2109 (18).
Skvortsov & Wang, G. Z. 1101 (31).
Smith 134 (22a); 136 (22a, 32); 137 (22a); 138 (32); 140 (32); 459 (5); 971 (33); 1050 (4a); 1083 (22a); 1168 (5); 1171 (22a); 2108 (18); 2649 (22a); 2844 (22a); 2847 (4a); 2849 (24); 4272 (31); 4694 (24); 6075 (5); 6080 (5); 6226 (33); 6230 (5); 6476 (22b); 6541 (22a); 6886 (32); 7108 (4b); 7232 (22a); 7305 (22a, 32); 7420:2 (32); 7582 (33); 8296 (22a); 8297 (4a); 10117 (18); 10118 (30); 10511 (24); 10524 (3); 10608 (28); 10747 (22a, 28); 10817 (12); 10825 (4a); 10928 (23); 10929 (24); 10961 (4b); 10982 (3); 11118 (22a); 11535 (18); 11801 (4a); 11986 (23).
Smith & Chin 8108 (22a).
Song, Z. G. 453 (22a).
Songming Pl. Exped. 51331 (13).
Soong, T. P. 38485 (30); 38763 (18); 38867 (18); 39018 (4a); 39065 (4b); 39460 (5); 46265 (22a).
Soong, Z. P. 39029 (26).
Soulie 96 (24); 161 (24, 26); 233 (4b); 350 (24); 449 (11, 24); 867 (4b); 1126 (5); 1127 (11); 1128 (13); 1256 (11); 1300 (22a); 1341 (4b); 2230 (26); 2564 (4a, 4b); 2964 (4b); 2969 (26); 3172 (17); 3636 (17); 3637 (26); 3638 (4a); 3639 (4a); 3640 (4a); 3641 (4a); 3929 (4b); 3978 (24).
South Hunan Exped. 3247 (22b); 3404 (22b).
Stainton et al. 726 (26); 1800 (27); 3894 (24); 6271 (26).
Steenis 20685 (14); 20726 (22a); 20863 (22a).
Stevens 47 (5); 83 (4b); 326 (4a); 496 (4a).
Steward 2450 (22b); 4706 (20b).

Steward & Cheo, C. C. 987 (20b).
Steward et al. 396 (17, 21, 22a).
Stewart 21523 (17).
Stotzner 55 (4a).
Strachey & Winterbottom 2 (4a).
Sugahara 8827 (22b).
Sun, B. X. 35 (20b); 83 (20b); 149 (20b); 351 (20b); 347 (18); 815 (18); 1081 (27, 30); 64142 (11); 64187 (20b); 64479 (20b); 64526 (20b); 64543 (20b); 64544 (20b); 64549 (20b); 64598 (20b); 64608 (20b); 64624 (20b); 64634 (20b); 64846 (20b).
Sun, C. X. 146 (32).
Sun, S. C., & Chang, K. 489 (30); 739 (18); 789 (18); 1125 (4b).
Sun, S. L. 1060 (4b); 1135 (4a); 2210 (18); 2415 (18).
Sung, C. S. 128 (31); 197 (31).
Sung, J. Z. 82 (6).
Suzuki 237 (14); 243 (22a); 5081 (22a); 5643 (22a); 6219 (20b).
Suzuki-Tokio 8709 (14); 8711 (22a); 9891 (32).
Tai, F. C., & Teng, C. M. 5019 (18); 5202 (4a).
Tai, L. Y. 24 (22a).
Taia, H. T. 52278 (24).
Taibaishan Exped. 334 (22b); 375 (17).
Tamazaki, T., et al. 1118 (14).
Tamura 21186 (16).
Tamura & Koyama 23190 (22a); 23270 (22a); 23291 (22a); 23295 (15); 23616 (14); 23677 (8); 23678 (15).
Tamura & Shimizu 20709 (16); 20727 (22a); 20728 (22a); 20742 (8); 20803 (8).
Tamura et al. 20954 (16); 21377 (22a); 21688 (14); 22001 (22a); 22034 (20b); 22043 (22a); 22152 (22a); 22136 (22a).
Tan, P. X. 62916 (20b); 63622 (20b).
Tanaka 501 (22a).
Tang, C. L. 838 (5).
Tang, H. K. 80 (19).
Tang, S. X. 230 (6).
Tang, T. 1060 (4a); 1087 (22a); 1111 (31, 32); 2182 (4b); 2269 (5); 21698 (6); 23354 (18).
Tang, Y. F., et al. 7000565 (20b).
Tao He Exped. 3705 (4b).
Tao, D. D. 10925 (24); 10936 (4a).
Tauo, Y. H. 51751 (22a).
Ten 18 (22a).
Teng, L. 7996 (5).
Teng, S. W. 90339 (24); 90435 (13); 90559 (5, 6).
Tianshui Exped. 151 (22a).
Tong, Y. H., et al. 024 (31); 142 (31); 143 (31).
Tonghua Exped. 74 (34).
Trippner 229 (4a); 343 (31).
Troong, P. C. 126 (5).
Tsai, H. T. 50355 (32); 51027 (18); 51048 (4b); 52278 (18); 52670 (18); 53119 (22a); 53788 (28); 53791 (4a, 4b); 53840 (26); 53866 (27); 53948 (28); 53957 (26); 53948 (24, 26); 54094 (22a); 54107 (22a); 54633 (20b); 54727 (20b); 56008 (20b); 56056 (5); 56155 (5); 56245 (5); 56248 (5); 56506 (4a); 57150 (11); 57170 (4b); 57559 (5); 57623 (11); 57623a (11); 57659 (22a); 57722 (26); 57732 (4b); 58043 (26); 58092 (26); 58244 (26); 58260 (4b); 58262 (4b); 58328 (22a); 58329 (22a, 24); 58457 (4b); 59104 (22a); 59535 (11); 59641 (17, 20b); 59706 (22a, 26); 59865 (12); 59868 (12, 22a); 59909 (17, 22a); 63041 (11); 63119 (22a); 63791 (4b); 588168 (20b).
Tsai, Y. W. 4274 (27).
Tsang, W. T. 28090 (20b).
Tsi, Z. H., et al. 539 (24).
Tsiang 11343 (19); 11708 (5).
Tsiang, H. C. 37986 (18).
Tsiang, H. C., & Hsiung, C. H. 34209 (30); 34587 (30).
Tsiang, H. L. 10801 (4b); 35383 (4a); 36404 (4a); 37718 (4a); 37783 (14).
Tsiang, H. L., & Tsiung, C. H. 36510 (3).
Tsiang, H. L., & Xiong, J. H. 34609 (4a); 34798 (4a).
Tsiang, Y. 5127 (18); 5380 (5); 6461 (5); 6582 (21); 7759 (11); 8312 (5); 8904 (6); 8954 (22b, 24); 10760 (21); 11372 (20b); 12405 (19); 12599 (5); 11522 (11); 21952 (22a).
Tsien, C. P., et al. 30758 (22a); 30921 (24); 30924 (18, 22a); 32055 (22b); 32107 (22a); 50357 (18).
Tsiu, H. P., et al. 80044 (4b).
Tsoong, K. K. 2022 (5); 2296 (19); 2489 (20a, 24); 2988 (5); 4050 (5); 4051 (20b, 21); 4597 (11).
Tsoong, P. C. 4093 (22b); 5722 (31); 5770 (26); 6145 (26); 8419 (31); 8529 (4a); 8936 (4a); 8997 (31); 10007 (22a).
Tsui, H. P., et al. 80–46 (32); 80–369 (22a); 80–455 (5).
Tsui, L. R. 1177 (33).
Tsui, Y. W. 1023 (31); 2033 (4b); 2067 (4b); 2339 (4b); 2378 (4a); 2671 (32); 2880 (4b); 4274 (24); 4360 (22a, 24); 4461 (4a); 5041 (4a); 5137 (24); 5165 (4a).
Tu, T. H. 47 (26, 30); 148 (18); 230 (24, 30); 282 (18); 522 (22a); 613 (4b); 2351 (4a); 4430 (26); 4431 (4b); 4495 (4a); 4616 (11); 4910 (14); 5608 (18, 24).
Tunghua Exped. 298 (4b).
Turpan Exped. 3890 (5).
Uni Univ. 20218 (13).
Waddell 32 (26).
Wallich 6325 (5); 6327 (6); 6328 (11); 6329A

(11); 2585 (14); 2654 (5); 3650 (26); 3651 (26); 3657 (26); 4914 (18); 4915 (4b); 4916 (22a).
Wollaston 99 (3).
Wu 247 (4b); 156 (4b); 4129 (17).
Wu & Yang 2884 (4a); 2963 (4b); 37398 (32).
Wu et al. 7728 (20b).
Wu'an Exped. 36 (22b).
Wu, C. C. 20258 (20b).
Wu, C. K. 8069 (24).
Wu, C. Y. 4129 (12); 5772 (11); 750293 (26).
Wu, C. Y., & Chen, C. J. 74 (22a); 96 (22a).
Wu, C. Y., et al. 49 (20b); 53 (18, 30); 73 (18); 76 (30); 100 (22a); 103 (18); 75782 (31); 75888 (26, 27); 750361 (26); 751440 (26); 751974 (27); 757974 (26).
Wu, J. Z. 254 (22b).
Wu, Q. R. 77 (33).
Wu, Q. R., et al. 96 (22a).
Wu, S. G. 1764 (4b); 2278 (26).
Wu, S. H. 119 (4b).
Wu, S. K. 2 (31); 1775 (20b); 2577 (20b); 8260 (22a); 8457 (4b); 8915 (4b).
Wu, W. Y., et al. 93 (9a).
Wu, Y. C. 451 (31).
Wu, Y. L. 16 (20b); 78–1 (22b); 001–1 (5).
Wu, Z. L. 32772 (22a); 33369 (22a).
Wudu Pl. Exped. 571 (11).
Wuling Exped. 1224 (6); 1890 (21); 2361 (21); 2627 (5); 3417 (5).
Wushang Exped. 1031 (14).
Wuyi Shan Exped. 428 (21); 830 (20b); 00540 (21); 790032 (21); 790835 (21).
Xang, Z. 4332 (31).
Xi, X. Y. 84210 (6).
Xia, G. C. 1715 (27).
Xian Univ. 206 (4b).
Xiao, C. L. 400 (5); 624 (5).
Xiaojin 6726 (22a).
Xiaolong Shan Exped. 744 (13).
Xiaowutai Shan Exped. 999 (4b); 1872 (22a); 1920 (4b); 2741 (4b); 3030 (4b); 5041 (22a); 10178 (4b); 74078 (4a).
Xie, S. B. 47 (22a).
Xinbin Pl. Exped. 90 (32).
Xing, J. Q. 16492 (6); 16962 (5); 17928 (5).
Xinjiang Branch Acad. Sin. 3577 (4a).
Xinjiang Exped. 176 (33); 321 (9b); 329 (33); 1171 (1); 1456 (5); 1548 (31); 1580 (33); 1646 (5); 1666 (33); 1854 (4a); 1901 (33); 1960 (33); 2105 (9a); 2127 (33); 2169 (4a); 2407 (5); 2478 (10); 2512 (4a); 2732 (5); 2922 (9a); 2938 (4a); 2944 (4a); 2970 (4a); 3026 (5); 3502 (9a); 3624 (1); 3625 (1); 3711 (4a); 3740 (9b); 3757 (9b); 3773 (4a); 4736 (6); 4746 (33); 4995 (1); 5404 (31); 5459 (4a); 5479 (4a); 6414 (33); 7285 (33); 7324 (4a); 7334 (5); 15506 (9b); 650070 (33).
Xinjiang Forest. Exped. 78 (9a).
Xinjiang Med. Pl. Exped. 432 (4a); 497 (5).
Xinping Exped. 32 (22a).
Xinxiang Teacher's College 6226 (22b).
Xion, R. Y., et al. 610070 (22a).
Xiong, J. 2417 (20b); 5946 (21); 6050 (20b); 6396 (22b); 6403 (20b); 6470 (20b).
Xiong, J. H., & Zhou, Z. L. 91142 (18); 92107 (5); 92459 (22a); 93500 (20b).
Xiong, J. H., et al. 30977 (18); 31494 (4b); 31533 (4b); 32638 (4b).
Xiong, Y. G. 113 (4b); 1412 (21); 4179 (21); 5352 (22b); 5648 (22b); 6863 (21); 7382 (21); 8406 (21); 8551 (20b); 9655 (21).
Xiu, Y. W. 177 (10).
Xizang Exped. 670 (3); 810 (22a); 860 (11, 24); 961 (22a); 1193 (11); 1199 (20a); 1216 (22a, 24); 1228 (17, 28); 1235 (22a, 28); 1551 (3); 1676 (24); 1724 (27); 1867 (4b); 2259 (22a, 28); 2415 (24); 2722 (24); 2912 (17); 3489 (31); 3803 (31); 4065 (12); 4113 (2); 751081 (3).
Xizang Med. Exped. 547 (3); 551 (26); 822 (26); 873 (27); 1180 (32); 3608 (4b); 3609 (17); 3785 (26); 4055 (3).
Xizang Med. Pl. Exped. 326 (22a); 380 (31); 411 (31); 553 (31); 720 (4a); 758 (31); 1086 (24); 1156 (4a); 1334 (11); 1537 (24); 1977 (17); 3902 (22a).
Xizang NWPBI Exped. 2674 (4a).
Xizang Veget Exped. 15548 (4b).
Xu, G. M. 83–225 (6).
Xu, G. Y. 23109 (22a); 23135 (4b); 23341 (26); 23477 (3).
Xu, S. G. 5310 (20b).
Xu, S. L. 556 (20b).
Xuing, J. H., & Zhou, Z. L. 93081 (6).
Yamazaki et al. 321 (8).
Yanbian Exped. 121 (34); 138 (31); 316 (31); 398 (5); 502 (31); 715 (31); 1160 (4a); 2687 (4a).
Yanbian Second Group 924 (4b).
Yang 5726 (21); 6011 (5).
Yang, B. C. 312 (20b).
Yang, B. M. 2080 (13); 2117 (6); 27079 (20b).
Yang, C. F. 361 (5).
Yang, C. G. 186 (31); 256 (4b); 483 (4b); 748 (4b); 1018 (5); 1369 (4b); 1371 (22a); 1489 (31); 1667 (4a, 4b); 1697 (5); 1719 (4b).
Yang, C. Y. 36292 (4b).
Yang, C. Y., & Li, Z. G. 36241 (33); 36431 (4a).
Yang, F. W. 335 (18).
Yang, G. F., & Zhao, R. L. 100 (31).
Yang, G. H. 54745 (18); 55114 (18); 55395 (30);

56132 (22a); 56509 (22a); 56560 (4a); 57442 (4b); 58730 (4b); 58825 (22a); 58956 (22b); 59344 (4b); 59439 (4b); 59692 (14); 59752 (4b); 59929 (13); 65091 (13); 65544 (5).
Yang, J. S. 6987 (26); 7008 (4b); 7510 (4a); 8495 (12).
Yang, J. X. 1172 (22b); 1173 (5); 1242 (5, 6); 1479 (22a); 2138 (6); 2328 (12).
Yang, J. X., & Hu, Z. X. 3404 (6); 3712 (5).
Yang, K. H. 58632 (11).
Yang, K. Z. 25 (5); 141 (5); 176 (11); 255 (21).
Yang, Q. Z. 185 (4a).
Yang, Y. B. 104 (4a); 7160 (4a); 7359 (5); 40037 (27); 40039 (4b).
Yang, Y. C. 1136 (17, 26, 31); 1339 (31); 1493 (4a); 7329 (26).
Yang, Y. C., & Zhang, Y. L. 384 (4a).
Yang, Z. B., et al. 1094 (5).
Yang, Z. H. 101229 (24); 101230 (13).
Yang, Z. T. 48 (1).
Yao, C. W. 2240 (18); 2362 (18); 3055 (26); 3066 (4b); 3289 (20b); 3353 (4b); 3367 (28); 3691 (18); 3942 (30, 31); 3943 (18); 3993 (26); 4352 (18, 24); 4353 (18); 4398 (22a); 4652 (4b); 4954 (5); 5112 (30).
Yao, Z. W. 2498 (13); 2979 (11).
Ye, T. 1657 (4a).
Yi, G. Q. 8 (31).
Yin, W. Q. 60–1265 (19); 2072 (20b).
Ying, S. S. 1571 (15).
Ying, T. S. 4111 (4a); 4238 (4a); 4507 (24); 4587 (24); 9563 (4b); 9857 (20b).
Ying, T. S., & Hong, D. Y. 336 (4b); 386 (24, 26); 412 (24); 567 (24); 906 (3); 1225 (24); 4507 (17); 650412 (17).
Ying, T. S., et al. 9857 (17).
You, D. Q. 6935 (20b).
Younghusband 196 (26).
Youngmongshan Exped. 522 (4a).
Yu 354 (33).
Yu, N. F. 60–1303 (5).
Yu, P. H. 360 (13); 1148 (11, 14); 1371 (17).
Yu, S. W. 63–134 (5).
Yu, S. Z. 49923 (4b).
Yu, T. T. 832 (30); 1274 (20b, 30); 2223 (26); 3042 (18); 3042a (18); 1251 (5); 1377 (5); 3432 (4a, 4b); 3552 (22a); 3683 (11); 3846 (4b); 6056 (4b); 6276 (26); 6632 (4a); 7139 (4a); 7616 (26); 7890 (3, 4b); 8774 (4b); 8885 (26); 9057 (26); 9238 (26); 9364 (4a); 9380 (4a, 4b); 9523 (17, 26); 9704 (26, 27); 9761 (4a, 4b); 9773 (26); 9955 (17); 10163 (12, 22a); 10222 (5); 10305 (11); 10393 (5); 10873 (4a); 11864 (26); 11876 (4a, 4b); 11961 (4a); 12233 (26); 12369 (4a); 12394 (26); 12609 (17); 12635 (4a, 4b); 12639 (4b); 13316 (5); 13489 (5); 13677 (20b, 24); 13817 (4a, 4b); 14353 (17); 14355 (28); 14357 (4b); 14421 (5); 14485 (5); 15212 (4b); 17121 (20b); 17586 (5); 17639 (19); 19588 (11); 19712 (3); 20808 (26); 22167 (26); 22482 (26); 22571 (3); 22701 (20b); 22732 (26); 27407 (3).
Yu, X. P. 181 (5); 250 (5).
Yue, J. S. 3287 (21); 4225 (21).
Yue, J. S., et al. 2035 (21); 5123 (21).
Yulong Shan Exped. 350 (22a); 769 (5); 817 (4b).
Yunnan Exped. 170 (7).
Yunnan Inst. Bot. 781 (22a).
Yunnan Univ. 51 (22a); 155 (20b); 817 (4b).
Yuo, J. S., et al. 3287 (21).
Zhai, P. 60–083 (20b).
Zhan, Q. S. 6374 (4b).
Zhang 423 (4b); 1683 (4b).
Zhang, A. L., & Yu, S. W. 100918 (20b).
Zhang, A. W. 4043 (33).
Zhang, C. W. 192 (22a).
Zhang, D. L., & Yu, J. W. 100859 (22a).
Zhang, G. D., et al. 707–233 (17).
Zhang, G. L., & Pu, X. 79–125 (22b).
Zhang, J. 2034 (32).
Zhang, J. W., & Fu, G. X. 1133 (4b); 1135 (29).
Zhang, L. C. 57 (4b).
Zhang, M. Q., et al. 420 (17).
Zhang, Q. Q. 328 (21).
Zhang, S. Y. 336 (20b).
Zhang, S. Z. 289 (14).
Zhang, T. J. 53323 (4b).
Zhang, W. X., et al. 53 (4b); 246 (22a).
Zhang, W. Z. 434 (22a); 2549 (4a).
Zhang, X. B. 363 (30); 364 (4b); 3055 (30).
Zhang, X. M. 178 (4b).
Zhang, X. Z. 134 (31, 33).
Zhang, Y. B. 7 (20b).
Zhang, Y. L. 1683 (4a); 1974 (32).
Zhang, Y. L., et al. 307 (31); 414 (22a); 546 (4a); 653 (4a); 666 (31); 1011 (31); 1087 (31); 1207 (31); 1974 (31).
Zhang, Y. M. 30 (4a).
Zhang, Y. T., & Lang, K. Y. 951 (17); 1067 (17); 1397 (27).
Zhang, Z. H. 1–830 (32); 1271 (4b); 3509 (4a).
Zhang, Z. W. 192 (22a).
Zhang, Z. W., et al. 3088 (4a); 3347 (33); 4443 (4a).
Zhang, Z. Y. 800 (4b); 7357 (22b); 8323 (13); 13975 (20b); 14229 (20b); 14757 (13); 16246 (17); 16376 (22b); 16440 (13); 16652 (22a); 16838 (14); 16918 (13); 17081 (13); 17181 (5).
Zhang, Z. Y., & H. F. Zhou 23514 (4a); 23476 (4a); 23763 (4a).

Zhang, Z. Y., et al. 125 (4b).
Zhangjiakuo Exped. 415 (22b).
Zhao, D. C., & Balarov 298 (31).
Zhao, H. D., & Jian, Z. F. 672 (4a).
Zhao, Q. 137 (4b).
Zhao, Q. S. 5502 (5); 6099 (5); 6210 (22a); 6324 (26); 6374a (4a); 6474b (4a); 6895 (4a); 7925 (4a); 8662 (22a); 8665 (5).
Zhao, Q. S., & Hu, G. 5845 (5).
Zhao, Q. S., et al. 5001 (20b); 5368 (5); 6738 (27); 6895 (4b); 7884 (24); 8099 (22a); 8246 (5); 8271 (14); 8272 (11, 17); 8276 (11, 22a).
Zhao, Y. T., et al. 543 (34); 4188 (34); 19723 (34).
Zhao, Y. Z. 2270 (28); 7714 (4a); 21295 (4a).
Zhao, Z. C. 88–1306 (6); 88–1426 (6); 88–1746 (22b); 88–1769 (22a); 88–1808 (22b); 88–1934 (22b); 88–1991 (22b).
Zhao, Z. L. 523 (5).
Zhao, Z. X. 818 (14).
Zhau, Y. Z. 77–205 (32).
Zhejiang Exped. 29342 (22b).
Zhejiang Museum 3625 (21).
Zheng, M. Z. 142 (4b).
Zhong, B. Q. 5589 (17); 6760 (22a); 6763 (22a).
Zhong, J. X. 10947 (21).
Zhong, Q. F. 439 (4b).
Zhong, S. Q., & Zhou, Y. 50 (5).
Zhongdian Exped. 791 (3); 1329 (23, 27); 3691 (31); 3692 (24); 633290 (26, 27); 633807 (28); 633945 (26).
Zhongdian Pl. Exped. 648 (22a); 791 (4b); 63–2819 (4b); 63–3721 (22a); 63–4285 (22a).
Zhongyang Exped. 428 (6).
Zhou et al. 583 (20b).
Zhou, G. S. 50 (32); 81240 (30).
Zhou, G. S., et al. 122 (4b).
Zhou, H. C. 1102 (6); 5105 (20b); 6604 (21).
Zhou, H. F., & Su, H. Y. 108976 (20b, 22a); 109908 (4b); 109918 (4b); 109962 (20b, 22a); 110151 (22a); 110674 (21); 110948 (20b, 21); 111052 (21); 25962 (6); 109589 (5); 109608 (6); 109736 (14); 110006 (13); 110071 (6); 110148 (14); 110348 (14); 110586 (6); 110893 (5); 111107 (14); 111143 (14); 111528 (14).
Zhou, J. 108 (4a).
Zhou, K. Y. 69 (4a).
Zhou, L. H. 2506 (4b).
Zhou, L. H., & Chang, H. T. 1507 (31).
Zhou, L. H., & Sun, L. N. 944 (31); 1974 (22a); 2310 (4a); 2333 (31).
Zhou, L. H., & Zhang, K. Z. 1267 (4b).
Zhou, X. M., & Sha, Q. 79 (31).
Zhou, Y. L. 1250 (4b).
Zhou, Y. L., et al. 2374 (31).
Zhu, C. Y. 67 (5).
Zhu, G. F. 43 (21).
Zhu, G. L. 5883 (4a).
Zhu, G. L., et al. 5771a (9a); 7204 (4a).
Zhu, G. Z., & Li, X. 75248 (5); 75287 (11); 75743 (22a); 75861 (5); 76503 (5).
Zhu, J. C. 532 (4b).
Zhu, S. F. 20120 (20b); 20310 (20b); 20430 (22a); 20431 (22a).
Zhu, T. Z. 657 (32).
Zhu, X. Y. 89138 (4b).
Zhu, Y. C. 148 (4b); 168 (4b); 1333 (22b).
Zhu, Y. C., & Balarov 39 (4b).
Zhu, Y. C., et al. 39 (4b); 148 (4a).
Zhu, Y. G., et al. 0431 (22a).
Zhuang, J. Y. 788 (26); 1105 (3).
Ziyun Shan Bot. Exped. 529 (20b).
Zuo, D. X., et al. 23496 (21).

INDEX TO SCIENTIFIC NAMES

Accepted names are in roman type; the main entry for each is in **boldface**. Synonyms are in *italics*.